Modulation Calorimetry

Springer
*Berlin
Heidelberg
New York
Hong Kong
London
Milan
Paris
Tokyo*

Yaakov Kraftmakher

Modulation Calorimetry

Theory and Applications

With 196 Figures and 27 Tables

 Springer

Professor Dr. Yaakov Kraftmakher
Bar-Ilan University
Department of Physics
52 900 Ramat Gan
ISRAEL
E-mail: krafty@mail.biu.ac.il

ISBN 3-540-21082-2 Springer-Verlag Berlin Heidelberg New York

Library of Congress Control Number: 2004102076

This work is subject to copyright. All rights are reserved, whether the whole or part of the material is concerned, specifically the rights of translation, reprinting, reuse of illustrations, recitation, broadcasting, reproduction on microfilm or in any other way, and storage in data banks. Duplication of this publication or parts thereof is permitted only under the provisions of the German Copyright Law of September 9, 1965, in its current version, and permission for use must always be obtained from Springer-Verlag. Violations are liable for prosecution under the German Copyright Law.

Springer-Verlag is a part of Springer Science+Business Media

springeronline.com

© Springer-Verlag Berlin Heidelberg 2004
Printed in Germany

The use of general descriptive names, registered names, trademarks, etc. in this publication does not imply, even in the absence of a specific statement, that such names are exempt from the relevant protective laws and regulations and therefore free for general use.

Typesetting and camera-ready copy by the author
Cover design: *design & production* GmbH, Heidelberg

Printed on acid-free paper   SPIN 10892962   57/3141/tr   5 4 3 2 1 0

Preface

Modulation calorimetry was invented by Orso Mario Corbino about one hundred years ago. Corbino developed the theory of modulation calorimetry and proposed two methods of measuring temperature oscillations in a wire sample through oscillations of its electrical resistance. Now the first method is known as the supplementary-current method, and the second one is known as the third-harmonic technique. Corbino's paper in Phys. Z. (1910) was entitled "Thermische Oszillationen wechselstromdurchflossener Lampen mit dünnem Faden und daraus sich ergebende Gleichrichterwirkung infolge der Anwesenheit geradzahliger Oberschwingungen." This title reflects the technique employed in the measurements. Corbino's second paper, published a year later, was entitled "Periodische Widerstandsänderungen feiner Metallfäden, die durch Wechselstrome zum Gluhen gebracht werden, sowie Ableitung ihrer thermischen Eigenschaften bei hoher Temperatur."

Modulation calorimetry and other methods based on periodically varying the temperature of the sample provide all advantages peculiar to modulation techniques. The excellent temperature resolution and sensitivity are very attractive, and in some cases this method is the best technique for measuring specific heat.

The idea of modulation calorimetry occurred to the present author when designing a high-temperature dilatometer. The main obstacle was the plastic deformation of the samples and external mechanical disturbances. To overcome this difficulty, the obvious approach seemed to be oscillating the temperature of the sample and measuring the corresponding changes in its length. With this approach, one directly measures the thermal expansivity and reduces the influence of a slow drift and of mechanical disturbances. A question that immediately arises is the determination of the amplitude of the temperature oscillations in the sample. This problem led me to the equivalent-impedance technique (1962). A short while later, I found out that modulation calorimetry was discovered long ago. Later, it became known to me that radio engineers knew the formulas for the equivalent impedance for a long time. However, calorimetric measurements were beyond their interests. Using the equivalent-impedance technique, the high-temperature specific heat of refractory metals was measured (1962–1964).

I thankfully remember my teacher, the late Professor P.G.Strelkov (1899–1968), and my students and collaborators: A.I.Akimov, I.M.Cheremisina, V.Y.Cherepanov, A.G.Cherevko, V.Y.Fridman, S.Y.Glazkov, O.M.Kanel', S.D.Krylov, E.B.Lanina, V.P.Nezhentsev, T.Y.Pinegina (Romashina), V.O.Shestopal, K.S.Sukhovei, G.G.Sushakova (Zaitseva), A.P.Tarasenko, V.L.Tonaevskii, and the late A.A.Varchenko.

I am grateful to my colleagues for useful discussions: the late Dr. A.Cezairliyan, Professor M.Deutsch, Dr. R.I.Efremova, the late Professor L.P.Filippov, Dr. V.A.Gruzdev, Dr. O.A.Kraev, Dr. K.D. Maglić, Professor A.A.Maradudin, Professor E.V.Matizen, Professor I.I.Novikov, Dr. F.Righini, the late Dr. G.Ruffino, Professor M.B.Salamon, Professor A.E.Sheindlin, Professor R.Taylor, Professor A.V.Voronel, Professor R.K.Wunderlich, Dr. M.M. Yakunkin.

I am indebted to the American Physical Society and to the authors for the permissions to reproduce graphs from papers published in the *Physical Review* and *Physical Review Letters*, namely:

Akutsu H, Saito K, Sorai M (2000) Phys Rev B 61:4346–4352 (Fig. 15.13).
Birge NO, Nagel SR (1985) Phys Rev Lett 54:2674–2677 (Fig. 7.15).
Charalambous M, Riou O, Gandit P, Billon B, Lejay P, Chaussy J, Hardy WN,
    Bonn DA, Liang R (1999) Phys Rev Lett 83:2042–2045 (Fig. 14.8).
Fominaya F, Villain J, Fournier T, Gandit P, Chaussy J, Fort A, Caneschi A (1999)
    Phys Rev B 59:519–528 (Fig. 15.12).
Fominaya F, Villain J, Gandit P, Chaussy J, Caneschi A (1997) Phys Rev Lett
    79:1126–1129 (Fig. 6.11).
Garnier PR, Salamon MB (1971) Phys Rev Lett 27:1523–1526 (Fig. 14.2).
Geer R, Stoebe T, Huang CC (1993) Phys Rev E 48:408–427 (Fig. 7.3).
Howson MA, Salamon MB, Friedmann TA, Rice JP, Ginsberg D (1990)
    Phys Rev B 41:300–306 (Fig. 1.7).
Inderhees SE, Salamon MB, Rice JP, Ginsberg DM (1991)
    Phys Rev Lett 66:232–235 (Fig. 14.7).
Jeong YH, Moon IK (1995) Phys Rev B 52:6381–6385 (Fig. 15.3).
Park T, Salamon MB, Choi EM, Kim HJ, Lee S-I (2003)
    Phys Rev Lett 90:177001-1–4 (Fig. 1.4).
Simons DS, Salamon MB (1971) Phys Rev Lett 26:750–752 (Fig. 14.5).
Steele LM, Yeager CJ, Finotello D (1993) Phys Rev Lett 71:3673–3676
    (Fig. 14.9b).
Terki F, Gandit P, Chaussy J (1992) Phys Rev B 46:922–929 (Fig. 9.4b).
Wunderlich RK, Lee DS, Johnson WL, Fecht H-J (1997)
    Phys Rev B 55:26–29 (Fig. 6.9).

I am grateful to Maria Grazia Ianniello (Roma) for the permission to reproduce the portrait and biographical data of O.M.Corbino from her article placed on the website <www.phys.uniroma1.it/docs/museo/Corbino/html> and to Dr. I.Dana for the translation of this article.

The first version of the book was published as a review paper in *Physics Reports*. Many thanks to the editor, Professor A.A.Maradudin, for his work for improving the review. The support by the Bar-Ilan University is gratefully acknowledged.

Bar-Ilan University,   Y.K.
Ramat-Gan,
November 2003

Table of Contents

1 Introduction...	1
1.1 History of Modulation Methods..	1
1.2 Features of Modulation Techniques.....................................	6
2 Theory of Modulation Calorimetry..	15
2.1 Basic Equation of Modulation Calorimetry...........................	15
2.1.1 Sample in a Temperature-Modulated Regime................	15
2.1.2 Basic Equation..	16
2.1.3 Complex-Form Presentation...	17
2.1.4 Temperature Oscillations in Resistively Heated Wires....	17
2.2 Thermal Coupling in the Calorimeter Cell............................	19
2.2.1 Sullivan-Seidel Analysis...	19
2.2.2 Thermal Conductance of the Sample..............................	20
2.2.3 Model Including a Substrate...	21
2.2.4 Choice of Modulation Frequency....................................	22
3 Modulation of Heating Power...	23
3.1 Direct Electric Heating and Resistive Heaters........................	23
3.1.1 Direct Heating...	23
3.1.2 Separate Resistive Heaters..	25
3.2 Modulated-Light Heating...	26
3.3 Other Methods...	28
3.3.1 Electron Bombardment...	29
3.3.2 Induction Heating...	30
3.3.3 Peltier Heating..	31
4 Measurement of Temperature Oscillations............................	33
4.1 Oscillations in Sample's Resistance.......................................	33
4.1.1 Supplementary-Current Technique.................................	33
4.1.2 Third-Harmonic Method...	35
4.1.3 Equivalent-Impedance Technique...................................	37
4.2 Photoelectric Detectors..	42
4.3 Pyroelectric Sensors..	45
4.4 Thermocouples and Resistance Thermometers......................	47
4.5 Lock-in Detection of Periodic Signals...................................	50
5 Brief Review of Methods of Calorimetry...............................	53
5.1 Adiabatic Calorimetry..	53
5.2 Drop Method..	56
5.3 Pulse and Dynamic Techniques...	58
5.3.1 Pulse Method..	58

5.3.2 Dynamic Calorimetry………………………………………..	59
5.4 Relaxation Techniques…………………………………………..	65
5.5 Rapid-Heating Experiments…………………………………….	67

6 Modulation Calorimetry I 73
   6.1 High Temperatures……………………………………………. 73
      6.1.1 Wire Samples…………………………………………….. 73
      6.1.2 Nonadiabatic Regime…………………………………….. 77
      6.1.3 Bulk Samples……………………………………………... 81
      6.1.4 Molten Metals……………………………………………. 81
      6.1.5 Active Thermal Shield…………………………………… 81
      6.1.6 Nonconducting Materials………………………………… 82
      6.1.7 Noncontact Calorimetry………………………………….. 84
   6.2 Medium Temperatures………………………………………… 87
   6.3 Low Temperatures…………………………………………….. 89
   6.4 Measurements Under High Pressures………………………… 91
   6.5 Measurement of Specific Heat and Thermal Diffusivity………. 96
   6.6 Sinku-Riko Calorimeter ACC-1………………………………. 98

7 Modulation Calorimetry II 101
   7.1 Modulation Microcalorimetry………………………………… 101
      7.1.1 Thin Deposited Films……………………………………. 101
      7.1.2 Freestanding Films………………………………………. 102
      7.1.3 Bath Modulation………………………………………… 104
      7.1.4 Nanocalorimetry at Low Temperatures…………………. 104
   7.2 Organic and Biological Materials…………………………….. 107
   7.3 Photoacoustic Techniques……………………………………. 113
   7.4 Specific-Heat Spectroscopy…………………………………... 115
      7.4.1 Measurement of Thermal Effusivity…………………….. 115
      7.4.2 Frequency-Dependent Effusivity……………………….. 118
      7.4.3 How to Extend the Frequency Band – a Proposal……… 121
      7.4.4 What is Worth Remembering…………………………… 123
   7.5 Modulated Differential Scanning Calorimetry……………….. 124
      7.5.1 Principle of Modulated Differential Scanning Calorimetry…….. 124
      7.5.2 What is Worth Remembering…………………………… 125

8 Modulation Dilatometry 129
   8.1 Brief Review of Dilatometic Techniques…………………….. 129
      8.1.1 Optical Methods…………………………………………. 130
      8.1.2 Capacitance Dilatometers………………………………. 131
      8.1.3 Dynamic Techniques……………………………………. 132
   8.2 Principle of Modulation Dilatometry………………………… 135
      8.2.1 Wire Samples……………………………………………. 135
      8.2.2 Differential Method…………………………………….. 137
      8.2.3 Bulk Samples……………………………………………. 139
      8.2.4 Interferometric Modulation Dilatometer……………….. 140
      8.2.5 Nonconducting Materials……………………………….. 142
   8.3 Extremely Small Periodic Displacements……………………. 143

9 Other Modulation Techniques... 147
9.1 Temperature Derivative of Resistance............................ 147
9.2 Direct Measurement of Thermopower........................... 151
9.3 Spectral Absorptance.. 153
   9.3.1 Calorimetric Techniques....................................... 153
   9.3.2 Compensation Method... 156
10 Noise Thermometry of Wire Samples............................. 159
10.1 Brief History of Noise Thermometry............................ 159
10.2 Noise Correlation Thermometer................................... 164
   10.2.1 Correlation Amplifier.. 164
   10.2.2 Compensation Method.. 165
   10.2.3 Temperature Derivative of Resistance................ 167
   10.2.4 Nyquist's Formula for High Current Densities... 168
11 Electronics for Modulation Measurements..................... 169
11.1 Electronic Instruments... 169
11.2 Examples of Modulation Set-Ups................................. 172
12 Accuracy of Modulation Measurements........................ 177
12.1 Requirements for Modulation Experiments................. 177
12.2 Comparison with Recommended Values..................... 180
   12.2.1 Specific Heat of Tungsten, Copper, and Iron..... 180
   12.2.2 Specific Heat of Molybdenum and Platinum..... 182
   12.2.3 Thermal Expansion of Tungsten and Platinum.. 183
   12.2.4 Electrical Resistivity of Platinum........................ 185
13 Studies at High Temperatures.. 187
13.1 Equilibrium Point Defects in Metals............................ 187
   13.1.1 Specific Heat... 189
   13.1.2 Thermal Expansion... 191
   13.1.3 Electrical Resistivity... 192
   13.1.4 Equilibrium Concentrations of Point Defects..... 193
13.2 Temperature Coefficient of Specific Heat.................... 196
13.3 Unexpected Premelting Anomaly................................. 199
13.4 Isochoric Specific Heat of Solds.................................. 201
   13.4.1 Temperature Fluctuations................................... 201
   13.4.2 Experimental... 202
   13.4.3 Determination of the C_p/C_v Ratio..................... 204
14 Phase Transitions... 207
14.1 First- and Second-Order Phase Transitions.................. 207
14.2 Ferro- and Antiferromagnets.. 211
14.3 Superconductors... 217
14.4 Thin Films and Confined Systems................................ 222
15 Relaxation Phenomena... 227
15.1 Formulas for Relaxation... 228
15.2 Glass Transitions.. 230
15.3 Point-Defect Equilibration in Metals............................ 232
   15.3.1 Enhancement of Modulation Frequency............ 232

    15.3.2 Relaxation Phenomenon in Tungsten and Platinum............ 234
    15.3.3 Equilibration Times... 235
    15.3.4 Two Proposals for Studies of Vacancy Equilibration........... 236
    15.3.5 Questions to be Answered by Rapid-Heating Experiments.... 237
    15.3.6 Determination of Vacancy-Related Enthalpy – a Proposal.... 239
  15.4 Other Examples of Relaxation in Specific Heat....................... 242
16 Five Student Experiments.. 245
  16.1 Pulse Calorimetry.. 245
  16.2 Principles of Modulation Calorimetry................................... 248
  16.3 Third-Harmonic Technique... 250
  16.4 Equivalent-Impedance Method... 252
  16.5 Determination of Boltzmann's Constant............................... 254
Conclusion... 259
References... 261
Index.. 283

List of Symbols

m	Mass of a sample
c	Specific heat
p_0, p	DC and AC components of heating power
Q, Q', Q''	Heat loss from a sample and its temperature derivatives
T_0	Mean temperature of a sample
C	Heat capacity; Capacitance
K	Thermal conductance; Heat transfer coefficient; Proportionality factor
R_0, R	Sample resistance at T_0; Resistance
R'	Temperature derivative of resistance
I_0, i	DC and AC components of current
V_0, v	DC and AC components of voltage
$\mathbf{Z}, Z$	Electrical impedance and its modulus
$V_{3\omega}$	Third-harmonic voltage
k_B	Boltzmann's constant
d	Diameter of a sample
C_p, C_v	Isobaric and isochoric specific heats
$\langle \Delta T_f^2 \rangle$	Spectral density of temperature fluctuations
T_c	Critical temperature
t	Time; $(T - T_c)/T_c$
D	Thermal diffusivity
e	Thermal effusivity
$T' = dT/dt$	Heating (cooling) rate
M	Mutual inductance
A, B	Real and imaginary parts of equivalent impedance; Constants
k	Extinction coefficient; Wave number
n	Refraction index
S	Surface area of a sample
$C(X)$	Complex specific heat ($X = \omega\tau$)
$C, \Delta C$	Non-relaxing and relaxing parts of specific heat
C_0	Specific heat under equilibrium
$\Delta a/a$	Relative change in the lattice parameter
N	Avogadro number
P	Electric power
G_F	Gibbs free energy of defect formation
H_F, S_F	Enthalpy and entropy of defect formation
q	Heating rate
R_λ	Spectral reflectivity

List of Symbols

Θ	AC component of temperature; Phase angle
Θ_0, Θ_1	Amplitude of temperature oscillations at fundamental frequency
Θ_2	Amplitude of second-harmonic component of temperature oscillations
$\beta = R'/R_0$	Temperature coefficient of resistance
ϕ, φ, ψ	Phase of temperature oscillations
ω	Angular frequency ($\omega = 2\pi f$)
α	Temperature coefficient of specific heat; Phase shift; Thermal expansivity; Critical index
$\langle \Theta_f^2 \rangle$	Spectral density of temperature oscillations caused by noise modulation
τ	Time constant; Relaxation time
κ	Thermal conductivity
ε	Hemispherical total emittance
σ	Stefan-Boltzmann constant
γ	Specific heat ratio ($\gamma = C_p/C_v$); Relative volume of point defects ($\gamma = V/\Omega$)
ρ	Electrical resistivity
ε_λ	Spectral emissivity
Ω	Angular frequency; Atomic volume
λ	Wavelength of light; Thermal wave length

1 Introduction

Modulation techniques are widely used in physical experiments. Modulation methods, employing selective amplifiers and/or lock-in detectors, are successfully used in NMR and ESR techniques, in studies of the Hall effect and I-V characteristics, in photometry, and in many other cases. However, the modulation principle was used for the first time just for measuring specific heat, many years before it was accepted as a tool for other physical measurements.

1.1 History of Modulation Methods

Orso Mario Corbino discovered modulation calorimetry about hundred years ago. The method consists in periodically changing the power dissipated in the sample and in measuring the temperature oscillations in it around a mean value. Clearly, these oscillations depend on the heat capacity of the sample. This is the basic distinction from other modulation techniques, where the modulation does not depend on the properties of the sample. Corbino developed the theory of modulation calorimetry and carried out the first modulation measurements of specific heat. He used the oscillations of the resistance of the sample to determine the temperature oscillations. A supplementary current of frequency equal to that of the temperature oscillations (Corbino 1910) or the third-harmonic technique (Corbino 1911) were proposed for this purpose. Fermi (1937) highly praised this work in a paper in memory of Corbino, and the paper was entitled "Un Maestro."

Orso Mario Corbino, Un Maestro

Born at Augusta in Sicily from a modest family of macaroni-maker artisans, Corbino got his degree in Physics at the University of Palermo when he was only 20 years old. In the period spent at Palermo, he devoted himself to the study of the properties of a circuit containing a dynamo and of some devices, which where commonly used at those times, but whose functioning was completely unclear, such as the induction coil and the arc of Duddell. In 1904, Corbino won the chair of Experimental Physics at the University of Messina. In 1908 he, luckily, escaped from the disastrous effects of the earthquake and transferred himself to Rome. In the roman period, Corbino started working intensely. In those years, he discovered the "Corbino effect", a variant of the Hall effect. In retrospect, however, the greatest merit of Corbino was to have laid down the favourable conditions for the birth of the School of Rome. He instituted the first chair of Theoretical Physics, which included at that time the atomic, nuclear, and quantum physics, and to which he called Fermi in 1926. He did not limit himself to insure to his young colleagues only institutional support, but he also followed closely their research with participation and love. He was a big teacher, with clear exposition, and a brilliant orator.

Orso Mario Corbino (1876-1937)

Development During Last Four Decades

In some further measurements, the temperature oscillations in the sample were detected by the oscillations of the thermionic current from it (Smith and Bigler 1922; Bockstahler 1925). Zwikker (1928) measured the specific heat of tungsten up to 2600 K. Considerable progress in modulation calorimetry was achieved, however, in the 1960s due to advances in experimental techniques. At the first stage, the method was used exclusively at high temperatures. Its most important feature was the smallness of the correction for heat losses. The samples, in the form of a wire or a rod, were heated by an electric current passing through them. The temperature oscillations were deduced from oscillations in the resistance of the samples or radiation from them. By the use of this method, the high-temperature specific heat of refractory metals was determined. Later, the method was employed in studies of phase transitions, where the main requirement became good temperature resolution. In these experiments, the samples also were heated by an electric current.

At the second stage, the modulation technique was applied to measurements at medium and low temperatures and for studying nonconducting materials. Even at low and medium temperatures, the traditional domain of adiabatic calorimetry, modulation calorimetry ensures better temperature resolution and higher sensitivity. Sometimes, small dimensions of the samples often are of great importance. Generally, absolute values obtainable by modulation calorimetry are less accurate than those obtained by adiabatic calorimetry. Even under very favourable conditions, the inaccuracy of modulation measurements is about 1–2%. Nevertheless, the excellent resolution of the method makes it preferable in many cases, e.g., in studies of phase transitions. The method is applicable also under high pressures.

Modulation calorimetry and other methods based on oscillating the temperature of the sample provide all the advantages peculiar to modulation techniques. The excellent temperature resolution and sensitivity are very attractive, and in some cases the method is the unique technique suitable for measuring specific heat.

The present author came to the modulation principle when designing a high-temperature dilatometer. The main obstacle was the plastic deformation of the samples and mechanical disturbances. The obvious approach to overcome these problems was to oscillate the temperature of the sample and to measure the corresponding changes of its length. With this approach, one directly measures the coefficient of thermal expansion and reduces the influence of a slow drift and irregular mechanical disturbances. A question that immediately arises is the determination of the temperature oscillations in the sample. This problem led me to modulation calorimetry and the equivalent-impedance technique (1962). In a short time, I became aware of the Corbino's discovery. Later, I also found that radio engineers knew the formulas for the equivalent impedance for a long time. However, calorimetric measurements were beyond their interests. Using the equivalent-impedance technique, the high-temperature specific heat of refractory metals was measured (1962–1964). In all the cases, a strong non-linear increase in specific heat was observed (Fig. 1.1). That time, these results gained no recognition, as well as previous observations on tantalum and molybdenum by Rasor and McClelland (1960) made by pulse calorimetry.

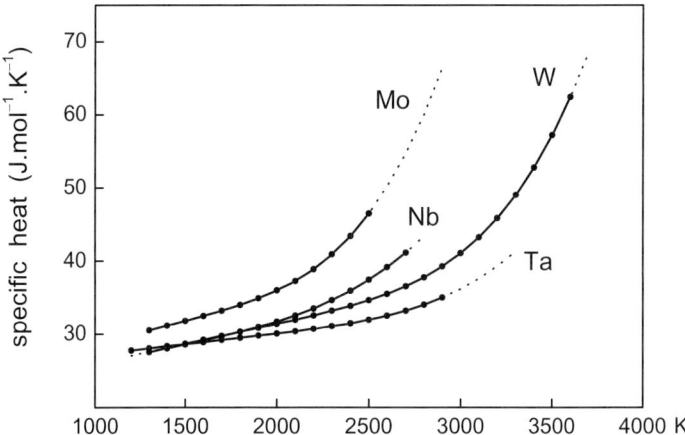

Fig. 1.1. Specific heat of refractory metals determined by equivalent-impedance technique and extrapolated to their melting points (Kraftmakher and Strelkov 1962; Kraftmakher 1963ab, 1964). The refractory metals manifest very strong non-linear increase in specific heat. The specific heat of Mo and W increases more than two times. The non-linear increase in specific heat was attributed to vacancy formation in crystal lattice (Sect. 13.1)

Meanwhile, it became possible to return to the idea of modulation dilatometry (1965). Simultaneously, modulation calorimetry was applied to studies of specific-heat anomalies near the Curie point of ferromagnets (1965–1967). Modulation techniques were proposed to directly measure the temperature derivative of resistance of the sample (1967) and thermopower (1970). The latter technique was introduced simultaneously by three groups and reinvented many times later.

The scientific community recognized the strong non-linear increase in high-temperature specific heat of refractory metals after well-known investigations by Ared Cezairliyan and his group at the National Bureau of Standards (1970–1971). A. Cezairliyan developed a subsecond dynamic technique for measurements of thermophysical properties of metals and alloys at high temperatures.

At the same time, L.P. Filippov and his group at Moscow University also employed modulation techniques. The general aim of these investigations was to apply the method of temperature waves to measurements of thermophysical properties of solid and liquid metals at high temperatures. Two monographs by Filippov (1967, 1984) summarize the results obtained by this group. Filippov (1960) was the first to use the third-harmonic method for studying thermal properties of a liquid surrounding a probe. Temperature oscillations in a heater immersed in a liquid under study depend on the product of the specific heat and thermal conductivity of the liquid.

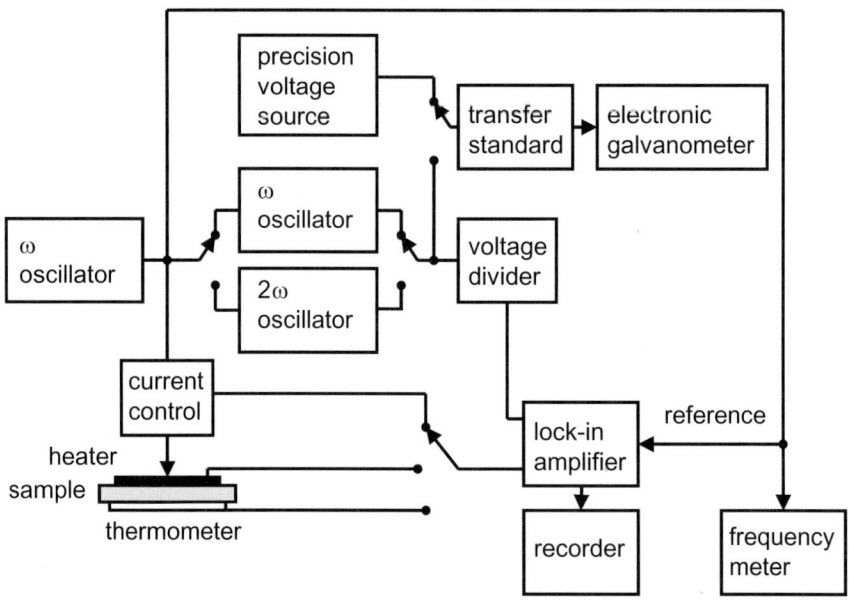

Fig. 1.2. Simplified diagram of the experimental set-up developed by Sullivan and Seidel for modulation measurements at low temperatures. The set-up was used to measure specific heat as a function of external magnetic field (Sullivan and Seidel 1968). The authors introduced the term 'AC calorimetry' now widely accepted

A very important milestone was put by P.F. Sullivan and G. Seidel, which rediscovered modulation calorimetry and applied it to low-temperature measurements (Fig. 1.2). Their famous papers (1966–1968) are known to all investigators using modulation calorimetry. The authors have clearly shown the advantages of modulation calorimetry. Accepting this idea, P. Handler, D.E. Mapother, and M. Rayl developed the method of modulated-light heating (1967), which became very popular. Sullivan and Seidel did not continue their work. M.B. Salamon (Illinois University, Urbana-Champaign) caught up the modulation technique and successfully employed it in numerous studies of phase transitions in solids. After his visit to Japan, modulation calorimetry began to develop also in this country (I. Hatta, A. Ikushima, H. Ikeda, K. Ema). In the United States, new groups employing modulation calorimetry have appeared (D. Finotello, C.W. Garland, C.C. Huang, G.S. Iannacchione).

It is impossible to list all laboratories employing modulation techniques. Today, the main objects of modulation calorimetry are high-temperature superconductors, liquid crystals, and glass-forming solids. Among recent achievements, it is necessary to mention the specific-heat spectroscopy (Birge and Nagel 1985). This technique makes it possible to search for relaxation phenomena, which appear when the modulation period becomes comparable to the relaxation time of a process contributing to the specific heat.

Another outstanding achievement is the nanocalorimetry capable of measurements on very small samples. A new branch of thermal analysis was founded, the modulated differential scanning calorimetry (MDSC). This technique combines advantages of DSC and of modulation calorimetry.

For a long time, there was no special term for modulation calorimetry. The term 'modulation method for measuring specific heat' was proposed in a paper describing the equivalent-impedance method for determining specific heat of wire samples (Kraftmakher 1962). However, many investigators became acquainted with modulation calorimetry only from the papers of Sullivan and Seidel (1966, 1967, 1968). The term 'AC calorimetry' introduced by Sullivan and Seidel is now generally accepted. A new term recently appeared, the 'temperature-modulated calorimetry' (Gmelin 1997).

Reading (2001) pointed out four areas of application of temperature modulation as follows:

- The determination of kinetic parameters.
- Calorimetric measurements.
- The separation of the sample responses that are dependent on rate of change of temperature from those principally dependent on temperature.
- The micro-thermal analysis.

Modulation calorimetry became very important and widely used technique, and a commercial modulation calorimeter designed by Sinku-Rico, Inc. already appeared (Fig. 1.3). The instrument employs modulated-light heating introduced by Handler et al. (1967). Modifications of this instrument were used in a wide temperature range, from 2 to 800 K. Determinations of thermal conductivity and thermal diffusivity are also possible.

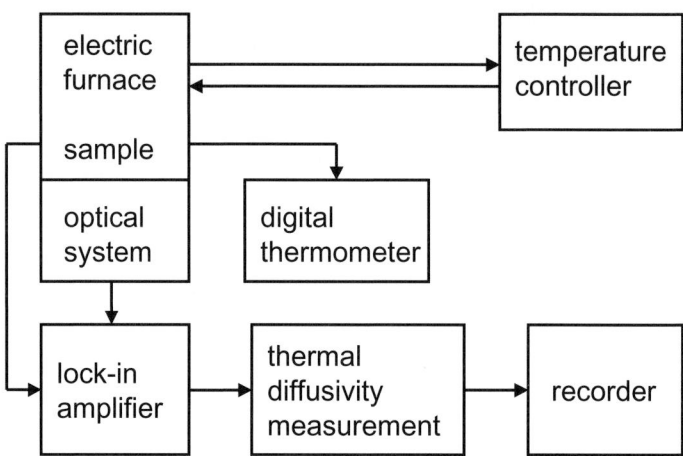

Fig. 1.3. Simplified block diagram of a version of thermal constants analyser manufactured by Sinku-Riko, Inc. (Sect. 6.6)

1.2 Features of Modulation Techniques

In modulation calorimetry, the heat capacity of the sample is calculated from the oscillations of the heating power and of the temperature of the sample. The use of periodic temperature oscillations provides important advantages. When the modulation frequency is sufficiently high, corrections for heat losses from the sample become negligible even at the highest temperatures. In this respect, the method is comparable to adiabatic calorimetry. Against the pulse method, the modulation technique has the advantage that selective amplifiers and lock-in detectors allow measurements of very small temperature oscillations. This feature is very important when good temperature resolution is crucial as, e.g., in studies of phase transitions. The theory of the measurements is simple and quite adequate to experimental conditions. Modifications of the method differ in the ways of modulating the heating power (direct electric heating, electron bombardment, use of separate heaters, radiation heating, induction heating, Peltier effect) and in the methods of detecting the temperature oscillations (through the resistance of the sample or radiation from it, and by the use of pyroelectric sensors, thermocouples, or resistance thermometers).

Modulation calorimetry is applicable in wide temperature ranges, from fractions of a kelvin up to melting points of refractory metals, and provides high sensitivity. In many cases, it is possible to assemble compensation schemes, whose balance does not depend on the amplitude of the oscillations in the applied power and to automatically record the quantities to be measured. The measurements can be controlled by a data-acquisition system and fully automated. Due to all these features, the method became very attractive and widely used. In treating the data, the mean temperature and the amplitude of the temperature oscillations are considered to be constant throughout the entire sample. In this respect, the modulation techniques differ from the method of temperature waves. As a rule, the measurements are carried out in a regime where the amplitude of the temperature oscillations in the sample is inversely proportional to its heat capacity.

The excellent resolution of modulation calorimetry appeared to be very useful in studies of high-temperature superconductors. The goal was to determine contributions that amount only to a few percent of the total specific heat. Small samples are generally employed in modulation calorimetry, but a problem arose when it was necessary to perform measurements on a microgram sample. One of the ways to solve the problem is the bath modulation (Graebner 1989). A more general approach consists in a miniaturisation of the calorimeter. At low temperatures, the sensitivity of such calorimeters is of the order of 10^{-12} J.K^{-1}.

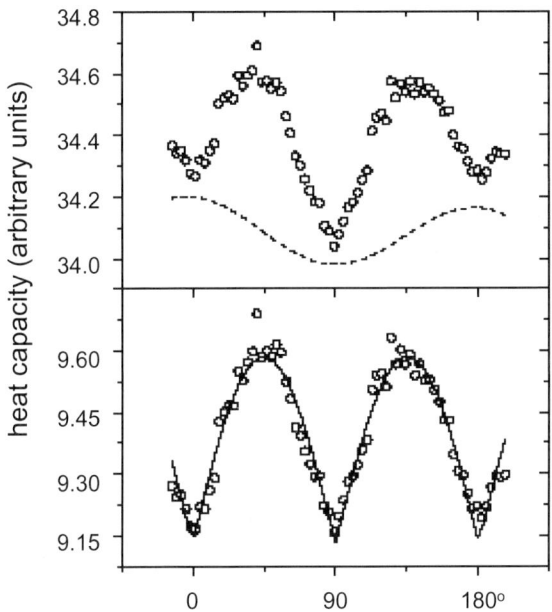

Fig. 1.4. Magnetic-field-angle dependence of the heat capacity of YNi$_2$B$_2$C single crystal at 2 K in 1-T magnetic field. The angle is measured with respect to the *a* axis. For details, see Park et al. 2003

Modulation calorimetry provides very attractive opportunities: temperatures from 0.1 K (Feng et al. 1988; Steinmetz et al. 1989) up to melting points of refractory metals; magnetic fields up to 30 T (Yu et al. 1988; Fortune et al. 1990); pressures up to 22 GPa (Wilhelm and Jaccard 2002ab); samples as small as 1 µg (Fominaya et al. 1997ab); resolution of the order of 0.01%. Temperature oscillations necessary for calorimetric measurements are of the order of 1 K at high temperatures, of 1 mK at room temperatures, and of 1 µK at liquid helium temperatures (Mehta and Gasparini 1997, 1998). One can measure specific heat as a function of an external parameter, such as magnetic field or pressure. Sullivan and Seidel (1967, 1968) proposed and confirmed this approach. Recently, Park et al. (2003) measured the specific heat of a single crystal of the superconductor YNi_2B_2C as a function of the orientation of an external magnetic field (Fig. 1.4). One of the recent achievements is the noncontact calorimetry successfully employed in space (R. Wunderlich et al. 1997, 2001; Egry 2000; Egry et al. 2001).

Sullivan and Seidel (1968) stressed the significant features of modulation calorimetry as follows:

- "The sample may be coupled thermally to a bath.
- The method is a steady-state measurement.
- Changes in heat capacity with some experimentally variable parameter may be recorded directly.
- Extremely small heat capacities may be measured with accuracy.
- The method possesses a precision an order of magnitude better than existing techniques."

Later, an important additional advantage was found, namely:

- The method is suitable for measurements at various modulation frequencies and search for relaxation phenomena in specific heat (Chap. 15).

In a review, Gmelin (1997) formulated the advantages of modulation calorimetry in more detail:

- Modulation measurements under quasi-adiabatic conditions are possible over a very wide temperature range.
- Heat losses of any type can be corrected or completely avoided.
- Extremely high sensitivity and temperature resolution are achievable.
- Experiments are possible with small and ultra-small samples.
- The method is best suited for studying samples under extreme external conditions as high pressure and high magnetic and electric fields.
- Sweep runs in heating and cooling modes are possible.
- The method provides short measuring time compared to many other calorimetric methods.
- Information on sample-specific relaxation phenomena is available.
- The method provides possibility for studying frequency-, pressure-, and magnetic field dependences under isothermal conditions.
- Due to use of selective and/or lock-in amplifiers, the method may be employed in mechanically or electrically noisy environment.

The disadvantages of modulation calorimetry were formulated as follows:

- Errors of absolute values in modulation calorimetry are in the range 1–10%.
- It is very difficult, or even impossible, to determine latent heats of first-order phase transitions.
- Strong limiting boundary conditions for the relaxation times must be fulfilled.

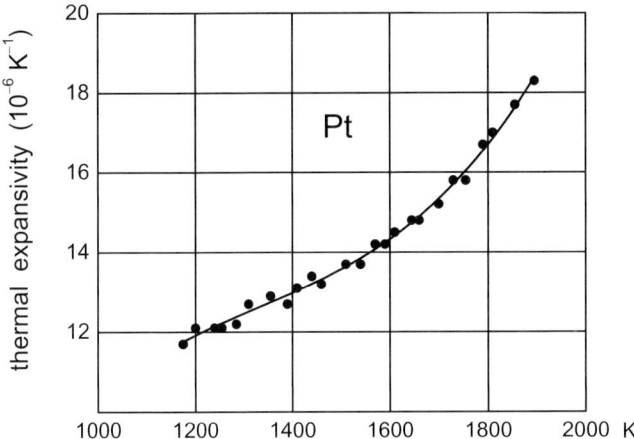

Fig. 1.5. Thermal expansivity of platinum (Kraftmakher 1967b). The advantages of direct measurements of thermal expansivity are clearly seen (Chap. 8)

- Large variety of partially very complex experimental accomplishments, often with insufficient description of details, makes it difficult to properly evaluate the results.

The modulation principle is also powerful for studying some other thermophysical properties. Measurements of oscillations in the length of the sample permit direct determination of thermal expansivity (Fig. 1.5). This technique enables one to circumvent the main difficulty of high-temperature dilatometry caused by the plastic deformation of the samples and to significantly improve the temperature resolution. The temperature derivative of resistance is available by registering oscillations of the resistance of the sample caused by the temperature oscillations. Direct measurements of this quantity more reliably reveal the behaviour of electrical resistivity, e.g., near phase transitions (Fig. 1.6). Measuring temperature oscillations by two thermocouples provides direct comparison of their thermopowers (Fig. 1.7). The modulation method was applied also to measurements of spectral absorptance.

High sensitivity and unique temperature resolution are characteristic of all modulation techniques. Both features are due to the periodic nature of the temperature changes. Employment of selective amplifiers and/or lock-in detectors reduces any influence of noise and interference. In contrast to pulse or dynamic calorimetry, the modulation technique is a steady-state method: the amplitude and the phase of the temperature oscillations in the sample, under constant experimental conditions, do not depend on time. Electronic equipment for modulation measurements (oscillators, amplifiers, oscilloscopes, lock-in detectors) is widely used in fundamental and applied studies and is readily accessible. It is therefore possible to perform the measurements using common scientific instruments, as well as data-acquisition systems and computers for controlling the measurements and processing the data.

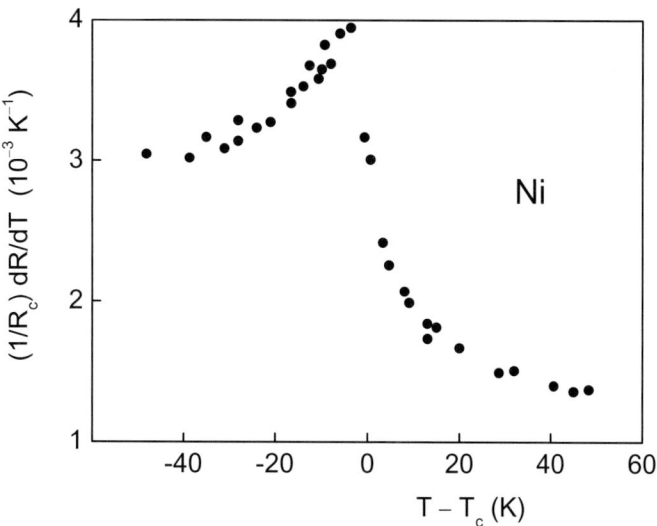

Fig. 1.6. Modulation measurements have shown that the temperature derivative of electrical resistance of nickel near its Curie point behaves like the specific heat (Kraftmakher 1967a). This approach has been also used in other studies of phase transitions in metals and solid electrolytes (Sect. 9.1 and Chap. 14)

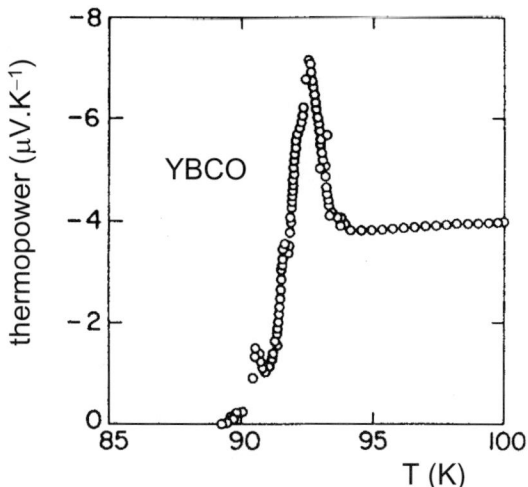

Fig. 1.7. Temperature dependence of thermopower for an $YBaCu_3O_{7-\delta}$ crystal (Howson et al. 1990) measured with modulation technique. This method was proposed more than thirty years ago and several times reinvented (Sect. 9.2 and Chap. 14)

Until today, only modulation calorimetry gained recognition by the scientific community. Other modulation methods are still waiting for a wider practical use.

Some modulation techniques were reinvented several times. This fact is a convincing confirmation of their merit. On the other hand, the principle of modulation is well known, and its application to studying various physical properties seems to be very natural and even ordinary. Probably, not all features of the modulation techniques are disclosed up to date.

Review papers and book chapters in this field are listed below. Table 1.1 briefly presents the long history of modulation calorimetry and related techniques.

Review Papers and Book Chapters on Modulation Calorimetry and Related Techniques

Filippov LP (1966) Methods of simultaneous measurement of heat conductivity, heat capacity and thermal diffusivity of solid and liquid metals at high temperatures. Int J Heat Mass Transfer 9:681–691

Filippov LP (1967) Measurement of Thermal Properties of Solid and Liquid Metals at High Temperatures, in Russian. University Press, Moscow

Kraftmakher YA (1973) Modulation method for measuring specific heat. High Temp – High Press 5:433–454

Kraftmakher YA (1973) Modulation methods for studying thermal expansion, electrical resistivity, and the Seebeck coefficient. High Temp – High Press 5:645–656

Kraftmakher YA (1978) Modulation method for studying thermal expansion, in: Peggs IA (ed) Thermal Expansion 6. Plenum, New York, pp 155–164

Hatta I, Ikushima AJ (1981) Studies on phase transitions by AC calorimetry. Jpn J Appl Phys 20:1995–2011

Filippov LP (1984) Measurement of Thermophysical Properties by Methods of Periodic Heating, in Russian. Energoatomizdat, Moscow

Kraftmakher YA (1984) Modulation calorimetry, in: Maglić KD, Cezairliyan A, Peletsky VE (eds) Compendium of Thermophysical Property Measurement Methods, vol. 1. Plenum, New York, pp 591–641

Garland CW (1985) High-resolution AC calorimetry and critical behaviour at phase transitions. Thermochim Acta 88:127–142

Kraftmakher YA (1988) Modulation calorimetry, in: Ho CY (ed) Specific Heat of Solids. Hemisphere, New York, pp 299–321

Kraftmakher YA (1989) Features of modulation dilatometry. Meas Tech 32:553–555

Kraftmakher YA (1992) Practical modulation calorimetry, in: Maglić KD, Cezairliyan A, Peletsky VE (eds) Compendium of Thermophysical Property Measurement Methods, vol. 2. Plenum, New York, pp 409–436

Kraftmakher Y (1992) Advances in modulation calorimetry and related techniques. High Temp – High Press 24:145–154

Huang CC, Stoebe T (1993) Thermal properties of 'stacked hexatic phases' in liquid crystals. Adv Phys 42:343–391

Kraftmakher Y (1994) Relaxation phenomena caused by equilibration of point defects. Int J Thermophys 15:983–991

Finotello D, Iannacchione GS (1995) High resolution calorimetric studies at phase transitions of alkylcyanobiphenyl liquid crystals confined to submicron size cylindrical cavities. Int J Mod Phys B 9:2247–2283

Kraftmakher Y (1996) Observation of equilibration phenomena in thermophysical properties of metals. Int J Thermophys 17:1137–1149

Hatta I (1997) History repeats itself: Progress in AC calorimetry. Thermochim Acta 300:7–13

Schick C, Höhne GWH (eds) (1997) Temperature modulated calorimetry. Thermochim Acta 304/305 (special issue)

Gmelin E (1997) Classical temperature-modulated calorimetry: A review Thermochim Acta 304/305:1–26

Hatta I (1997) Potentiality of an AC calorimetric method in the study of phase transitions. Thermochim Acta 304/305:27–34

Birge NO, Dixon PK, Menon N (1997) Specific heat spectroscopy: Origin, status and applications of the 3ω method. Thermochim Acta 304/305:51–66

Jeong Y-H (1997) Progress in experimental techniques for dynamic calorimetry. Thermochim Acta 304/305:67–98

Minakov AA (1997) Low-temperature AC calorimetry: Possibilities and limitations. Thermochim Acta 304/305:165–170

Finotello D, Qian S, Iannacchione GS (1997) AC calorimetric studies of phase transitions in porous substrates. Superfluid helium and liquid crystals. Thermochim Acta 304/305:303–316

Hatta I, Nakayama S (1998) First-order phase transitions studied by temperature-modulated calorimetry. Thermochim Acta 318:21–27

Kraftmakher Y (1998) Equilibrium vacancies and thermophysical properties of metals at high temperatures. Physics Reports 299:79–188

Menczel JD, Judovits L (eds) (1998) Temperature-modulated differential scanning calorimetry. J Therm Anal Calorim 54:409–704 (special issue)

Reading M (1998) A personal perspective on the rise of MTDSC. J Therm Anal Calorim 54:411–418

Kraftmakher Y (2000) Lecture Notes on Equilibrium Point Defects and Thermophysical Properties of Metals. World Scientific, Singapore

Wunderlich B (2000) Temperature-modulated calorimetry in the 21st century. Thermochim Acta 355:43–57

Jeong Y-H (2001) Modern calorimetry: going beyond tradition. Thermochim Acta 377:1–7

Reading M (2001) The use of modulated temperature programs in thermal methods. J Therm Anal Calorim 64:7–14

Kraftmakher Y (2002) Modulation calorimetry and related techniques. Physics Reports 356:1–117

Wunderlich B (2003) The thermal properties of complex, nanophase-separated macromolecules as revealed by temperature-modulated calorimetry. Thermochim Acta 403:1–13

Table 1.1. Brief presentation of the long history of modulation techniques

Item	Reference
Theory, supplementary-current and 3ω methods	Corbino 1910, 1911
Use of thermionic current	Smith and Bigler 1922
Specific heat of tungsten up to 2600 K	Zwikker 1928
3ω technique, probe inside a medium	Filippov 1960; Rosenthal 1961, 1965
Equivalent-impedance method	Kraftmakher 1962
Specific heat of tungsten, 1500–3600 K	Kraftmakher and Strelkov 1962
Photoelectric detectors	Lowenthal 1963
Molten metals, high temperatures	Akhmatova 1965, 1967
Electron-bombardment heating	Filippov and Yurchak 1965
Modulation dilatometry	Kraftmakher and Cheremisina 1965
Analysis of resistively heated wires	Holland and Smith 1966
Measurement at various frequencies	Smith 1966
Low temperatures, thermal coupling	Sullivan and Seidel 1966, 1967, 1968
Modulated-light heating	Handler, Mapother, Rayl 1967
Modulation measurement of dR/dT	Kraftmakher 1967a; Salamon et al. 1969
Induction heating	Filippov and Makarenko 1968
Nonconducting materials	Glass 1968
Modulation measurement of thermopower	Freeman and Bass 1970; Hellenthal and Ostholt 1970; Kraftmakher and Pinegina 1970
Microcalorimetry, low temperatures	Zally and Mochel 1971, 1972
Temperatures down to 0.3 K	Manuel et al. 1972
High pressures, low temperatures	Chu and Knapp 1973
Nonadiabatic regime	Varchenko and Kraftmakher 1973
High pressures, high temperatures	Filippov et al. 1976
Liquid crystals	Schantz and Johnson 1978
Organic liquids	Smaardyk and Mochel 1978
Organic and biological materials	Tanasijczuk and Oja 1978
Active thermal shield	Kraftmakher and Cherepanov 1978
Improvement of modulated-light technique	Ikeda and Ishikawa 1979
Observation of temperature fluctuations	Kraftmakher and Krylov 1980ab
Modulation frequencies up to 10^5 Hz	Kraftmakher 1981
Specific-heat spectroscopy, 0.01–6000 Hz	Birge and Nagel 1985, 1987
Freestanding thin films	Pitchford et al. 1986
Dilatometry of nonconducting materials	Johansen 1987
Modulated-bath calorimetry	Graebner 1989
Noncontact calorimetry	Monazam et al. 1989
Modulated differential scanning calorimetry	Gill et al. 1993; Reading et al. 1993
Nanocalorimetry, low temperatures	Fominaya et al. 1997ab; Riou et al. 1997
Noncontact calorimetry in space	R. Wunderlich et al. 1997, 2001; Egry 2000; Egry et al. 2001
Photopyroelectric method, up to 10^5 Hz	Chirtoc et al. 2001; Bentefour et al. 2003
Pressures up to 22 GPa, 0.3–10 K	Wilhelm and Jaccard 2002ab
3ω calorimeter, up to 30 kHz	Jung et al. 2003
Integrated circuit calorimeter	Merzlyakov 2003
Magnetic-field-angle dependent heat capacity	Park et al. 2003

2 Theory of Modulation Calorimetry

The theory of modulation calorimetry is simple and straightforward. When the power heating the sample is modulated by a sine wave and equals $p_0 + p\sin\omega t$, the temperature of the sample starts to oscillate around a mean value T_0. Clearly, the amplitude of the temperature oscillations depends on the heat capacity of the sample. This is the basic point discovered by O.M.Corbino about a century ago.

2.1 Basic Equation of Modulation Calorimetry

The basic theory of modulation calorimetry was developed by Corbino (1910, 1911). Corbino considered a wire sample heated by an AC current passing through it and derived an expression for the temperature oscillations in the sample. This analysis is also valid for other methods of modulation of the heating power.

2.1.1 Sample in a Temperature-Modulated Regime

For a short interval Δt, during which the quantities involved remain constant, the heat balance equation takes the form

$$(p_0 + p\sin\omega t)\Delta t = mc\Delta T + Q(T)\Delta t, \qquad (2.1)$$

where m, c and T are the mass, specific heat and temperature of the sample, $Q(T)$ is the power of heat losses from the sample, $\omega = 2\pi f$ is the angular modulation frequency, and ΔT is the change of the temperature during the interval Δt. The equation has a very simple meaning: heat input = heat accumulated in the sample + heat losses. By assuming $T = T_0 + \Theta$, $\Theta \ll T_0$, and taking $Q(T) = Q(T_0) + Q'\Theta$ (where $Q' = dQ/dT$ is called the heat transfer coefficient), one obtains

$$mc\Theta' + Q(T_0) + Q'\Theta = p_0 + p\sin\omega t, \qquad (2.2)$$

where $\Theta' = d\Theta/dt$. The steady-state solution to this equation is

$$Q(T_0) = p_0, \qquad (2.3\text{a})$$

$$\Theta = \Theta_0\sin(\omega t - \phi), \qquad (2.3\text{b})$$

$$\Theta_0 = (p/mc\omega)\sin\phi = (p\cos\phi)/Q' = p/(m^2c^2\omega^2 + Q'^2)^{1/2}, \quad (2.3c)$$

$$\tan\phi = mc\omega/Q', \quad (2.3d)$$

where ϕ is a phase shift between the oscillations of the power dissipated in the sample and the temperature oscillations. It is possible to present these results as a frequency dependence of Θ_0 and $\tan\phi$ (Fig. 2.1).

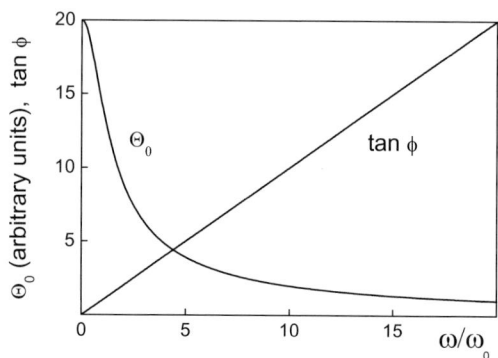

Fig. 2.1. Frequency dependence of the amplitude and the phase of temperature oscillations. The modulation frequency ω_0 is a frequency, for which $\tan\phi = 1$. An adiabatic regime of the measurements is achieved when $\tan\phi \geq 10$

An analogy exists between a sample subjected to modulated heat input and an integrating RC circuit fed by a voltage containing both DC and AC components. The condition $\tan\phi \gg 1$ ($\sin\phi \cong 1$) is a criterion of the so-called adiabatic regime. More rigorously, it should be called quasiadiabatic: the adiabaticity relates only to the AC component of the heating power. When the phase shift ϕ is close to 90°, corrections for heat losses become negligible.

2.1.2 Basic Equation

An adiabatic regime means that the oscillations of the heat losses due to the temperature oscillations in the sample are much smaller than the oscillations of the heating power. Although the heat transfer coefficient grows rapidly with temperature, the regime of the measurements can be kept adiabatic by increasing the modulation frequency. Under adiabaticity conditions, the heat capacity of the sample equals

$$mc = p/\omega\Theta_0, \quad (2.4)$$

which is the basic equation of modulation calorimetry. More generally,

$$mc = (p/\omega\Theta_0)\sin\phi. \quad (2.5)$$

This relation shows how to determine the heat capacity under a nonadiabatic regime. In such cases, it is sufficient to determine the phase shift between the oscillations of the power dissipated in the sample and the temperature oscillations. The temperature oscillations can also be presented as a polar diagram $\Theta_0(\phi)$ (Fig. 2.2).

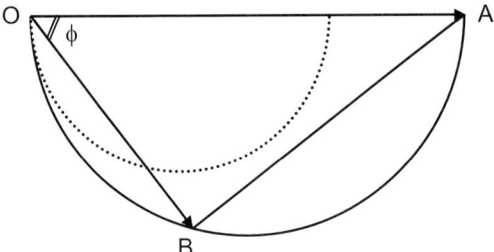

Fig. 2.2. Polar diagram $\Theta_0(\phi)$. The vector **OA** corresponds to $\omega = 0$, $\phi = 0$, and equals p/Q'. The vector **OB** represents the temperature oscillations in a nonadiabatic regime. In an adiabatic regime, it should be nearly perpendicular to **OA**. The dotted circle corresponds to temperature oscillations when the sample is immersed in a dense medium (Rosenthal 1965)

2.1.3 Complex-Form Presentation

The temperature oscillations written in a complex form show both the amplitude and the phase of the temperature oscillations:

$$\Theta = p/(Q' + imc\omega) = \Theta_0 \exp(-i\phi), \qquad (2.6a)$$

$$\Theta_0 = (p/mc\omega)\sin\phi, \qquad (2.6b)$$

$$\tan\phi = mc\omega/Q'. \qquad (2.6c)$$

Equation (2.6a) is a complex-form presentation of Eqs. (2.3b)–(2.3d). Under adiabatic conditions ($Q' \ll mc\omega$),

$$\Theta = -ip/mc\omega = \Theta_0 \exp(-i\pi/2), \qquad (2.7a)$$

$$\Theta_0 = p/mc\omega. \qquad (2.7b)$$

2.1.4 Temperature Oscillations in Resistively Heated Wires

When an electric current heats a wire sample, it is necessary to take into account the temperature dependence of the resistance of the sample as it was done by Corbino (1910). Hence, $R = R_0 + R'\Theta$, where R_0 and R' are the resistance of the sample and its temperature derivative at the mean temperature. When a DC current I_0 superimposed by a small AC component $i\sin\omega t$ passes through the sample, the electric power dissipated in it equals

$$P = I_0^2 R_0 + I_0^2 R'\Theta + 2I_0 i R_0 \sin\omega t$$
$$+ 2I_0 i R'\Theta \sin\omega t + i^2 R_0 \sin^2\omega t + i^2 R'\Theta \sin^2\omega t. \tag{2.8}$$

The last three terms are insignificant because $i \ll I_0$ and $R'\Theta \ll R_0$. In addition, these terms correspond to power oscillations of higher frequencies, while selective amplifiers tuned to the fundamental frequency and/or lock-in amplifiers usually serve for the measurements. The heat-balance equation for the oscillating part of the heating power is given by

$$mc\Theta' + (Q' - I_0^2 R')\Theta = 2I_0 i R_0 \sin\omega t, \tag{2.9}$$

and the solution to it differs from Eqs. (2.3b)–(2.3d) only by the phase shift:

$$\Theta = \Theta_0 \sin(\omega t - \varphi), \tag{2.10a}$$

$$\Theta_0 = (2I_0 i R_0 / mc\omega)/\sin\varphi, \tag{2.10b}$$

$$\tan\varphi = mc\omega/(Q' - I_0^2 R'), \tag{2.10c}$$

where φ is the phase shift between the AC component of the heating current and the temperature oscillations. It was assumed here that the current passing through the sample does not depend on the resistance of the sample because of high internal resistance of the source. On the other hand, when the internal resistance of the source is negligibly small, then

$$\Theta = \Theta_0 \sin(\omega t - \psi), \tag{2.11a}$$

$$\Theta_0 = (2U_0 U / R_0 mc\omega) \sin\psi, \tag{2.11b}$$

$$\tan\psi = mc\omega/(Q' + I_0^2 R'), \tag{2.11c}$$

where U_0 and U are the DC and AC components of the voltage applied to the sample, and ψ is the phase difference between the AC component and the temperature oscillations. Usually, $I_0^2 R'$ is several times smaller than Q'. The difference between the phase shifts ϕ, φ and ψ is of no importance when the measurements are performed in an adiabatic regime ($mc\omega \gg Q'$), and there is no need to take into account the temperature dependence of the resistance of the sample. This difference becomes important only in a nonadiabatic regime.

Under direct electrical heating, axial and radial temperature gradients may appear in the samples. However, Holland and Smith (1966) have shown that for thin samples and under proper modulation frequencies the gradients are sufficiently small. The authors gave expressions for the magnitude and phase of all harmonic components of the temperature oscillations. Radial and end effects were treated as corrections to the solutions for a long thin wire. It turned out that corrections may be required if the temperature coefficient of the resistance of the sample is very large. Second, the surface temperature oscillations may differ from those in the interior. Fortunately, such corrections may become meaningful only in special, rather exotic cases.

2.2 Thermal Coupling in Calorimeter Cell

When using separate heaters and thermometers, one has to take into account thermal coupling in a calorimeter cell, as well as the finite thermal conductance of the sample.

2.2.1 Sullivan-Seidel Analysis

Sullivan and Seidel (1968) considered a heater (heat capacity C_h, temperature T_h), a sample (C_s, T_s), and a thermometer (C_t, T_t) interconnected by thermal conductances K_h and K_t (Fig. 2.3).

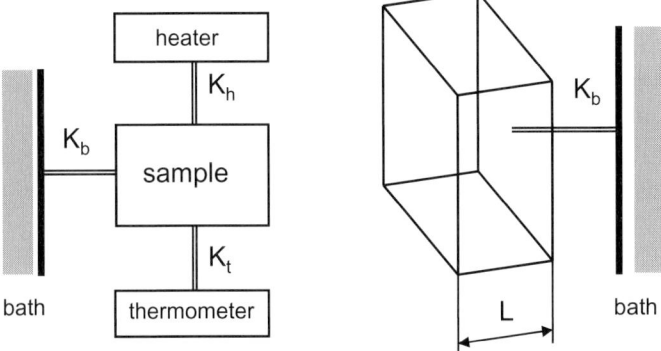

Fig. 2.3. Presentation of calorimeter cell consisting of a heater, a sample, and a thermometer. As a rule, the modulation period is much longer than the equilibration time inside the cell and much shorter than that between the sample and the bath (Sullivan and Seidel 1968)

An AC power applied to the heater provides an oscillating heat input to the sample. The heat flows through the sample and out to the heat sink (bath) via a thermal link K_b. First, the thermal conductance of the sample was assumed to be infinite. The heat-balance equations for the system are as follows ($T' = dT/dt$):

$$C_h T_h' = p_0 + p\sin\omega t - K_h(T_h - T_s), \qquad (2.12a)$$

$$C_s T_s' = K_h(T_h - T_s) - K_b(T_s - T_b) - K_t(T_s - T_t), \qquad (2.12b)$$

$$C_t T_t' = K_t(T_s - T_t). \qquad (2.12c)$$

The temperature oscillations are sufficiently small to consider the parameters C and K to be constant. The steady-state solution to these equations consists of two terms, a constant term depending upon K_b, and an oscillatory term inversely proportional to the heat capacity of the calorimeter cell:

$$T_t = T_b + p_0/K_b + (pB/\omega C)\sin(\omega t - \phi), \tag{2.13}$$

where $C = C_s + C_h + C_t$, B is a complicated expression involving quantities from Eqs. (2.12a)–(2.12c), and ϕ is a phase shift close to 90° under conditions listed below.

If (i) the heat capacities of the heater and of the thermometer are much smaller than that of the sample, (ii) the sample, the heater, and the thermometer come to equilibrium in a time much shorter than the modulation period, and (iii) the modulation period is much shorter than the sample-to-bath relaxation time τ_s, then, to first order in $1/\omega^2\tau_s^2$ and $\omega^2(\tau_h^2 + \tau_t^2)$,

$$B = [1 + 1/\omega^2\tau_s^2 + \omega^2(\tau_h^2 + \tau_t^2)]^{-\frac{1}{2}}, \tag{2.14a}$$

$$\cot\phi = 1/\omega\tau_s - \omega(\tau_h + \tau_t), \tag{2.14b}$$

where the relaxation times are defined as $\tau_s = C/K_b$, $\tau_h = C_h/K_h$, and $\tau_t = C_t/K_t$.

2.2.2 Thermal Conductance of the Sample

To take into account the finite thermal conductance of the sample, Sullivan and Seidel (1968) considered a sample in the form of a slab heated uniformly on one side by a sine heat flux (Fig. 2.3). The other side of the slab is coupled to the bath through the thermal conductance K_b. The thermal diffusivity of the sample is $D = \kappa/\rho c$, where κ is the thermal conductivity, c is the specific heat, and ρ is the density of the sample. The heater at $x = 0$ and the thermometer at $x = L$ were assumed to be coupled to the sample with very short relaxation time. Under the conditions that $K_s \gg K_b$ (where K_s is the thermal conductance of the sample in the direction of heat flow), the temperature oscillations in the slab at $x = L$ were found to be

$$\Theta_0 = p/(1 + 1/\omega^2\tau_s^2 + \omega^2\tau_i^2 + 2K_b/3K_s)^{\frac{1}{2}}\omega C_s, \tag{2.15}$$

where τ_i is an internal time constant, which was calculated to be $L^2/90^{\frac{1}{2}}D$.

The finite thermal conductance of the sample thus results in a correction term, $\omega^2\tau_i^2$, of the same form as that arising from the finite thermal conductance between the sample and the thermometer and heater. Therefore, the amplitude of the temperature oscillations measured at $x = L$ is

$$\Theta_0 = p/(1 + 1/\omega^2\tau_s^2 + \omega^2\tau^2 + 2K_b/3K_s)^{\frac{1}{2}}\omega C, \tag{2.16}$$

where various time constants are lumped into $\tau = (\tau_h^2 + \tau_t^2 + \tau_i^2)^{\frac{1}{2}}$. As a rule, τ_s is two to three orders of magnitude larger than τ, so the condition $\omega^2\tau_s^2 \gg 1 \gg \omega^2\tau^2$ is easily achievable by a proper choice of the modulation frequency. The term $2K_b/3K_s$ is small owing to the small thickness of the sample. It is difficult to fulfil the basic requirements of modulation calorimetry only in one case, when samples of low thermal conductivity are studied under high pressures, and a dense material is used as pressure-transmitting medium.

2.2.3 Model Including a Substrate

When measuring specific heat under high pressures, Eichler and Gey (1979) considered separate heat links to the bath from a sample, a heater, and a thermometer (Fig. 2.4). These heat links should be taken into account when a medium of high thermal conductivity surrounds the calorimeter cell.

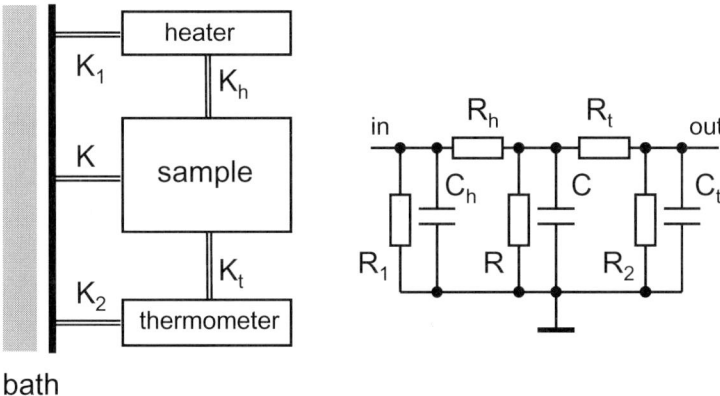

Fig. 2.4. Presentation of calorimeter cell containing a substrate, a sample, a heater, and a thermometer, and its equivalent electrical circuit (Eichler and Gey 1979). The resistances R are inversely proportional to the thermal conductances K

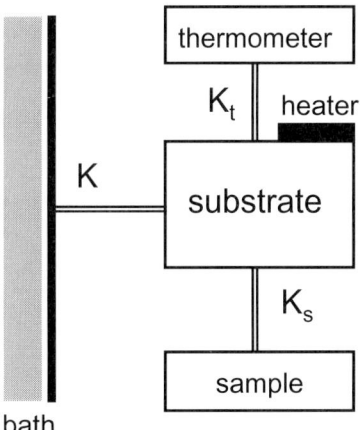

Fig. 2.5. Model of modulation calorimeter analysed by Velichkov (1992). The thermometer is assumed to have good contact with the substrate

Velichkov (1992) analysed a model of a modulation calorimeter containing a substrate, a heater, a sample, and a thermometer (Fig. 2.5). Assuming the thermometer to have good thermal contact to the substrate and negligible heat capacity, the author obtained mathematical solutions concerning the stationary amplitude and phase of the temperature oscillations measured by the thermometer. He presented the results as $\omega\Theta_0$ and $\tan\phi$ versus the modulation frequency. Velichkov (1992) stressed the possibility to perform the measurements in a nonadiabatic regime, where $\tan\phi < 10$. Earlier, such a regime was explored, and it was shown that measurements in a nonadiabatic regime provide about the same accuracy as in an adiabatic regime (Varchenko and Kraftmakher 1973). The nonadiabatic regime is considered in Sect. 6.1.2.

2.2.4 Choice of Modulation Frequency

Modulation measurements are usually performed under adiabatic conditions, so that the expression for calculating the heat capacity has the simple form given by Eq. (2.4). When the conditions of the measurements are adequate to the theoretical model, the quantity $\omega\Theta_0$ should not depend on the modulation frequency. To check this prediction, one plots a corresponding graph (Fig. 2.6).

The decrease of the quantity $\omega\Theta_0$ at low frequencies shows that the criterion of adiabaticity is not satisfied, while the decrease at high frequencies is due to the thermal inertia of the temperature sensor, a thermocouple or a resistance thermometer. Usually, the operating frequency is chosen inside an interval, where the quantity $\omega\Theta_0$ remains constant. However, a nonadiabatic regime of the measurements is also applicable when necessary. In modulation measurements of other thermophysical properties, the criterion of adiabaticity is of no importance, and the choice of the modulation frequency depends on other considerations. In any case, the temperature sensor must have low thermal inertia allowing correct measurements of the temperature oscillations.

It is worth remembering that solutions of basic problems related to propagation of heat in solids, including temperature waves, were given by Carslaw and Jaeger (1959).

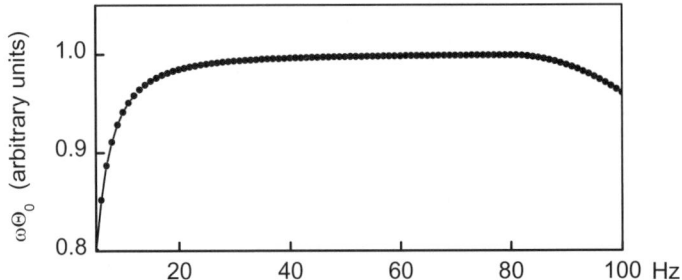

Fig. 2.6. Frequency dependence of $\omega\Theta_0$, schematically. Usually, the modulation frequency is chosen in a range, where this quantity is constant

3 Modulation of Heating Power

The choice of a method to periodically heat the sample depends on its shape and electrical conductivity, and on the temperature range of the measurements. At high temperatures, direct electric heating or electron bombardment are preferable for conducting samples. In studies of nonconducting samples, separate resistive heaters or modulated-light heating should be used. Various methods of the heating provide different accuracy. For example, the modulated-light heating is usable, where there is no need to accurately determine absolute values of the specific heat. No data on heating-power oscillations are necessary for modulation measurements of thermal expansivity, temperature derivative of resistance, and thermopower.

3.1 Direct Electric Heating and Resistive Heaters

Heating by a current passing through a sample or by separate resistive heaters ensures high accuracy of the measurements of the power heating the sample, which is necessary for the determination of absolute values of the specific heat.

3.1.1 Direct Heating

As a rule, conducting samples are heated by passing an electric current through them. Several methods are usable for modulating the heating power (Figs. 3.1 and 3.2):

• Heating by an AC current. The modulation frequency is twice the frequency of the current. The amplitude of the power oscillations equals the effective electric power. If the current is the only means to heat the sample in a wide temperature range, the amplitude of the temperature oscillations depends strongly on the temperature. When a sample of resistance R is heated by an AC current $I\sin\omega t$, the electric power dissipated in the sample equals $P = I^2 R \sin^2 \omega t = \frac{1}{2} I^2 R(1 - \cos 2\omega t)$. The amplitude of the oscillations in the power heating the sample thus equals the mean applied power. In an adiabatic regime, the temperature oscillations created in the sample are $\Theta = -\Theta_0 \sin 2\omega t$, $\Theta_0 = I^2 R / 2mc\omega$. The supplementary-current method and the third-harmonic technique are applicable for measuring the temperature oscillations.

• Heating by a DC current I_0 with a small AC component $i\sin\omega t$ superimposed ($I \ll I_0$). The modulation frequency equals that of the AC component of the heating current. The mean temperature and the amplitude of the temperature oscillations in the sample are controlled independently. The oscillations of the power dissipated in the sample equal $2I_0iR\sin\omega t$. An important advantage is that the equivalent-impedance method is applicable permitting the heat capacity of the sample to be calculated from parameters of a bridge or potentiometer circuit, which are independent of the amplitude of the AC component of the heating current.

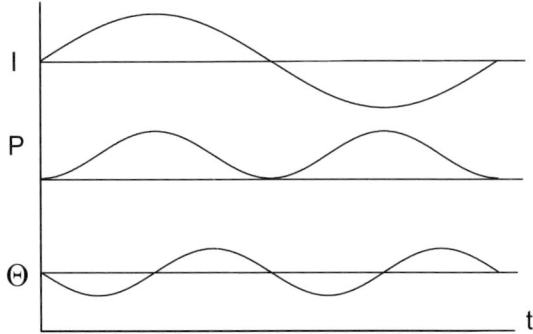

Fig. 3.1. Diagram of the current, power, and temperature oscillations when a sample is heated by AC current

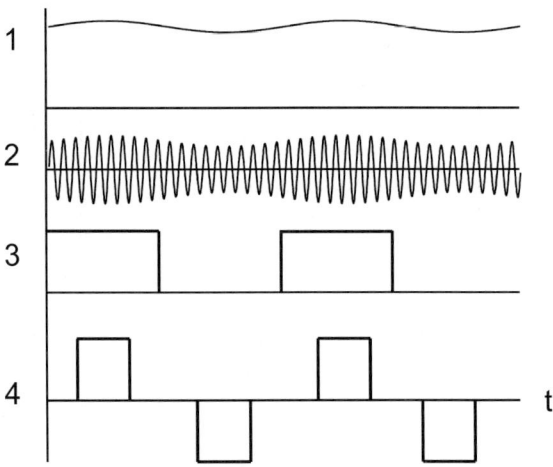

Fig. 3.2. Waveforms of heating current for modulation measurements: 1 – DC current with a small AC component, 2 – modulated high-frequency current, 3 – unipolar square-wave pulses, 4 – pulses of special waveform

- Heating by a high-frequency current modulated by the necessary low frequency. Such heating is useful when the temperature of the sample and the temperature oscillations are measured by a thermocouple connected electrically to the sample. In this case, one easily separates the low-frequency signal related to the temperature oscillations from a high-frequency voltage that may appear in the thermocouple circuit. This method is also useful in studies of relaxation phenomena in specific heat, where temperature oscillations of two frequencies are simultaneously created in the sample, and the specific heats corresponding to the two frequencies are compared.

- Heating by square-wave unipolar pulses. When the modulation period is sufficiently short, the temperature oscillations are of triangular shape, and their period equals that of the pulses. Pulses of special waveform allow avoiding such a coincidence (Jin et al. 1984).
- To obtain a purely sine modulation of the heating power, Schmiedeshoff et al. (1987) used a heating current given by $A(\sin\omega t + 1)^{1/2}$.

3.1.2 Separate Resistive Heaters

Separate resistive heaters for periodic heating are used mainly at low and medium temperatures and in studies of nonconducting samples. The main requirements for the heater are its small heat capacity and good thermal coupling to the sample. Microresistors or thin deposited films often serve as the heaters (Fig. 3.3). Employing an AC current, one accurately measures the oscillations of the applied power. In all the cases of electric heating, it is easy to precisely determine the AC component of the power applied to the sample.

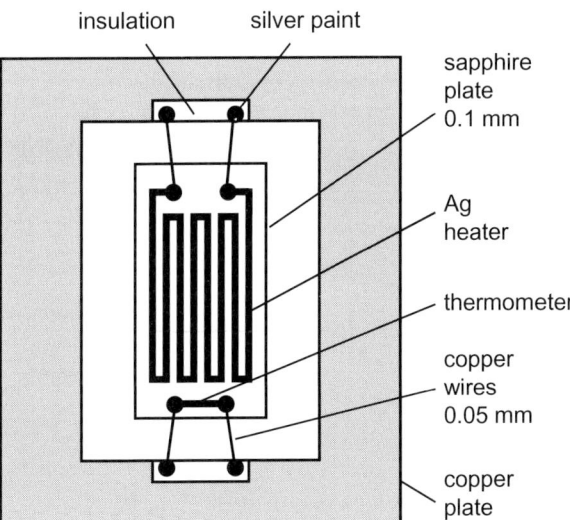

Fig. 3.3. Example of low-temperature calorimeter cell with thin-film heater (Kämpf et al. 1981). A sapphire plate, about 0.1 mm thick, is linked to a cryostat by four copper wires 0.05 mm thick. On the backside of the plate, a very thin Ag heater and a carbon resistance thermometer are prepared

3.2 Modulated-Light Heating

This elegant method, invented for measurements of the specific heat of nickel near the Curie point (Handler et al. 1967), became very popular. A sample in the form of a foil or a thin slab is placed in a furnace, which controls its mean temperature (Fig. 3.4). A light passed through a chopper provides the AC heating power for the sample. A thermocouple and a lock-in amplifier measure temperature oscillations created in the sample, of about 10 mK. A photocell provides the reference voltage for the lock-in amplifier. The output DC signal of the lock-in amplifier is proportional to the input AC voltage from the thermocouple. In an adiabatic regime, this DC signal is inversely proportional to the heat capacity of the sample. It is fed to the Y-input of a plotter. A second thermocouple, with the hot junction inside the furnace, drives the X-input. During the measurements, the temperature of the furnace changes gradually. The choice of the modulation frequency depends on the thickness of the sample and its thermal diffusivity.

Connelly et al. (1971) described this technique in more detail. A nickel single crystal of 3×3×0.1 mm size was placed into a copper block inside a furnace. Helium gas at about 50 kPa ensured the necessary heat exchange. An incandescent lamp powered by a stabilised source provided a radiation constant within 0.4% for several hours. Control of the radiation by a thermopile permitted introducing corrections if necessary. A strip thermocouple of cross-section 70×5 μm measured the temperature oscillations in the sample. The increase in the mean temperature of the sample above the furnace temperature amounted to 0.5 K. In the frequency range 25–60 Hz, the quantity $\omega\Theta_0$ remained constant within 1%. The rate of heating the furnace was adjustable from 0.06 to 0.4 K.min^{-1}.

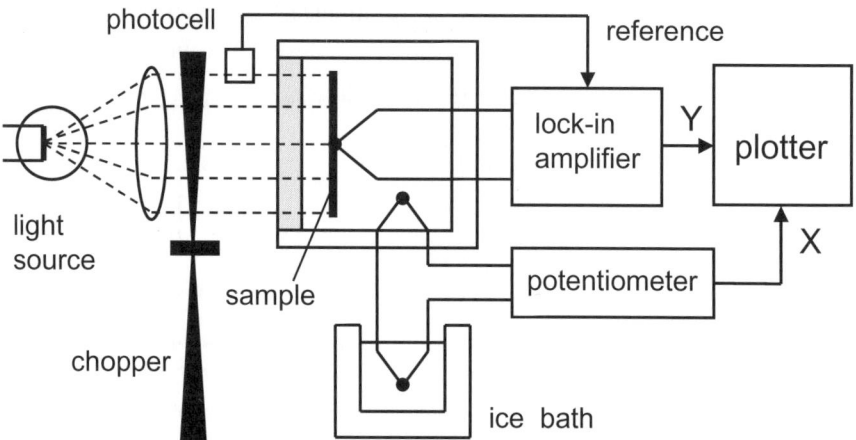

Fig. 3.4. Diagram of set-up employing modulated-light heating (Handler et al. 1967). A thermocouple and a lock-in amplifier measure small temperature oscillations in a thin sample generated by absorption of modulated light. A photocell provides reference voltage for the lock-in amplifier. The mean temperature of the sample changes gradually

In most measurements utilising the modulated-light heating, the amplitude of the heating-power oscillations has not been determined, but precautions have been taken to make it independent of the temperature. For this purpose, one covers the samples with a thin layer of graphite or PbS, whose spectral absorptance is high and does not depend on temperature. To evaluate the oscillations of the heating power caused by modulated light, Salamon (1970) employed the relation between the mean power supplied to the sample and the temperature increment of the sample ΔT: $P = K\Delta T$, where K is the thermal conductance between the sample and the bath. Ikeda and Ishikawa (1979) also used this approach. They considered the thermal conductance to contain two parts, the conductance of the gas surrounding the sample, and the heat flow through other paths such as thermocouples.

The modulated-light heating makes it easy to build a fully automated measuring set-up. Stokka and Fossheim (1982a) described such a calorimeter for the range 2–380 K (Fig. 3.5), with a possibility to apply to the samples uniaxial pressures up to 0.1 GPa. A thermocouple and a lock-in amplifier measure the temperature oscillations in the sample caused by the absorbed light. The second thermocouple measures the temperature difference between the sample and a copper block, while a platinum or germanium thermometer and an automatic AC bridge serve for determining the temperature of the block. A microcomputer provides a flexible program of the measurements. A scanning voltmeter measures the signals corresponding to the mean temperature and the temperature oscillations. All the data are registered by a tape recorder and then processed by a larger computer. Nowadays, automation of modulation measurements became much simpler by using data-acquisition systems and computers.

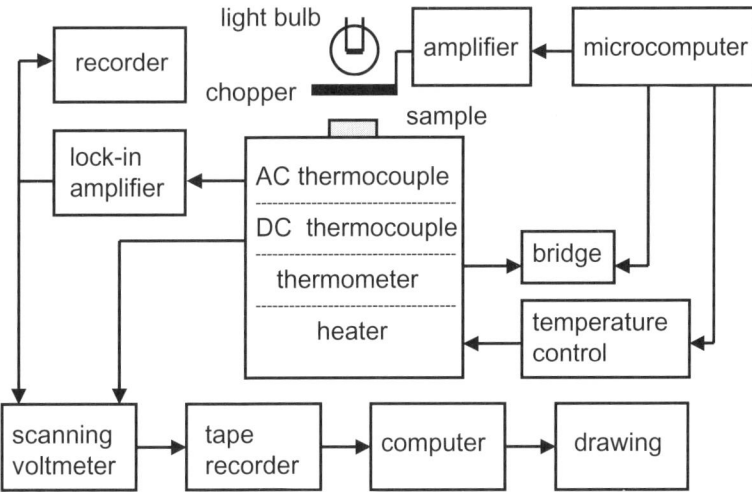

Fig. 3.5. Simplified diagram of automated calorimeter employing modulated-light heating (Stokka and Fossheim 1982a). The calorimeter operates in the range 2–380 K, with a possibility to apply uniaxial pressures up to 0.1 GPa

Regan et al. (1991) used a modulated-light calorimeter with a fibre light guide, and Garfield et al. (1998) described this calorimeter in detail. The optical fibre passes through a vacuum feed-through at one end of a stainless-steel tube. An infrared light-emitting diode serves as the light source. Four contact blocks are glued to a copper heat sink so as not to give electrical contact but to ensure good thermal contact (Fig. 3.6). A platinum thermometer provides good sensitivity in the temperature range 30–300 K. A small heater is wrapped around the heat sink to control temperature. The samples for the measurements have dimensions $1 \times 1 \times 0.05$ mm. Helium gas at a pressure of approximately 1 kPa provides a thermal link between the sample and the heat sink. A flattened thermocouple junction is in the form of a cross. The two thermocouples serve for the detection of the mean temperature and the temperature oscillations in the sample. The thermocouple wires held rigidly the sample. The sample assembly is glued onto the heat sink with GE varnish. A lock-in amplifier measures the AC signal from the thermocouple. The authors pointed out that the calorimeter is not usable for scanning over wide temperature ranges, but only a temperature region around a phase transition.

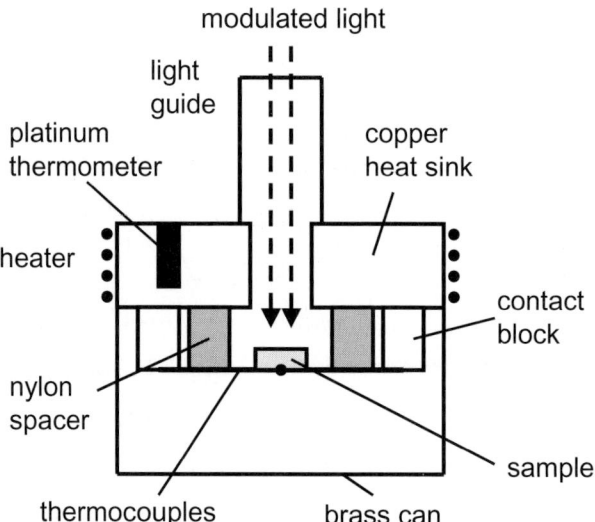

Fig. 3.6. Simplified diagram of calorimeter cell with a light guide described by Garfield et al. (1998). An infrared light-emitting diode provides the AC heating

3.3 Other Methods

Other methods of modulating the heating power include electron bombardment, induction heating, and Peltier heating. The employment of these techniques is very limited.

3.3.1 Electron Bombardment

An advantage of electron-bombardment heating is the possibility to use samples of irregular shape. At the same time, it is easy to determine the AC component of the heating power. The power supplied to the sample could be modulated as follows:

- In the saturation regime, the accelerating voltage is modulated, while the electron-beam current remains constant.
- The temperature of the cathode is modulated or a control electrode is utilised to modulate the electron current, while the accelerating voltage is constant.
- The accelerating voltage is periodically switched on and off, so that square-wave pulses of the heating power are applied to the sample. A drawback of this method is that the power oscillations increase along with the mean heating power.

The power dissipated in the sample equals IU, where I is the electron-beam current, and U is the accelerating voltage. A thermocouple can be either welded to the sample or placed into a thin-wall insulating capillary passing through the sample. Welding ensures better thermal coupling, but a part of the anode current may branch off into the thermocouple circuit. The second method requires longer modulation periods.

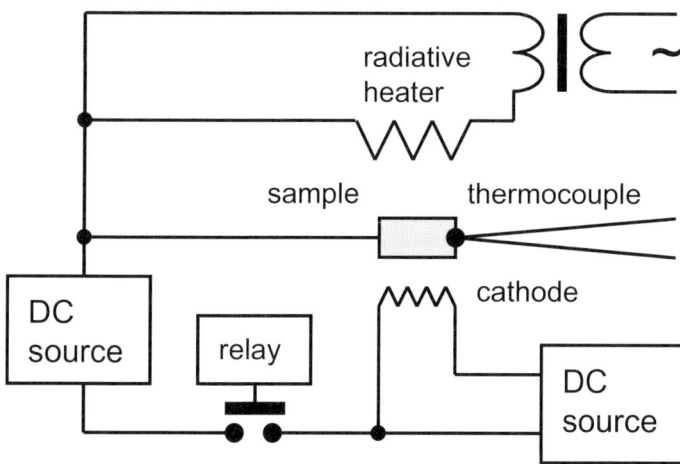

Fig. 3.7. Electron-bombardment heating (Varchenko et al. 1978). With this method, samples of irregular shape are also acceptable. A radiative heater controls the mean temperature of the sample

An additional radiative heater was employed to independently control the mean temperature of the sample and the amplitude of the temperature oscillations (Varchenko et al. 1978). The electron beam was modulated either by a relay powered by an infra-low frequency generator (Fig. 3.7) or by a control electrode between the cathode and the sample. A plotter recorded the temperature oscillations. With this technique, the specific heat of iron was measured in the range 600–1250 K (Fig. 3.8). The sample was 5 mm in diameter and 2 mm thick. A Pt/Pt+10%Rh thermocouple welded to the sample measured the temperature. The modulation period, 5 s, was inside an interval satisfying the adiabaticity criterion. On the other hand, the length of the temperature wave in the sample was one order of magnitude longer than its thickness. The temperature oscillations amount to 0.5 K, while the mean temperature was constant throughout the sample within 1 K.

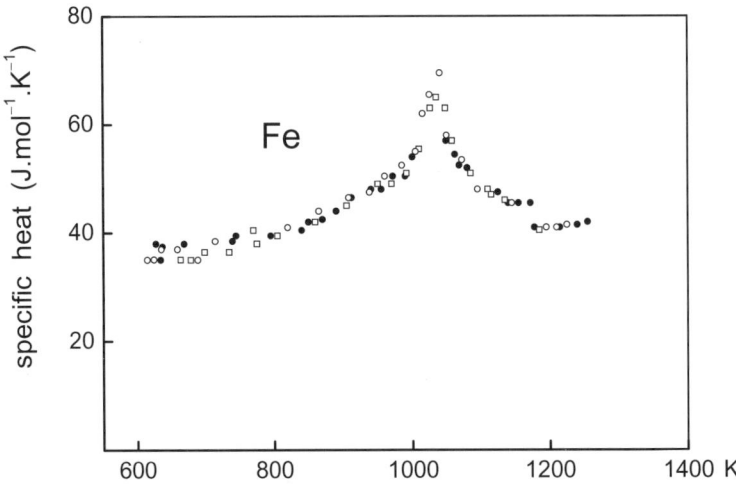

Fig. 3.8. Specific heat of iron from modulation measurements (Varchenko et al. 1978). Various symbols denote data from different runs

3.3.2 Induction Heating

Filippov and Makarenko (1968) introduced the induction heating method. The sample is placed in an induction furnace fed by a high-frequency current modulated by a low frequency (Fig. 3.9). A micropyrometer measures the mean temperature of the sample, while a photomultiplier detects the temperature oscillations. A blackbody cavity in the sample allows one to avoid corrections for the spectral emittance. A separate coil serves for measuring the AC magnetic field, which should be known for evaluating the power dissipated in the sample. The electrical conductivity of the sample, which also must be known, is measured by the four-probe method.

Using induction heating, it is difficult to accurately determine the power dissipated in the sample. Induction heating was employed in noncontact calorimetry (Fecht and Johnson 1991; Fecht and Wunderlich 1994; R.Wunderlich and Fecht 1993, 1996; R.Wunderlich et al. 1993), including modulation measurements in space (R.Wunderlich et al. 1997, 2001; Egry 2000; Egry et al. 2001).

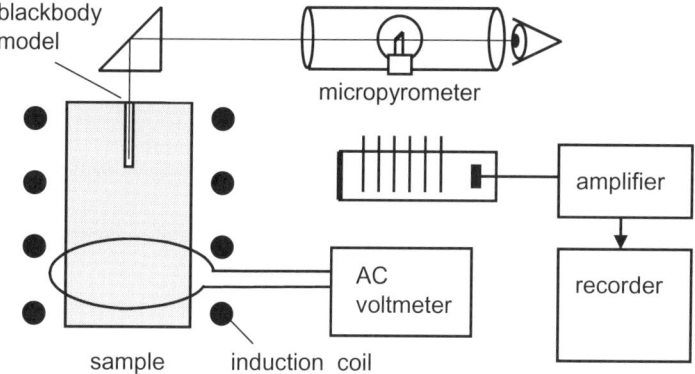

Fig. 3.9. Simplified diagram of arrangement for induction heating (Filippov and Makarenko 1968). The mean temperature of the blackbody model is measured with a micropyrometer

3.3.3 Peltier Heating

This method of heating has an advantage that no DC temperature increment in the sample is introduced. McNeill (1962) employed AC Peltier heating in measurements of thermal diffusivity of thermoelectric materials. Johansen (1987) used a Peltier element for periodically heating the sample in a modulation dilatometer. Moon et al. (2000b) and Jung et al. (2002) described a Peltier microcalorimeter (Fig. 3.10). At 320 K, the frequency range of the calorimeter is 0.001 to 1 Hz.

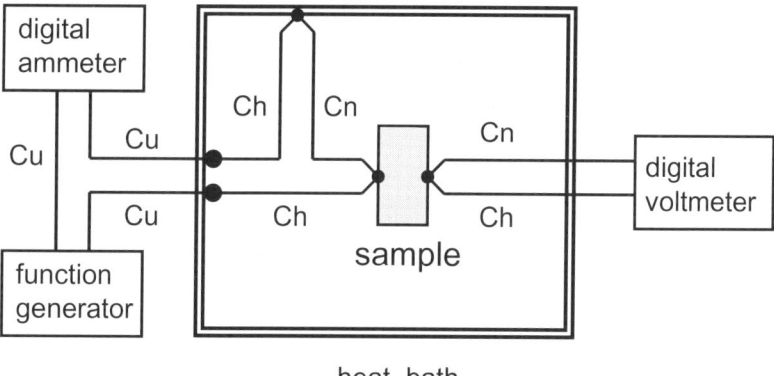

Fig. 3.10. Simplified diagram of microcalorimeter employing Peltier heating (Moon et al. 2000b; Jung et al. 2002). Ch, Cn, and Cu denote chromel, constantan, and copper, respectively. Wires of either diameter 25 μm or 12 μm are used

In conclusion, methods of modulating the heating power necessary for modulation measurements in different temperature ranges are presented (Fig. 3.11). With appropriate samples, direct electric heating is usable over all the temperature range of modulation measurements.

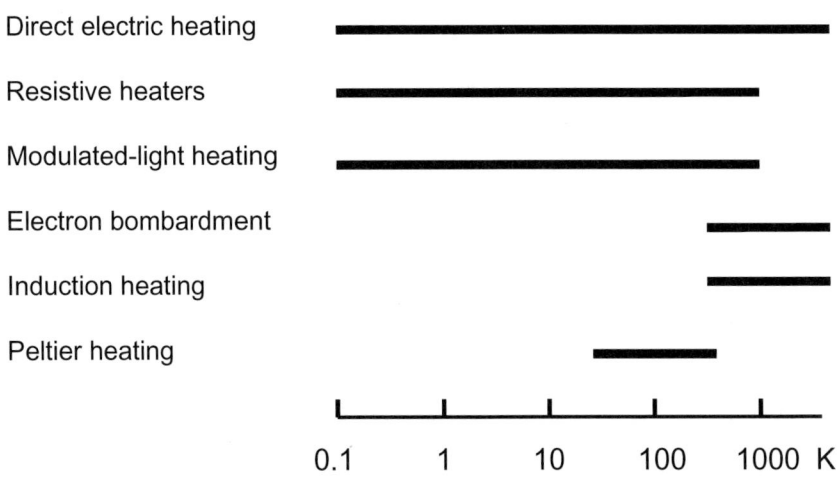

Fig. 3.11. Methods of modulating the heating power in modulation measurements usable in different temperature ranges

4 Measurement of Temperature Oscillations

In modulation measurements, various methods are applicable to detect temperature oscillations. First, they were determined from oscillations in the resistance of the sample, thermionic current, or radiation from it. Later, separate temperature sensors, thermocouples and resistance thermometers, served for this purpose. Oscillations of the thermionic current are usable only in special cases, and this technique is now out of use. The choice of an adequate method to measure the temperature oscillations is very important. It depends on the temperature range, shape and electrical conductivity of the sample, and on the modulation frequency.

4.1 Oscillations in Sample's Resistance

An important advantage of this method is that high modulation frequencies are possible because the problem of thermal inertia of a temperature sensor is avoided. In addition, the measurements are possible at relatively low temperatures, where the radiation from the sample is negligible. The only drawback of the method is the necessity to know the temperature dependence of the resistance of the sample and of its temperature derivative. The method is inapplicable when these quantities exhibit anomalies. Three techniques for measuring temperature oscillations are known: (i) the supplementary-current method, (ii) the third-harmonic technique, and (iii) the equivalent-impedance method. In all the cases, the temperature derivative of the resistance of the sample, $R' = dR/dT$, enters the expressions for the determinations of the temperature oscillations and the heat capacity of the sample.

4.1.1 Supplementary-Current Technique

This technique was discovered by Corbino (1910). He observed a DC voltage drop across the sample heated by an AC current when a small supplementary current of a frequency equal to that of the temperature oscillations passed through the sample. In these observations, the second harmonic of the heating current played role of the supplementary current. The title of the Corbino's paper explicitly tells its contents: "Thermische Oszillationen wechselstromdurchflossener Lampen mit dünnem Faden und daraus sich ergebende Gleichrichterwirkung infolge der Anwesenheit geradzahliger Oberschwingungen." Under adiabatic conditions, a current $I\sin\omega t$ passing through the sample causes temperature oscillations in it $\Theta = -\Theta_0\sin 2\omega t$.

The resistance of the sample is $R_0 - R'\Theta_0\sin2\omega t$, where R' is the temperature derivative of the resistance. When a supplementary current $i\sin(2\omega t - \alpha)$ passes through the sample ($i \ll I$), a voltage across it equals

$$V = [I\sin\omega t + i\sin(2\omega t - \alpha)] \times (R_0 - R'\Theta_0\sin2\omega t). \tag{4.1}$$

The DC component of this voltage equals $(iR'\Theta_0/2)\cos\alpha$. It reaches a maximum when the supplementary current has the same phase as the temperature oscillations ($\alpha = 0$). Clearly, the supplementary current should be several times smaller than the heating current.

Supplementary Current of High Frequency

Gerlich et al. (1965) employed a supplementary current of a frequency much higher than that of the temperature oscillations (Fig. 4.1). A 1-Hz current passes through the sample setting up temperature oscillations at 2 Hz. The oscillations in the resistance of the sample are detected by means of a 320-Hz current. A potentiometer balances the main component of the 320-Hz voltage across the sample. A lock-in amplifier measures the remaining modulated 320-Hz voltage. The 320-Hz oscillator provides also the necessary voltage for the potentiometer and the reference voltage for the amplifier. A 2-Hz signal from the lock-in amplifier proceeds, through a filter, to the Y-input of an oscilloscope. The X-input of the oscilloscope is connected to the oscillator of the frequency 1 Hz. The Lissajous figure is photographed, and the specific heat of the sample is evaluated from its shape and the temperature derivative of the resistance of the sample.

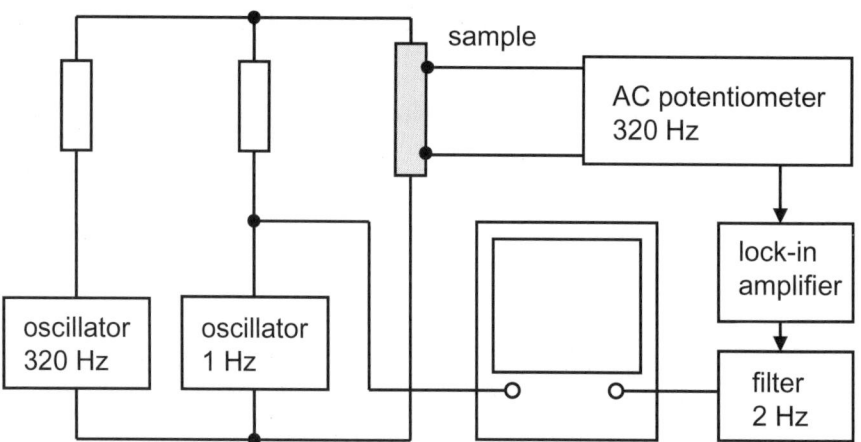

Fig. 4.1. Simplified diagram of set-up utilising supplementary current of frequency much higher than that of temperature oscillations (Gerlich et al. 1965). With this technique, specific heat of germanium, silicon, and Ge-Si alloys was measured in the range 300–1000 K

Frequency Conversion

The frequency-conversion method employs a supplementary current of a frequency close to that of the temperature oscillations (Kraftmakher and Tonaevskii 1972). This results in a difference-frequency component of the voltage across the sample, which is proportional to the oscillations in the resistance of the sample and to the supplementary current. In this case, the supplementary current is $i\sin\omega_1 t$, where $\omega_1 = 2\omega \pm \Omega$ ($\Omega \ll \omega$), and the voltage drop across the sample contains a component of the frequency Ω:

$$V_\Omega = (iR'\Theta_0/2)\sin\Omega t. \qquad (4.2)$$

An advantage of the method is that the difference-frequency signal appears only due to the oscillations in the resistance of the sample. The method was used to directly determine the temperature coefficient of the specific heat, where it was necessary to measure a weak second-harmonic component of the temperature oscillations in the presence of a much stronger fundamental signal (Sect. 13.2). The frequency-conversion method may turn out a good solution to the problem of measuring high-frequency temperature oscillations in specific-heat spectroscopy.

4.1.2 Third-Harmonic Method

This method was also discovered by Corbino (1911). Corbino found that when an AC voltage is applied to a sample, the current through it contains a third-harmonic component proportional to the temperature oscillations in the sample and the temperature derivative of the resistance of the sample. He measured this component by compensating the fundamental-frequency current and observing the Lissajous figure by means of a cathode-ray tube (that time, the Braun tube). Corbino (1911) has shown how to evaluate the heat capacity of the sample, and this is seen from the title of his paper: "Periodische Widerstandsänderungen feiner Metallfäden, die durch Wechselströme zum Gluhen gebracht werden, sowie Ableitung ihrer thermischen Eigenschaften bei hoher Temperatur."

When a current $I\sin\omega t$ passes through a sample, the temperature oscillations in it, under adiabatic conditions, are $\Theta = -\Theta_0\sin 2\omega t$. If the sample is fed through a sufficiently high resistance, the voltage across the sample equals

$$V = (I\sin\omega t)\times(R_0 - R'\Theta_0\sin 2\omega t) =$$
$$= IR_0[\sin\omega t - (\beta\Theta_0/2)\cos\omega t + (\beta\Theta_0/2)\cos 3\omega t], \qquad (4.3)$$

where $\beta = R'/R_0$ is the temperature coefficient of resistance. The third-harmonic voltage across the sample, under adiabatic conditions, is

$$V_{3\omega} = (\beta\Theta_0 IR_0/2)\cos 3\omega t. \qquad (4.4)$$

From this relation, the amplitude of the temperature oscillations in the sample is available. In calorimetric measurements, the third-harmonic component of the voltage across the sample directly relates to the heat capacity of the sample:

$$V_{3\omega} = I^3 R_0 R'/8mc\omega = U^3 R'/8R_0^2 mc\omega, \tag{4.5}$$

where I and U are the amplitudes of the fundamental-frequency current through the sample and of the voltage across it.

Filippov (1960) measured the third-harmonic signal by a bridge circuit with the sample as one of its arms (Fig. 4.2). The bridge is balanced at the fundamental frequency, and the third-harmonic output signal is observed by means of a selective amplifier (not shown in the figure) and an oscilloscope. A variable capacitor shunting one of the arms of the bridge balances the out-of-phase fundamental-frequency voltage caused by the temperature oscillations, i.e., $(\beta\Theta_0 I R_0/2)\cos\omega t$.

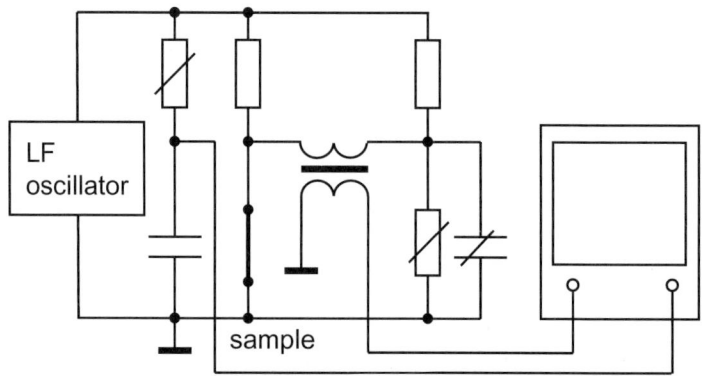

Fig. 4.2. Measurement of the third-harmonic signal (Filippov 1960; Rosenthal 1961, 1965). The variable capacitor balances the quadrature signal of fundamental frequency caused by the temperature oscillations in the sample

Rosenthal (1961, 1965) studied the temperature oscillations in thin wires over a wide frequency range. He presented the data as a polar diagram $\Theta_0(\phi)$. Holland (1963) employed the third-harmonic method for measuring the specific heat of titanium over the range 600–1345 K. At 1155 K, titanium undergoes a phase transition from a hexagonal close-packed low-temperature form to a body-centred cubic form. The sample was a 0.25-mm filament, 40 cm long. Potential probes, 25-μm tungsten wires, were placed 10 cm from each end of the filament. The mean temperature of the sample was computed from the electric power dissipated in the sample. In the temperature range of the measurements, the hemispherical total emissivity of titanium is nearly constant. A Kelvin bridge served for balancing the fundamental voltage, together with any third harmonic originating in the source of the heating current. The amplified output signal proceeded to the moving coil of an electrodynamometer, which operated as a lock-in detector (Fig. 4.3). A frequency tripler provided the current for its fixed coil. The mean torque on the moving coil is proportional to the averaged product of the currents in the two coils. When the current in the fixed coil is sinusoidal, only a current of the same frequency in the moving coil produces a torque with a nonzero time average. Smith (1966) used the third-harmonic technique to measure the specific heat of a germanium whisker in the range 600–900 K, and Skelskey and Van den Sype (1970) in measurements of the specific heat of gold.

Birge and Nagel (1985, 1987) and Birge (1986) used the third-harmonic method in measurements of frequency-dependent specific heat of supercooled liquids over the range 0.01–6000 Hz. Jung et al. (1992) described a fully automated calorimeter employing this technique up to 10 kHz. Recently, Jung et al. (2003) extended the frequency range of the measurements up to 30 kHz. Measurements with different modulation frequencies are aimed at searching for relaxation phenomena in specific heat. The method, often referred to as the 3ω technique, was also used in many other studies (Dixon and Nagel 1988; Dixon 1990; Jeong et al. 1991; Jeong and Moon 1995; Menon 1996; Ema and Yao 1997).

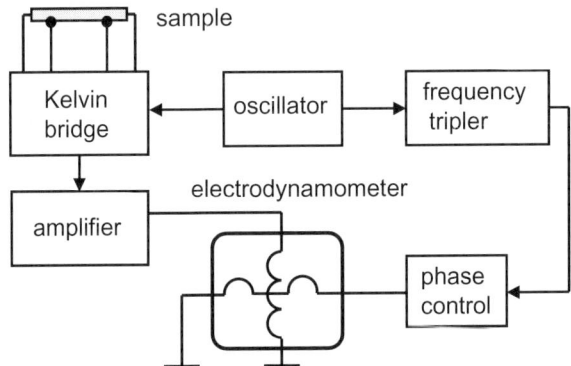

Fig. 4.3. Block diagram of set-up with an electrodynamometer for measuring the third-harmonic signal (Holland 1963; Smith 1966). The electrodynamometer operates as a lock-in detector, providing exact multiplication of two signals

4.1.3 Equivalent-Impedance Technique

Radio engineers found long ago that a temperature-sensitive resistor, through which a DC current is flowing, displays equivalent impedance depending, in particular, on its heat capacity (Griesheimer 1947; Jones 1953; Van der Ziel 1958). However, a time elapsed before thermophysicists realised the advantages of specific-heat measurements based on this principle.

Compensation Circuit

When a current $I = I_0 + i\sin\omega t$ ($i \ll I_0$) heats a wire sample, its resistance follows the relation

$$R = R_0 + R'\Theta = R_0 + R'\Theta_0\sin(\omega t - \varphi) \qquad (4.6)$$
$$= R_0 + R'\Theta_0\cos\varphi \sin\omega t - R'\Theta_0\sin\varphi \cos\omega t.$$

The voltage across the sample equals

$$IR = (I_0 + i\sin\omega t) \times (R_0 + R'\Theta_0\cos\varphi \sin\omega t - R'\Theta_0\sin\varphi \cos\omega t)$$
$$= I_0R_0 + I_0R'\Theta_0\cos\varphi \sin\omega t - I_0R'\Theta_0\sin\varphi \cos\omega t + iR_0\sin\omega t \quad (4.7)$$
$$+ iR'\Theta_0\cos\varphi \sin^2\omega t - iR'\Theta_0\sin\varphi \sin\omega t \cos\omega t.$$

Excluding the DC terms and neglecting the small AC terms, the AC voltage across the sample is

$$V = iR_0\sin\omega t + I_0R'\Theta_0\cos\varphi \sin\omega t - I_0R'\Theta_0\sin\varphi \cos\omega t. \quad (4.8)$$

This voltage contains two components, one in phase with the AC component of the heating current, and the other in quadrature. The latter, $-I_0R'\Theta_0\sin\varphi \cos\omega t$, reflects the temperature oscillations in the sample. This term can be balanced by means of an adjustable mutual inductance (Kraftmakher 1962). The sample is included in a bridge circuit fed by a DC current with a small AC component (Fig. 4.4).

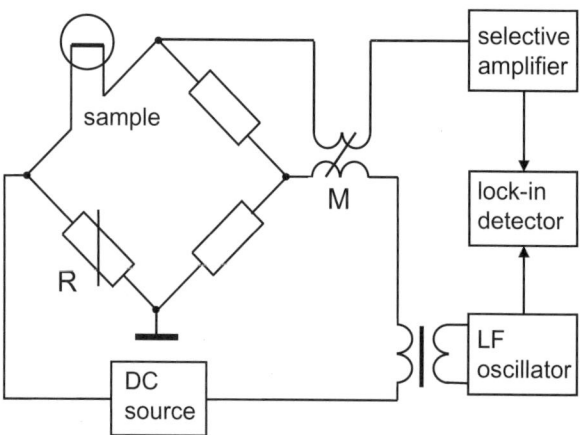

Fig. 4.4. Compensation circuit with variable mutual inductance for balancing the quadrature-lagging voltage generated by the sample due to the temperature oscillations in it (Kraftmakher 1962). That time, a homemade lock-in detector was used

A variable resistor R balances the in-phase voltage, so only the quadrature-lagging component remains at the output of the bridge. The mutual inductance M provides the additional compensation voltage. The current feeding the bridge passes through its primary winding. The voltage generated by the secondary winding balances the quadrature-lagging output voltage. The balance does not depend on the amplitude of the AC component of the current feeding the bridge.

Equivalent-Impedance Parameters

For the calorimetric measurements, a better approach is to utilise the equivalent impedance of the sample, which are directly related to its heat capacity. The impedance of the sample describes the amplitude and phase relations between the AC components of the current through the sample and the voltage across it. The impedance may be written as a complex quantity $Z = R_0 + A - iB$, where the terms A and B depend on the temperature oscillations in the sample. The impedance is obtainable from Eqs. (4.8) and (2.10b) by dividing the AC voltage across the sample by the AC component of the current:

$$Z = R_0 + (2I_0^2 R_0 R'/mc\omega)\sin\varphi \cos\varphi - i(2I_0^2 R_0 R'/mc\omega)\sin^2\varphi. \quad (4.9)$$

The ratio B/A equals $\tan\varphi$, and at high frequencies, i.e., in an adiabatic regime, the additional real part of the equivalent impedance, A, is much smaller than the imaginary part, B (Fig. 4.5). As a rule, measurements are performed under conditions $A \ll B$, $A \ll R_0$, and $\sin^2\varphi \cong 1$, so that $B = 2I_0^2 R_0 R'/mc\omega$.

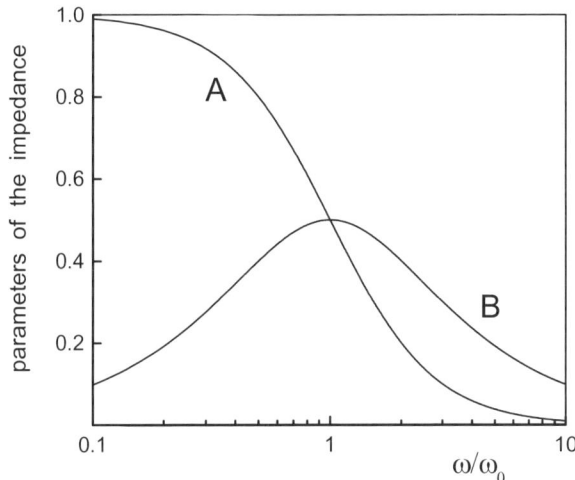

Fig. 4.5. Real and imaginary parts of equivalent impedance caused by temperature oscillations. $\omega/\omega_0 = B/A = \tan\varphi$, where ω_0 is a frequency, for which $\varphi = 45°$

The impedance of the sample is analogous to that of an electric circuit represented by a resistor and a capacitor connected in series: $Z = R^* - i/\omega C^*$. The resistance R^* equals $R_0 + A$, and the capacitance C^* equals $mc/2I_0^2 R_0 R'$. In an adiabatic regime,

$$mc = 2I_0^2 R_0 R' C^*. \quad (4.10)$$

The equivalent capacitance C^* does not depend on the frequency of the temperature oscillations. However, a large value of the equivalent capacitance C^* makes this approach impractical. To avoid this difficulty, the equivalent impedance was represented by a resistor R and a capacitor C connected in parallel. The impedance of this combination equals

$$Z_1 = R/(1 + \omega^2 R^2 C^2) - i\omega R^2 C/(1 + \omega^2 R^2 C^2). \quad (4.11)$$

At sufficiently high modulation frequencies, $\omega^2 R^2 C^2 \ll 1$. Hence, $R = R_0$, $B = \omega R^2 C$, and the main relation for the equivalent-impedance technique is given by

$$mc = 2I_0^2 R'/\omega^2 RC. \tag{4.12}$$

The equivalent-impedance parameters, R and C, are directly related to the heat capacity of the sample. Under adiabaticity conditions, the equivalent capacitance C is inversely proportional to the modulation frequency squared. This means that the term $\omega^2 R^2 C^2$ decreases with increasing frequency. When deriving the above expressions, the temperature derivative of the resistance of the sample was assumed to be positive.

Bridge Circuit

The heat capacity of the sample is thus available from the parameters of the equivalent impedance, R and C. A bridge circuit, one arm of which is shunted by a variable capacitor (Fig. 4.6), serves for the measurements.

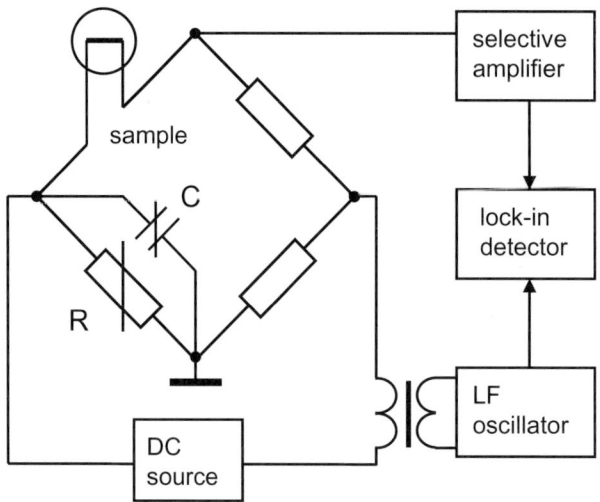

Fig. 4.6. Bridge circuit for measuring specific heat of wire samples (Kraftmakher 1962). Because of temperature oscillations in the sample, an adjustable capacitor is necessary to balance the bridge. Under balance, the parameters R and C are directly related to the heat capacity of the sample

A selective amplifier tuned to the modulation frequency ensures high sensitivity even when the temperature oscillations in the sample are small. For very small temperature oscillations, it is necessary to employ a lock-in amplifier. Using a dual-channel lock-in amplifier, one of the detectors is set to be sensitive to the in-phase component of the input voltage, and the other to the quadrature component. Signs of the output DC voltages of the detectors show either an increase or decrease of the resistance R and capacitance C are needed to complete the compensation. This method allowed us to determine the specific heat of tungsten at temperatures up to 3600 K (Kraftmakher and Strelkov 1962).

Kraev (1967) found that the above bridge circuit is suitable for measuring specific heat even when only an AC current passes through the sample. In this case, the variable capacitor compensates for the additional voltage of fundamental frequency generated by the sample. According to Eq. (4.3), this quadrature-lagging voltage equals $-(\beta\Theta_0 IR_0/2)\cos\omega t$. The third-harmonic signal at the output of the bridge remains uncompensated, but a selective amplifier solves this problem. Kraev (1967) measured the specific heat of tungsten up to 3000 K. A drawback of this approach is a strong temperature dependence of the amplitude of the temperature oscillations.

Potentiometer Circuit

The bridge circuit is impracticable in measurements at relatively low temperatures where cold-end effects become significant. In this case, one has to use very long wire samples or to heat the current leads, to which the sample is welded. As a good alternative, a potentiometer circuit is usable for measuring the heat capacity of a central portion of the sample (Fig. 4.7). A DC current with a small AC component heats the sample. The potential probes are much thinner than the sample and cause no significant temperature changes at the points where they are welded to the sample. To satisfy this requirement more reliably, it is possible to heat the probes by passing an additional current through them. A resistor R_2 is placed in series with the sample. The resistors R_1, R, and R_3, and the capacitors C_1 and C constitute a compensation circuit. The voltages across the resistors R_2 and R_3 are equalised by adjustment of the resistor R_1 and capacitor C_1. A selective amplifier serves as a null detector. The circuit R_1C_1 makes it possible to obtain proper amplitude of the AC current in the compensation circuit and its phase strictly coinciding with that of the AC component of the heating current. The amplifier is then switched to measure the equivalent impedance of the central portion of the sample. As in the case of the bridge circuit, the parameters of the equivalent impedance, R and C, do not depend on the AC component of the heating current. An oscilloscope indicates the balance, and a lock-in amplifier is also applicable. The resistance R_1 is much larger than R, so that the amplitude and the phase of the current in the compensation circuit are independent of R and C. Equation (4.12) remains applicable, but all the quantities correspond to the central portion of the sample.

The equivalent-impedance method allows very accurate determinations of the quantities I_0, R, and C. Errors in these quantities are of the order of 0.1%. The mass of the sample is measured with an error smaller than 1%. The total accuracy of the specific heat depends mainly on the reliability of the accepted values of R'. Most favourable is a case, where the temperature dependence of the resistance of the sample is nearly linear, so that the temperature derivative R' weakly depends on temperature. For tungsten and molybdenum, the temperature dependence of this derivative at high temperatures is of about 1% per 100 K. On the other hand, the method is inapplicable in studies of phase transitions accompanied by anomalies in the electrical resistivity.

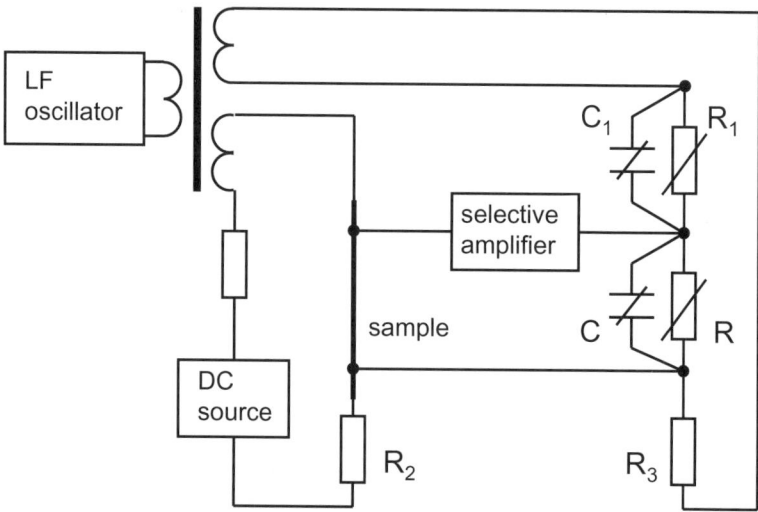

Fig. 4.7. Potentiometer circuit eliminates the influence of cold-end effects (Kraftmakher 1966b). Only a central portion of the sample is involved in the measurements. The equivalent impedance of the central portion is compared with the impedance of an adjustable RC circuit

4.2 Photoelectric Detectors

Many workers determined the temperature oscillations by measuring the radiation from the sample. With this method, Lowenthal (1963) measured the specific heat of tungsten, tantalum, molybdenum, and niobium in the 1200–2400 K range. An AC current heated the samples, and a photomultiplier served for the measurement of the temperature oscillations. The dependence of the photomultiplier's current on the temperature of the sample was assumed to be $I = TA^n$, with n weakly dependent on temperature. Hence, the amplitude of the temperature oscillations obeys the relation

$$\Theta_0 = TV/nV_0, \qquad (4.13)$$

where V_0 and V are the DC and AC components of the output voltage of the photomultiplier.

Filippov et al. (1964) and Filippov and Yurchak (1965) considered the dependence of the photomultiplier's current on the temperature of the sample to be $I = B\exp(-A/T)$. In this case,

$$\Theta_0 = T^2V/AV_0. \qquad (4.14)$$

Akhmatova (1965, 1967) employed comparative measurements in studies of specific heat of molten metals. The samples were placed in a niobium capillary heated by an AC current. The oscillations of the radiation from the empty and filled capillary kept at the same mean temperature were compared. This method provides the ratio of the heat capacities of the molten metal and the capillary.

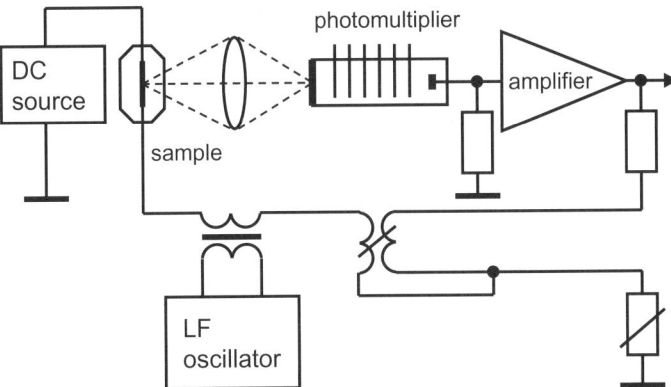

Fig. 4.8. Compensation circuit employing a photomultiplier. A variable mutual inductance balances the main output voltage (Kraftmakher 1966c). A variable resistor is necessary for compensating for the minor part of the output voltage, whose phase coincides with the AC component of the heating current

In our measurements, a compensation circuit with an adjustable mutual inductance was assembled to balance the AC output voltage of a photomultiplier (Fig. 4.8). The balance does not depend on the amplitude of the AC component of the current heating the sample (Kraftmakher 1966c). In these measurements, the temperature of the sample was determined from its electrical resistance, and the necessary calibration of the current of the photomultiplier was done. A variable resistor in the compensation circuit balances the in-phase component of the output voltage of the photomultiplier, which appears because the phase shift between the AC component of the heating current and the temperature oscillations in the sample is somewhat smaller than 90°.

Samples with Blackbody Models

However, neither of the above methods suffices because the quantities *A, B,* and *n* are temperature dependent (*A* and *B* depend on the effective wavelength and on the spectral emissivity of the sample). The calibration of a photosensor through the electrical resistance of the sample is also ambiguous. Fortunately, it is possible to circumvent these difficulties and to gain an important additional advantage by the use of samples with blackbody cavities. For all such samples maintained at equal mean temperatures, the relation between oscillations of the radiation from the cavities and temperature oscillations in the sample is the same. This enables one to directly compare temperature oscillations in various samples, without calibration of the photosensor. With a reference sample of known specific heat, comparative measurements of specific heat are possible (Kraftmakher 1992). Tungsten and platinum may serve as reference materials for such measurements. Their specific heat was determined using all the calorimetric methods already known. Tungsten is a good reference material for temperatures up to 3000 K, and platinum up to 1500 K. The high-temperature specific heat of other metals and alloys is not so well known, and comparative measurements of their specific heats may appear to be useful.

Compensation Circuit

To enhance the accuracy of such measurements, a compensation circuit, whose balance is independent of the AC component of the heating power, was proposed (Kraftmakher 1992). In this circuit, a DC current I_0 with a small AC component heats a sample with a blackbody cavity (Fig. 4.9). To exclude cold-end effects, only a central portion of the sample restricted by thin potential probes is involved in the measurements. The mean temperature and the amplitude of the temperature oscillations are assumed to be constant throughout the central portion. In an adiabatic regime, the heat capacity of the sample is given by

$$mc = 2I_0 U/\omega\Theta_0, \qquad (4.15)$$

where m and U are the mass of the central portion of the sample and the AC voltage across it.

The blackbody cavity is projected onto a photodiode, so that the image of the cavity exceeds the photosensitive area. The AC output voltage of the photodiode is proportional to the temperature oscillations in the sample: $V_1 = K_1\Theta_0$, where K_1 is a proportionality factor. Owing to the blackbody cavity, this factor depends only upon the mean temperature of the sample. The potential probes are connected to the input of an integrating RC circuit. The output voltage of this circuit is amplified and then fed to the load resistor of the photodiode. The AC voltage at the capacitor C equals $U/\omega RC$ ($\omega^2 R^2 C^2 \gg 1$). The compensation voltage is $V_2 = K_2 U/\omega RC$, with K_2 a proportionality factor. Under compensation, $V_1 = V_2$, i.e.,

$$mc = 2K_1 I_0 RC/K_2 = KI_0 RC. \qquad (4.16)$$

The coefficient $K = 2K_1/K_2$ is the same for all samples kept at a given mean temperature. Only the variable capacitor C provides the compensation. Therefore, for two samples one obtains

$$m_1 c_1/m_2 c_2 = I_{01} C_1/I_{02} C_2, \qquad (4.17)$$

where I_{01} and I_{02} are the DC currents providing equal mean temperatures of the samples, and C_1 and C_2 are the values of the capacitance corresponding to the compensation.

This method allows comparative measurements on any conducting samples provided with blackbody cavities. In addition, the total radiation from the cavity is utilised instead of radiation in a narrow spectral band. This circumstance compensates for the decrease in the radiant flux due to the small size of the cavity. The radiation from the cavity and from a standard strip lamp is in turn projected, by the same optical system, onto a photodiode or another photosensor. The standard lamp serves to determine the mean temperature of the samples. Both samples and the standard lamp are located in a vacuum chamber and are replaced by means of a turn-plate. No corrections are thus needed for the reflectance and absorption of the radiation in the window of the chamber.

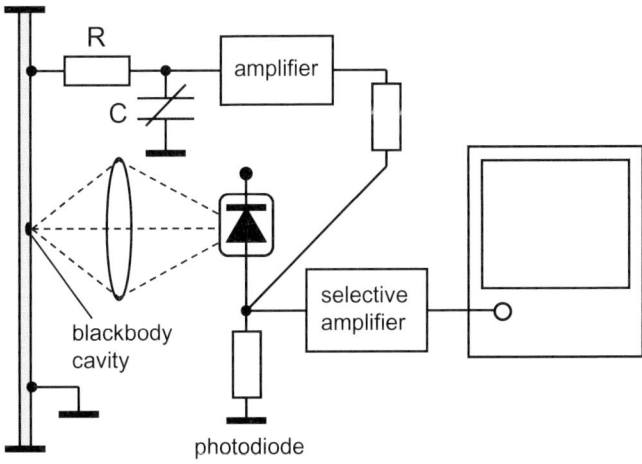

Fig. 4.9. Use of samples with blackbody cavities (Kraftmakher 1992). Under compensation, $mc = KI_0RC$, where K depends only on the mean temperature of the sample. Temperature oscillations in different samples are compared directly

4.3 Pyroelectric Sensors

Many investigators employed pyroelectric sensors for detecting temperature oscillations. Coufal (1983) reported on the application of a pyroelectric thin-film transducer made out of polyvinylidene difluoride (PVDF). A 9-μm PVDF foil with nickel electrodes was attached to the sample. The sensitivity of the transducer was typically 0.1 V.K^{-1}, while the response was flat in the range 10^{-1} to 10^7 K.s^{-1}. The measurements were carried out using a modulated light source and a laser with pulse width of 12 ns.

Mandelis and Zver (1985) developed a theory of this approach referred to as photopyroelectric technique. Mandelis et al. (1985), Zammit et al. (1998), Marinelli et al. (1990, 1992, 1994ab, 1996, 1998), Glorieux et al. (1994), Mercuri et al. (1994), Schoubs et al. (1994) also employed the method.

In the back detection configuration, this technique consists of a sample with one side periodically heated by a modulated laser beam, while a pyroelectric transducer detects the induced temperature oscillations on the opposite side (Fig. 4.10). From the data obtained with various modulation frequencies, specific heat, thermal diffusivity, and thermal conductivity are available (Marinelli et al. 1994b). For instance, the thermal properties of antiferromagnets Cr_2O_3, FeF_2, and Cr were studied close to the Néel points.

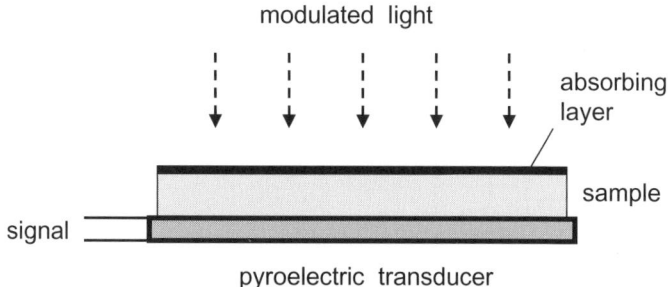

Fig. 4.10. Calorimeter cell with pyroelectric sensor in the back detection configuration. A thin-film pyroelectric transducer senses temperature oscillations in a sample subjected to modulated-light heating (Marinelli et al. 1992)

Marinelli et al. (1994a) used the front detection photopyroelectric configuration to study the non-linear heat transport processes in the vicinity of the smectic-A–nematic phase transition of 9CB liquid crystal (Fig. 4.11).

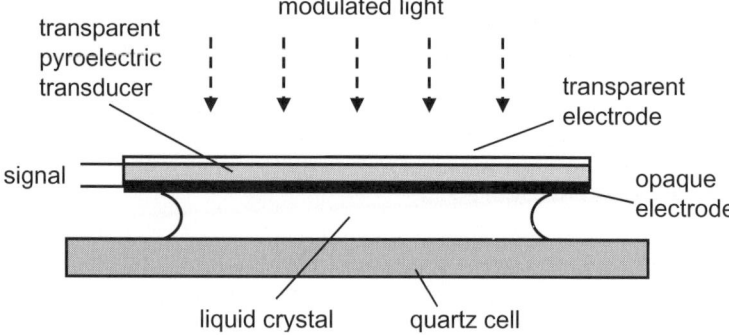

Fig. 4.11. Front detection photopyroelectric configuration used for studying non-linear heat transport processes in the vicinity of the smectic-A–nematic phase transition (Marinelli et al. (1994a)

The liquid crystal is contained in a quartz cell 0.1 mm thick. A $LiTaO_3$ pyroelectric transducer, 0.3 mm thick, acts as a cover for the cell. One of its electrodes is optically transparent in the visible. The second electrode is optically opaque and thermally transparent metallic thin film. An acousto-optically modulated light from a 5-mW He-Ne laser is partially absorbed by the opaque electrode. The signal from the transducer is analysed by a dual-channel lock-in amplifier. The non-linear effects near the phase transition were detected via the second-harmonic component of the signal from the transducer. This component depends on the temperature derivative of the specific heat and of the thermal conductivity. Measurements of the temperature derivative of specific heat are considered in Sect. 13.2. Using the photopyroelectric technique, Marinelli et al. (1996) studied the critical behaviour of the specific heat, thermal conductivity, and thermal diffusivity of $RbMnF_3$ close to the Néel temperature.

Christofides (1993) and Thoen and Glorieux (1997) reviewed the photopyroelectric techniques. Recently, the modulation frequency accessible by this method was extended up to 100 kHz (Chirtoc et al. 2001; Bentefour et al. 2003).

4.4 Thermocouples and Resistance Thermometers

At medium and low temperatures, thermocouples and resistance thermometers are the most reliable tools for measuring the mean temperature of a sample and the temperature oscillations in it.

Thermocouples

Thermocouples are widely used with modulated-light heating or separate electrical heaters. In some cases, the thermocouples were formed by deposited thin films, 10–100 nm thick. A good thermal coupling and low thermal inertia are thus achievable allowing measurements over a wide frequency range. With thermocouples, it is possible to measure temperature oscillations as small as 0.1 mK (Bonilla and Garland 1974). When using thermocouples, one has to ensure good thermal coupling of the junction to the sample and to make the additional heat capacity to be negligible. Craven et al. (1974) measured the low-temperature specific heat of TTF-TCNQ on thin platelets, typically 10×0.3×0.02 mm. The samples, selected for size and uniformity, were mounted on a pair of thermocouples formed by spot-welded 25-µm chromel and alumel wires, flattened to 5 µm in the junction region. Pulses of light from a quartz-iodine lamp heated the sample. One of the junctions of the crossed thermocouples detected the temperature oscillations of the system, while the other formed the cold junction of a thermocouple with its hot junction in an ice bath. Garfield and Patel (1998) described a method of preparing thin thermocouples.

Overend et al. (1996) employed a chromel-alumel thermocouple when measuring the specific heat of single-crystal samples of $YBa_2Cu_3O_{7-\delta}$ in magnetic fields up to 8 T. The samples were mounted on crossed thermocouple junctions made by spot welding of 25-µm thermocouple wires (Fig. 4.12). The samples were attached to the thermocouple junction using GE varnish. An infrared light-emitting diode and a fibre light guide provided the periodical heating of the sample. The end of the fibre was held approximately 5 mm behind the sample. This distance was adjusted, so that the diverging light beam covered the whole crystal and ensured homogeneous heating. Helium gas at a pressure of 50 kPa provided the necessary thermal link between the sample and a copper block, whose temperature was measured with a platinum thermometer. The modulation frequency was 17 Hz, and the amplitude of the temperature oscillations measured by one of the thermocouples was about 15 mK. The second thermocouple measured the enhancement of the mean temperature of the sample due to the modulated-light heating. This technique allowed the specific heat to be determined with a resolution of 10^{-4}.

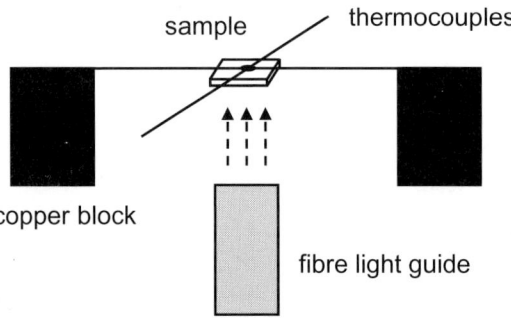

Fig. 4.12. Arrangement for measuring specific heat of small samples using modulated-light heating and a thermocouple (Overend et al. 1996). Similar arrangement was used in many other works

Williams and Wickramasinghe (1986) developed a thermocouple sensor with dimensions approaching 0.1 µm. The temperature resolution of this ultra small probe was of the order of 10^{-4} K. Majumdar et al. (1995) applied a similar small thermocouple probe (Fig. 4.13) for thermal imaging of surfaces. The spatial resolution of 0.5 µm was demonstrated though the probes were about 25 µm in diameter. Forster and Gmelin (1996) described a miniature W/Au+0.07%Fe thermocouple, which was used in a scanning tunnel microscope set-up to measure temperature profiles with sub micrometer and sub microsecond resolution. The construction of the tip resembles that described by Williams and Wickramasinghe (1986).

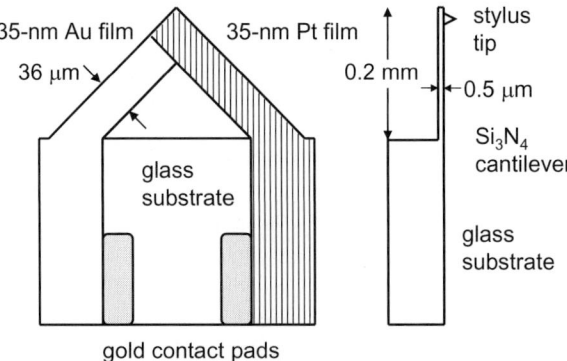

Fig. 4.13. Schematic diagram of thin-film thermocouple probe designed by Majumdar et al. (1995). The probe was used in atomic force microscope to simultaneously obtain thermal and topographical images of surfaces with sub micrometer spatial resolution

Gmelin et al. (1998) reviewed scanning thermal microscopy, i.e., measurements of thermal parameters in the sub micrometer range. This technique, derived from scanning tunneling microscopy and atomic force microscopy, provides also tools useful for modulation nanocalorimetry.

Resistance Thermometers

The simplest way to employ a resistance thermometer consists in including it in a DC bridge or potentiometer circuit. Since the resistance of the thermometer oscillates according to the temperature oscillations, the AC output voltage of the bridge immediately provides the necessary data. As a rule, a lock-in amplifier measures this voltage. The reference voltage for the amplifier is taken from the source of the modulation. A drawback of this simple technique is an additional noise at low frequencies. To circumvent this difficulty, Zally and Mochel (1971, 1972) utilised an AC bridge (Fig. 4.14). A resistive heater fed by an AC current of a frequency $f_0/2$ provides AC heating of a frequency f_0. A current of a high frequency f feeds the bridge including a thermometer. The output voltage of the bridge has the same frequency f but is modulated by the frequency f_0, according to the temperature oscillations. A lock-in amplifier measures this voltage. Mixing the signals of the two frequencies creates a double-sideband signal at frequencies $f+f_0$ and $f-f_0$, which serves as the reference for the lock-in amplifier. The DC output voltage of the amplifier is proportional to the amplitude of the temperature oscillations in the calorimeter cell.

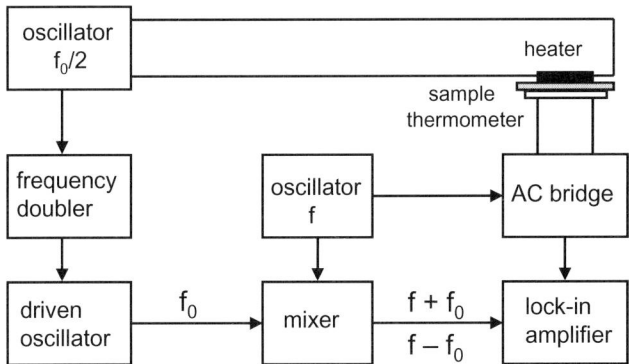

Fig. 4.14. Simplified diagram of calorimetric set-up designed by Zally and Mochel (1971, 1972). The frequency conversion enhances the sensitivity of the temperature measurements

To calibrate a resistance thermometer, one needs a precisely calibrated thermometer, usually commercially available. Metallic thermometers, such as platinum thermometer, satisfy the relation $T = \Sigma A_n R^n$. However, they are impracticable at temperatures below about 20 K because of low sensitivity. Dissolved impurities and defects of crystal structure cause a residual resistance, and the resistance of the thermometer becomes less sensitive to temperature changes. At low temperatures, semiconductors provide much higher sensitivity, which increases with decreasing temperature. For such thermometers, a more appropriate relation is $1/T = \Sigma A_n (\ln R)^n = A_0 + A_1 \ln R + A_2 (\ln R)^2 +...$ Thermistors or usual carbon resistors are applicable also at room temperatures. To check the possible influence of self-heating, it is useful to calibrate the thermometer with different measuring currents.

Very small temperature oscillations, of the order of 1 μK, are measurable at liquid helium temperatures (Mehta and Gasparini 1997, 1998; Mehta et al. 1999; Kimball et al. 2000). Measurements of extremely small temperature oscillations at high temperatures are described in Sect. 13.4.

4.5 Lock-in Detection of Periodic Signals

Lock-in detection is an excellent technique for measuring weak periodic signals in the presence of signals of other frequencies or noise. The method rests on the exact knowledge of the frequency of the expected signal. This periodic signal is measured by a detector controlled by a reference voltage taken from the oscillator governing the process under study (Fig. 4.15). The frequency of the expected signal therefore strictly coincides with that of the reference, while the phase shift between them, under constant experimental conditions, remains constant.

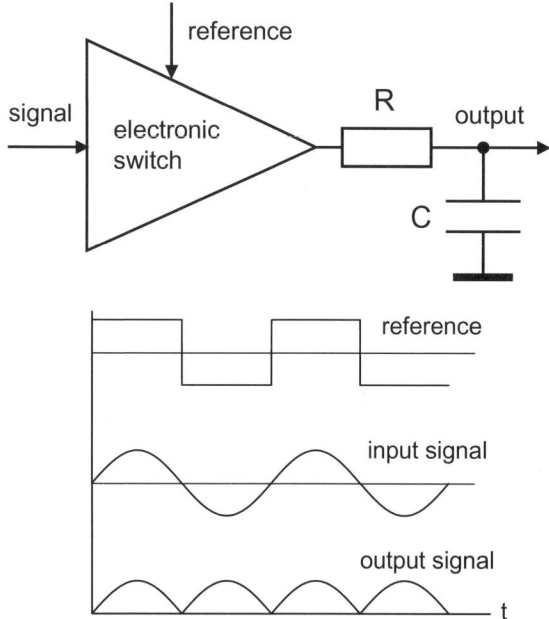

Fig. 4.15. Principle of the lock-in detection. An electronic switch controlled by reference voltage and an integration *RC* circuit provide the necessary operation. Averaged output voltage is proportional to the AC input signal of frequency coinciding with that of the reference and to the cosine of the phase shift between them. When the signal contains no component of the reference frequency, the averaged output voltage remains zero

In the modulation techniques, the reference voltage is supplied either by the source of the modulated power or by a special sensor, e.g., a photocell when using the modulated-light heating. The operation of a lock-in detector may be explained as follows. An electronic switch controlled by the reference voltage periodically changes the polarity of the signal fed to an integrating RC circuit. The output signal of the detector is averaged over a time sufficiently long to suppress irregular pulses due to signals of other frequencies or noise. The effective bandwidth of a lock-in detector is inversely proportional to the averaging time. The detector is thus always tuned to the expected signal and has a readily adjustable bandwidth. A DC voltage at the output of a lock-in detector appears only when the signal contains a component of the reference frequency. The DC voltage is proportional to this component and to the cosine of the phase shift between it and the reference. The detector incorporates an adjustable phase shifter to achieve the maximum output voltage. When the signal contains no component of the reference frequency, the averaged output voltage remains zero. The reason for this is the absence of fixed phase relations between the reference and signals of other frequencies or noise.

Amplifiers employing lock-in detection are called lock-in amplifiers. Owing to the narrow effective bandwidth, they provide high sensitivity and noise immunity. Some types of lock-in amplifiers incorporate two detectors governed by references with phases shifted by 90°. This makes it possible to measure the signal and its phase shift relative to the reference. Dual-channel lock-in amplifiers are necessary when the signal to be measured or balanced contains both in-phase and quadrature components, e.g., in the equivalent-impedance technique. A simplified diagram of a lock-in amplifier is shown here (Fig. 4.16).

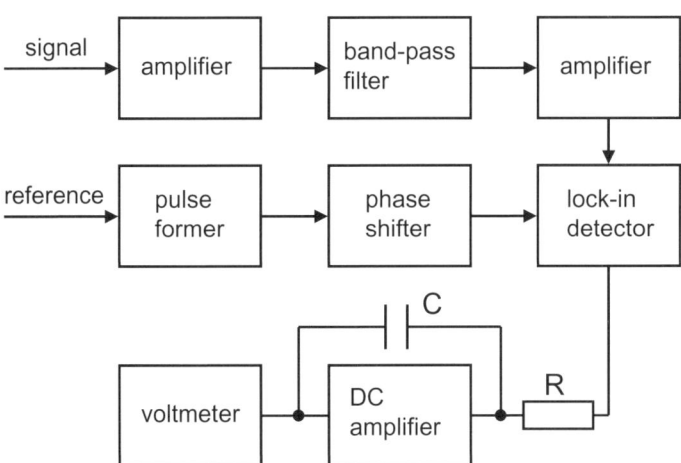

Fig. 4.16. Simplified diagram of lock-in amplifier. In addition to a lock-in detector, the device incorporates amplifiers, a band-pass filter, and a variable phase shifter

A lock-in detector is also efficient for measuring small phase changes. In this case, the phase shift between the signal and the reference is set at 90°. Under such conditions, the output voltage of the detector is zero, but the sensitivity to changes in the phase of the signal is the best. Monitoring the phase of the temperature oscillations appeared to be very useful in many cases of modulation measurements.

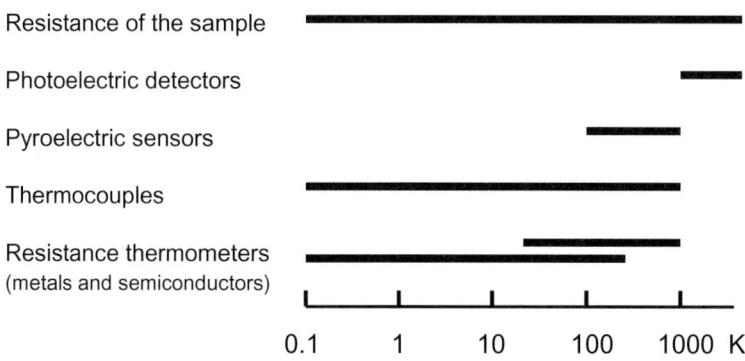

Fig. 4.17. Methods of measuring temperature oscillations in modulation calorimetry. Below about 20 K, only semiconductor thermometers are usable for temperature measurements

Methods of measuring temperature oscillations in modulation calorimetry for different temperature ranges are summarized here (Fig. 4.17). Clearly, the choice of a method of measuring the temperature oscillations in the sample depends on how the heating power is modulated. Recommendations for such a choice are given below (Fig. 4.18).

	Direct heating	Modulated light	Electron bombardment	Resistive heaters	Induction heating	Peltier heating
Resistance of the sample	●	●		●		●
Photoelectric detectors	●		●		●	
Pyroelectric sensors		●		●		●
Thermocouples	●	●	●	●	●	●
Resistance thermometers		●	●	●		●

Fig. 4.18. Compatibility of methods of periodically heating and of measuring the temperature oscillations. Thermocouples are the most universal tools for the temperature measurements

5 Brief Review of Methods of Calorimetry

Calorimetric measurements seem, at first glance, to be simple and straightforward. By definition, one has to supply some heat to the sample and to measure the corresponding increment in its temperature. However, no simple solution for this problem exists in a wide temperature range. First, the accuracy of temperature measurements in various temperature ranges is very different. Second, it is impossible to completely avoid uncontrollable heat exchange between the sample and its surroundings when the temperature of the sample is far from room temperature. During many years of development, the following methods have been proposed to solve the problem:

- All possible precautions are undertaken to reduce the unwanted heat exchange between the sample and its surroundings (adiabatic calorimetry).
- The enthalpy of the sample is measured instead of the specific heat: the sample heated up to a high temperature drops into a calorimeter usually kept at room temperature, and the heat released from the sample is measured (drop method).
- Shortening the time of the measurements minimises the influence of any uncontrollable heat exchange (modulation, pulse, and dynamic techniques).
- The heat exchange between the sample and its surroundings is taken into account and involved in the measurements of specific heat (relaxation method).

5.1 Adiabatic Calorimetry

Adiabatic calorimetry reduces to a minimum any heat exchange between the calorimeter and its environment. For this purpose, the calorimeter is surrounded by the so-called adiabatic shield, whose temperature is kept equal to that of the calorimeter during the entire experiment (e.g., Kraftmakher and Strelkov 1960). The calorimeter and the shield are placed in a vacuum chamber, so that only the radiative heat exchange occurs (Fig. 5.1). An electrical heater heats the calorimeter, while a resistance thermometer or a thermocouple measures its temperature. The heat supplied to the calorimeter and the temperature increment are thus accurately known. To reduce the heat flow through the electrical leads, they are thermally anchored to the adiabatic shield. With a temperature difference between the calorimeter and the shield ΔT, the power of the radiative heat exchange is proportional to $T^3 \Delta T$. Adiabatic calorimetry, being an excellent technique at low and medium temperatures, fails therefore at high temperatures.

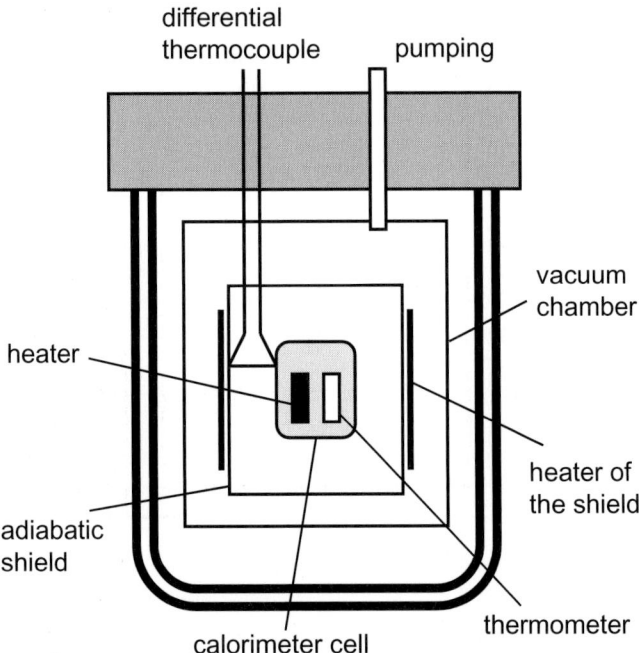

Fig. 5.1. Simplified diagram of adiabatic calorimeter for low temperatures. Temperature of the adiabatic shield is automatically kept equal to that of calorimeter cell during the entire experiment

Continuous operation with enhanced heating rate reduces the role of heat losses. Braun et al. (1968) developed an adiabatic calorimeter of continuous heating for the range 300–1900 K (Fig. 5.2) and investigated many metals. The calorimeter is capable of measurements of specific heat of solids (with an error of 2%) and liquids (3%), and latent heats of phase transitions in solids (0.5%) and of melting (1.5%). Three modes of operation are feasible:

- The applied power is constant, and the heating rate is inversely proportional to the heat capacity of the sample.
- The heating rate is kept constant, so that the applied power is proportional to the heat capacity.
- With the heater switched off, the temperature difference between the calorimeter and the thermal shield is monitored.

Schnelle and Gmelin (1995) described a high-resolution adiabatic calorimeter for small samples operating in the 20–300 K range. Three measuring modes are possible: (i) the sample is heated with a constant power; (ii) the temperature of the adiabatic shield is increased by a programmed rate, and the temperature difference between the shield and the sample is kept at zero; (iii) a single heat pulse heats the sample (the classical Nernst method).

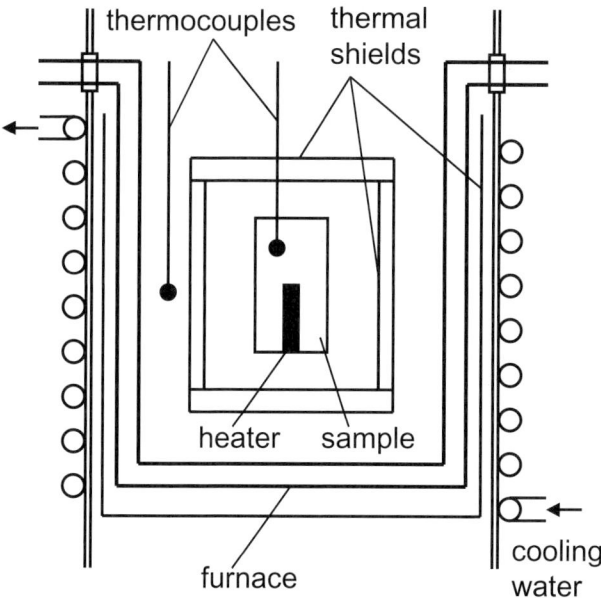

Fig. 5.2. Simplified diagram of adiabatic calorimeter for high temperatures designed by Braun et al. (1968). The calorimeter employs continuous heating and operates in the range 300–1900 K

Buckingham et al. (1973) designed a high-precision calorimeter of continuous heating. The aim was to measure specific heats near phase transitions, where the rate of change of specific heat and of thermal relaxation time may become very large. In this apparatus (Fig. 5.3), the sample is suspended in vacuum inside a reference stage (stage 1). Two servo-controlled shields (stage 2 and stage 3) surround the stage 1 providing effective thermal isolation of the sample. The operating temperature of the stage 3 is a fraction of kelvin below that of the stage 2. The stage 1 is heated in such a manner that its temperature T_1 rises linearly with time. The sample is servo-controlled to keep its temperature T_0 the same as T_1. Under such conditions, the power delivered to the sample, P_0, is given by $P_0 = C_0 T'$, where $T' = dT/dt$, and C_0 is the heat capacity of the sample. The method permits negative scans as well as positive, and any dependence on the heating (cooling) rate is easily disclosed. Each stage is equipped with a manganin heater and thermistors used as thermometers. The outer stage (stage 3) has a thermoelectric heat sink for operation below room temperature. Close to room temperature, the stability of the thermistors is equivalent to a drift less than 0.3 mK over a period of two months. The temperature drift of the stage 1 is less than 10^{-6} K.min^{-1} for periods of 2 to 3 hr. At the same time, the internal relaxation times characterizing the equilibration within the stages 1, 2 and 3 are 33, 60, and 120 s, respectively. To operate in the cooling mode, the temperature difference between the stages 1 and 2 should be adjusted to cool the stage 1 at a desired rate with no power dissipated in its heater. The servo loop linking the calorimeter and the stage 1 now operates to decrease the cooling rate of the calorimeter until it equals to that of the stage 1. The power dissipated in the calorimeter is now proportional to a constant minus the heat capacity of the sample. The apparatus operates also in the conventional heat-pulse mode.

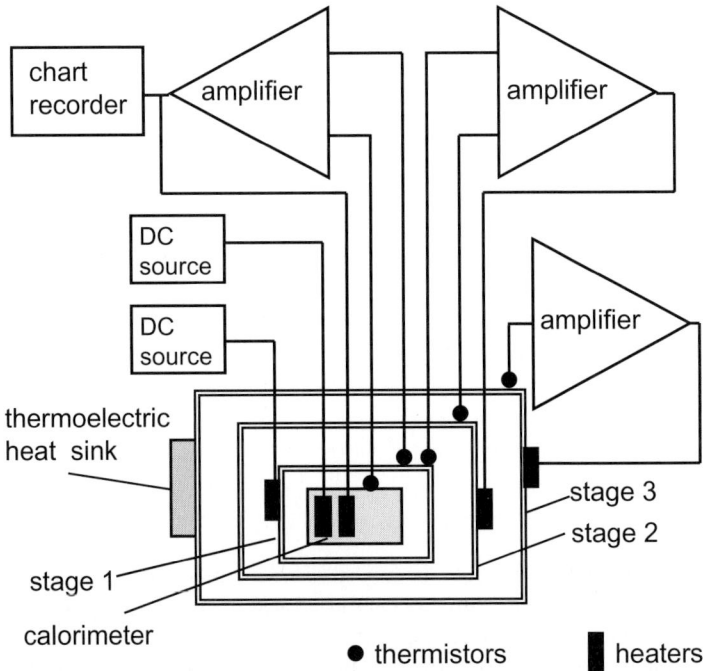

Fig. 5.3. Simplified diagram of high-precision adiabatic calorimeter of continuous heating (Buckingham et al. 1973). The temperature drift of the stage 1 is less than 10^{-6} K.min^{-1} for periods of 2 to 3 hr

Moon and Jeong (1996) also reported on an adiabatic calorimeter operating both in the heating and cooling mode in range 100–400 K.

Kagan (1984) reviewed adiabatic calorimetry for medium and high temperatures. Stewart (1983), Gmelin (1987, 1999), and Suga (2000) reviewed calorimetric techniques for low temperatures. High-temperature calorimetry was reviewed by Rogez (1998).

5.2 Drop Method

This technique was developed for measurements at high temperatures. A sample placed in a furnace is heated up to a selected temperature measured by a thermocouple or an optical pyrometer (Fig. 5.4).

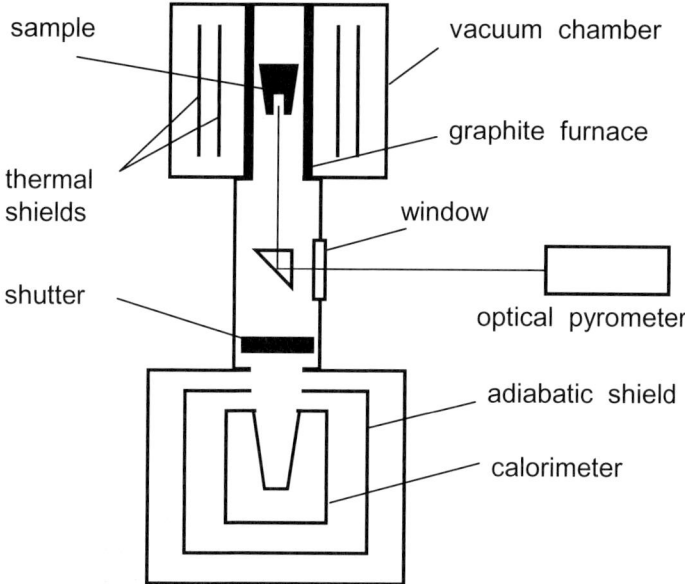

Fig. 5.4. Simplified diagram of set-up for drop calorimetry (Glukhikh et al. 1966). The heat released from the sample is measured with an adiabatic calorimeter

Then it drops into a calorimeter kept at a temperature convenient for measurements of the heat released from the sample, usually room temperature. A resistance thermometer measures the increment in the temperature of the calorimeter, which is proportional to the enthalpy of the sample. The calorimetric measurements are thus carried out under conditions most favourable for reducing any unwanted heat exchange. The price for this gain is that the result of the measurements is the enthalpy, instead of the specific heat. The specific heat is obtainable as the temperature derivative of the enthalpy.

When the specific heat is weakly temperature dependent, the method is quite adequate. The situation becomes complicated when the specific heat varies in narrow temperature intervals but the corresponding changes in the enthalpy are too small to be determined precisely. An additional problem arises when a first-order phase transition occurs at intermediate temperatures, and thermodynamic equilibrium in the sample after cooling is doubtful. Drop calorimetry was proposed when there was no alternative at high temperatures. Nevertheless, it remains useful until today, especially for measurements on nonconducting materials (for details see Ditmars 1984).

Levitation Calorimetry

A modification of the drop technique is levitation calorimetry, where the sample is levitated and heated by a high-frequency electromagnetic field (for reviews see

Chekhovskoi 1984, 1992; Frohberg 1999). No container is thus needed, and the temperature range of the measurements can be extended above the melting points of refractory materials. For instance, Arpaci and Frohberg (1984) measured the specific heat of tungsten up to 4000 K.

It is much easier to achieve electromagnetic levitation under microgravity conditions (Sect. 6.1.7).

5.3 Pulse and Dynamic Techniques

The influence of any unwanted heat exchange between a sample and its surroundings is proportional to the time of the measurements. A decrease of this time is an efficient method even at very high temperatures. This approach is applicable whenever it is possible to heat the sample and to measure its temperature rapidly. Conducting samples heated by an electric current or by electron bombardment are well suited for such measurements. The temperature of the sample is measured through its resistance or radiation. When the heat losses cannot be completely avoided, they are taken into account. Pulse calorimetry employs small increments in the temperature of the sample, and the result of one measurement is the specific heat at a single temperature. A furnace or electric heating provides the initial temperature of the sample. Dynamic calorimetry consists in rapidly heating the sample over a wide temperature interval. The heating power and the temperature of the sample are measured continuously during the run. When the heat losses from the sample are significant, they are measured during the cooling period and taken into account. These data are sufficient to evaluate the specific heat of the sample in the whole temperature range of the measurements.

5.3.1 Pulse Method

Rasor and McClelland (1960a) developed an apparatus with a sample placed in a graphite furnace to achieve the initial temperature of the experiment (Fig. 5.5). A photomultiplier senses the temperature increment caused by passing through the sample a short pulse of electric current. By means of a four-channel oscilloscope, the heating current I, the voltage drop across a central portion of the sample U, the temperature of the sample T, and the heating rate $T_h' = (dT/dt)_h$ are recorded simultaneously. With heat losses neglected, the heat capacity of the sample is

$$mc = P/T_h', \qquad (5.1)$$

where $P = IU$.

Rasor and McClelland (1960b) measured the specific heat of molybdenum, tantalum, and graphite at high temperatures, up to 3920 K for graphite. They observed, for the first time, a strong non-linear increase in the specific heat of refractory metals. That time, this observation gained no recognition, and the authors did not publish their results during about five years.

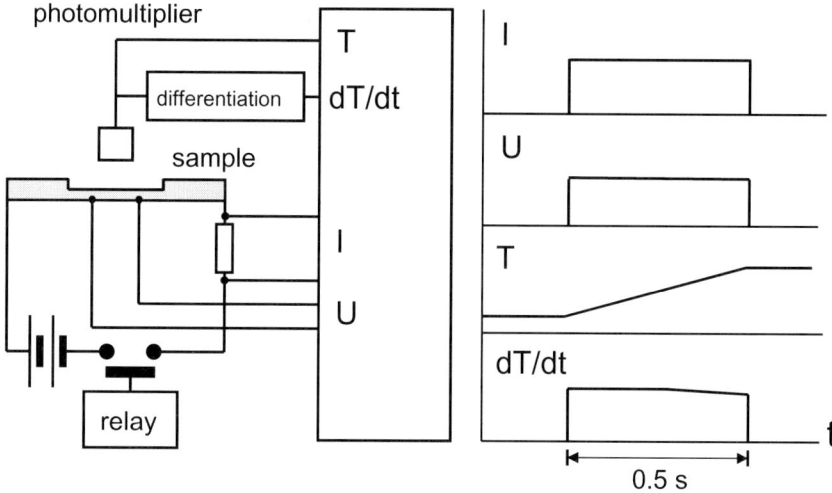

Fig. 5.5. Pulse calorimetry at high temperatures (Rasor and McClelland 1960a). The authors observed, for the first time, a strong non-linear increase in the specific heat of Mo, Ta, and graphite

5.3.2 Dynamic Calorimetry

In the last three decades, dynamic calorimetry became a main technique for calorimetric studies of electrical conductors at high temperatures. This technique was developed in several laboratories, and many experimental data have been obtained.

Subsecond Technique Developed at NBS

Cezairliyan and co-workers at the National Bureau of Standards (now the Institute of Standards and Technology, Gaithersburg) developed a convenient and accurate subsecond technique for temperatures up to 3600 K (for reviews see Cezairliyan 1984, 1988, 1992). The temperature of the samples with blackbody models was measured by an optical pyrometer (Fig. 5.6).

The heat-balance equations for the sample during the heating and cooling periods are given by

$$mcT_h' + Q = P, \quad (5.2a)$$

$$mcT_c' + Q = 0, \quad (5.2b)$$

where P is the electric power applied to the sample during the heating period, Q is the power of the heat losses from the sample ($P > Q$), $T_h' = (dT/dt)_h$ is the heating rate, and $T_c' = (dT/dt)_c$ is the cooling rate after ending the heating. From the above relations, the heat capacity of the sample is

$$mc = P/(T_h' - T_c'), \tag{5.3}$$

where T_h' and T_c' relate to the same temperature of the sample. This technique was successfully employed in studies of many high-melting-point metals and alloys. The specific heat of tungsten was determined up to 3600 K (Cezairliyan and McClure 1971).

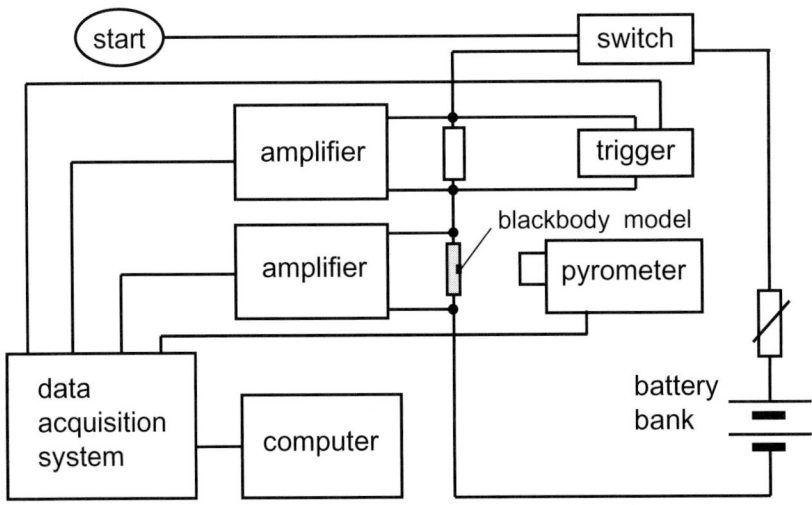

Fig. 5.6. Simplified diagram of set-up for dynamic calorimetry developed by Cezairliyan and co-workers. In one run, specific heat, electrical resistivity, and thermal radiation properties are measured in a wide temperature range. Many high-melting-point metals and alloys have been studied

To determine the temperature of a sample without blackbody cavities, it is necessary to know the spectral emissivity of its surface, ε_λ. Spectral radiometry and laser polarimetry are two independent techniques for such measurements. The most direct method of determining normal spectral emissivity utilises measurements of the radiance from the surface of the material under study and that from a blackbody cavity at the same temperature. The method is best realised with a sample incorporating a blackbody cavity. However, it is not applicable to samples, which either are too small or are inaccessible for the inclusion of a blackbody cavity. A method resolving this problem is the laser polarimetry (Cezairliyan et al. 1996, 1998). The objectives of the study were (i) to extend the laser polarimetry method to measurements on rapidly heated samples, and (ii) to validate this technique by performing experiments, where the normal spectral emissivity of the samples is simultaneously measured by conventional spectral radiometry.

With the conventional spectral radiometry method, normal spectral emissivity is available as the ratio of radiance, at a given wavelength, from the surface of the sample to that from a blackbody cavity at the same temperature. After the spectral emissivity is determined, the true temperature is measurable by an optical pyrometer. The laser polarimetry is based on the determination of spectral reflectivity, R_λ. For opaque materials, $\varepsilon_\lambda = 1 - R_\lambda$. Spectral reflectivity is given by

$$R_\lambda = [(n - n_0)^2 + k^2]/[(n + n_0)^2 + k^2], \qquad (5.4)$$

where n is the real part of the index of refraction, n_0 is the refractive index of the transparent ambient medium at the material surface, and k is the extinction coefficient. To obtain n and k, the target surface is illuminated at a non-normal angle of incidence with laser radiation of known polarization. A polarimeter measures the four Stokes parameters of the reflected radiation, which provide the complete description of the state of polarization of the radiation. From these data, the quantities n, k, and n_0 are obtainable. Conventional designs for determining the polarization state of radiation generally contain moving parts and therefore are not suitable for rapid measurements. A new polarimeter design with stationary optical elements and four detectors was adopted for millisecond-resolution measurements. The samples were in the form of thin-wall tubes with a small blackbody cavity fabricated in the wall at the middle of the sample.

Two pyrometers were used in the measurements. The first pyrometer viewed the blackbody cavity in the sample, while the second one viewed the surface of the sample on the side opposite to the cavity. The first pyrometer operated at 0.65 µm, while the second one at 0.6 and 0.65 µm to bracket the wavelength of the He-Ne laser polarimeter (0.633 µm). The laser target was close to the target of the second pyrometer. Signals from the two pyrometers and the polarimeter were digitised using analogue-to-digital converters. A data-acquisition system collected the data at the rate of 2 kHz. Measurements were performed on tubular samples made of molybdenum and of tungsten. The uncertainty in normal spectral emissivity determined by either of the two techniques was estimated to be not more than 2%. Using this approach, specific heat of molybdenum was measured in the range 2000–2800 K (Cezairliyan et al. 1998).

Dynamic Method at NRML

Matsumoto and Ono (2001) described measurements by the pulse method with a brief steady state, which was developed at the NRLM (National Research Laboratory of Metrology, Tsukuba). The method is based on a fast resistive self-heating of a ribbon-shaped sample up to a selected high temperature, and then keeping it in a steady state for about 0.5 ms. The true temperature of the sample is determined using a high-speed pyrometer and an ellipsometer. The specific heat of four refractory metals (Nb, Mo, Ta, and W) was measured in the range from about 1500 K to temperatures close to the melting point of each metal. The specific-heat data for the four metals agreed with the values obtained earlier at the NBS.

The authors have given a list of major sources of uncertainties in hemispherical total emissivity and specific heat as follows:

- Sources due to the uncertainty in the measured quantities: current; voltage; sample geometry; sample weight; temperature.

- Sources due to the deviation from the ideal measurement conditions: heat loss due to the longitudinal thermal conduction; voltage difference between the probes due to the temperature non-uniformity in the sample; remaining temperature drift of the sample at the plateau; thermal expansion of the sample.

Among the sources listed, the main contributions to the error in specific heat were introduced by measurements of the emissivity, which is necessary for evaluating the true temperature (1% at 2200 K), by the temperature non-uniformity (0.9%), and by the thermal expansion of the sample (1%). The errors are given as standard deviations. The authors did not correct their results for the thermal expansion "because reliable data on thermal expansion of the sample are not always available."

Dynamic Calorimetry at IMGC

A variant of dynamic calorimetry was developed at the IMGC (Instituto di Metrologia "G. Colonnetti", Torino). This technique was described in detail in a review paper by Righini et al. (2000). The experimental set-up includes a battery bank, a switch, a standard resistor, and a high-speed pyrometer (Fig. 5.7). A rapid heating of a sample placed in a chamber is achieved by the passage of a current pulse (several thousand amperes) of subsecond duration. The experimental quantities (current, voltage across the central portion of the sample, temperature) are measured with a submillisecond time resolution. The most accurate experiments were performed on tubular samples with the following typical dimensions: length, 70–100 mm; diameter, 6–10 mm; and wall thickness, 0.5–1 mm. A rectangular hole in the middle of the sample defined a blackbody cavity with emissivity greater than 0.98. The upper clamp is fixed, while the lower clamp is movable to permit the thermal expansion of the sample. The measurements are possible either in vacuum or in a controlled gas atmosphere. Temperature is measured by a high-speed pyrometer designed at the IMGC.

The measurements may also be performed on strip samples with simultaneous determinations of the normal spectral emissivity of the strip during the pulse experiment. The strips have the following typical dimensions: length, 80–100 mm; width, 8–10 mm; and thickness, 0.5–1 mm. The measurement of the normal spectral emissivity requires an additional apparatus consisting of a small integrating sphere inserted in the chamber. The radiance temperature on one side of the strip is measured by a high-speed pyrometer. The other side of the strip is placed outside a hole of the integrating sphere (Fig. 5.8). A modulated laser beam strikes the side of the strip facing the sphere. The reflected light is collected by the integrating sphere and measured by a silicon detector. A lock-in technique is used to discriminate between the reflected modulated light and the continuous thermal radiation from the sample. The reflectance of the sample is measured in comparison to that of a known $BaSO_4$ reference. The quantity that is determined is the hemispherical spectral reflectivity; the normal emissivity is obtainable in accordance with Kirhhoff's law for opaque materials. Using this technique, Righini et al. (1999) determined the specific heat of niobium in the 1100–2700 K range.

Righini et al. (2000) claimed that pulse calorimetry may be considered the most accurate measuring technique for the determination of heat capacity of electrical conductors at high temperatures. Cezairliyan and Righini (1996) reviewed high-speed optical pyrometry.

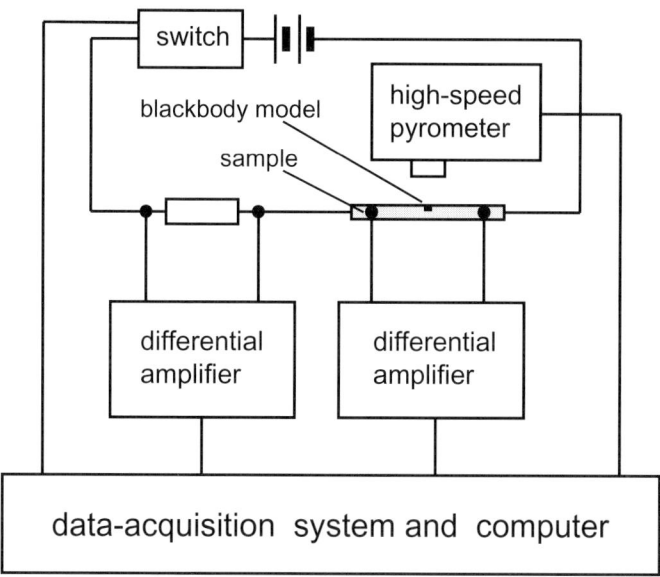

Fig. 5.7. Simplified diagram of dynamic set-up developed at IMGC (Righini et al. 2000). A current pulse of several thousand amperes heats the sample

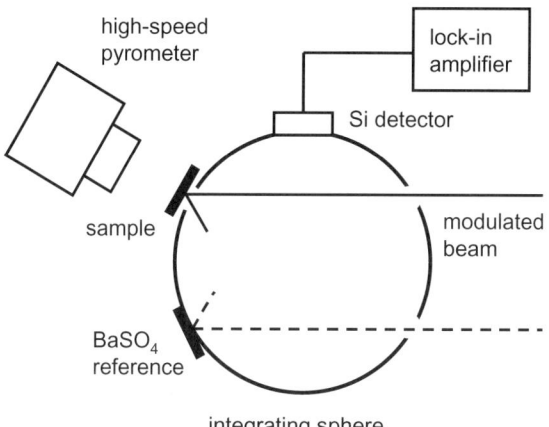

Fig. 5.8. Determination of the spectral emissivity of the surface of the sample during dynamic experiment. The integrating sphere averages the intensity of light reflected in all directions (Righini et al. 2000)

Dynamic Measurements at INS

An apparatus designed at the INS (Institute of Nuclear Sciences, Beograd) and described by Dobrosavljević and Maglić (1989) employs wire samples and heating rates up to 1500 K.s^{-1}. The set-up consists of a vacuum chamber, an electric power circuit, measuring and control devices, and a computer (Fig. 5.9).

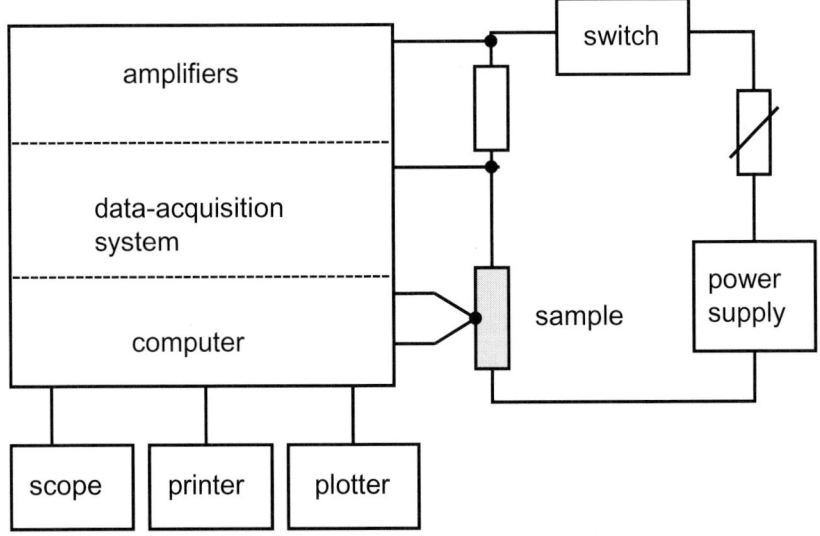

Fig. 5.9. Simplified diagram of apparatus for dynamic measurements designed at INS (Dobrosavljvić and Maglić 1989)

The sample, 2 mm in diameter, has a total length of about 200 mm. The central portion of the sample is 20 mm long. Three 0.05 or 0.1-mm thermocouples are spot-welded at the centre and symmetrically at 10-mm separations on both sides. The thermoelectrodes serve also as potential leads to measure the voltage drop across the central portion of the sample. The data are collected at a 1-kHz sampling rate. Contact temperature measurements in the presence of an electric current pose a problem because it is impossible to position both thermoelectrodes exactly along the same equipotential line. Comparison of the last thermocouple reading during the heating period and the first reading after ending the current enables one to introduce necessary corrections. In the 300–1900 K range, the maximum uncertainties were estimated to be 3% in the specific heat and 1% in the electrical resistivity. For experimental results by this technique see Dobrosavljević et al. (1989), Dobrosavljević and Maglić (1989, 1991), Maglić and Dobrosavljević (1992), Maglić et al. (1994, 1995/96, 1997), Perović et al. (1996), Vuković et al. (1996), Milošević et al. (1999), Maglić and Pavičić (2001), Pavičić and Maglić (2002), and Maglić (2003).

Ronchi et al. (1999) reported on laser-pulse apparatus for thermophysical measurements at high temperatures.

5.4 Relaxation Techniques

This technique employs measurements of the cooling (or heating) rate of a sample, whose temperature differs from that of the surroundings (Bachmann et al. 1972; Denlinger et al. 1994). This rate depends on the temperature difference, on the heat capacity of the sample, and on the heat transfer coefficient, i.e., the temperature derivative of the heat losses from the sample. The relaxation method is applicable even to very small samples. One of the methods employs a calorimeter, in which a sample under study and a reference one are placed in turn. The temperature dependence of the heat losses from the calorimeter remains the same, and it is easy to calculate the ratio of the specific heats of the two samples.

Denlinger et al. (1994) described a microcalorimeter for measuring heat capacity of thin films in a wide temperature range. The addenda of the calorimeter, including substrate, is 4×10^{-6} J.K^{-1} at room temperature and 2×10^{-9} J.K^{-1} at 4.3 K. The device is capable of measuring heat capacities of samples as small as a few micrograms. Semiconductor processing techniques were used to create a device with an amorphous silicon-nitride membrane as the sample substrate. The authors employed the relaxation method, but they pointed out that the microcalorimeter may be used with any of calorimetric techniques, including modulation calorimetry. The microcalorimeter includes a platinum thin-film thermometer for temperatures above 40 K and either a thin-film amorphous Nb-Si or a boron-doped polycrystalline silicon thermometer for lower temperatures. A DC current is supplied to the heater, causing the temperature of the calorimeter cell to rise to a temperature $T_0 + \Delta T$. After turning off the heater, the temperature decays exponentially with the time constant τ. The internal time constant (through the sample and between the sample, the heater, and the thermometer) must be much shorter than τ. This requirement is met when the sample is conducting or when a conducting layer is deposited either under or over the sample. The authors were probably the first to apply the modern technology of miniaturisation of semiconductor devices for calorimetric techniques. The authors reported measurements from 4.3 to 360 K and claimed that the microcalorimeter is suitable for the 1.5–800 K range.

Step Method

Using the step method, the calorimeter is first brought to a temperature $T = T_0 + \Delta T$, slightly higher than that of the surroundings, T_0. After ending the heating, the temperature of the calorimeter decays exponentially:

$$T = T_0 + \Delta T \exp(-t/\tau), \qquad (5.5)$$

where $\tau = C/K$ is the relaxation time, C is the sum of the heat capacities of the sample and of the calorimeter itself, and K is the heat transfer coefficient.

The determinations of specific heat thus include measurements of the relaxation time and of the heat transfer coefficient. Under steady-state conditions, $P = K\Delta T$, where P is the power applied to the calorimeter, which is necessary to increase its temperature by ΔT. The temperature increment is small, so that the linear relation is valid.

Sweep Method

With the sweep method, a wide temperature range is covered in one run. For this purpose, the steady-state temperature of the calorimeter is determined beforehand as a function of the heating power. Then the heat capacity of the sample becomes available from the measured cooling curve. The relaxation technique was employed by Zinov'ev and Lebedev (1976) in measurements on tungsten in the range 2400–3600 K. After heating a wire sample to the highest temperature, the heating current was switched off and the cooling was monitored by means of a photomultiplier and an oscilloscope. The heat capacity obeys the relation

$$mc = -Q/T_c', \qquad (5.6)$$

where Q is the power of heat losses from the sample, and T_c' is the cooling rate at this temperature. The results obtained by Zinov'ev and Lebedev (1976) well coincide with those from the modulation measurements (Kraftmakher and Strelkov 1962).

Hatta (1979) designed a relaxation calorimeter employing light heating (Fig. 5.10). Ema et al. (1993) described a calorimeter operating in both relaxation and modulation modes. Shepherd (1985) considered a case of poor thermal coupling between the sample and the holder. In this case, the cooling curve obeys the dependence $\Delta T = A_1 \exp(-t/\tau_1) + A_2 \exp(-t/\tau_2)$. Usually, the contribution of the second term (the τ_2 effect) is small.

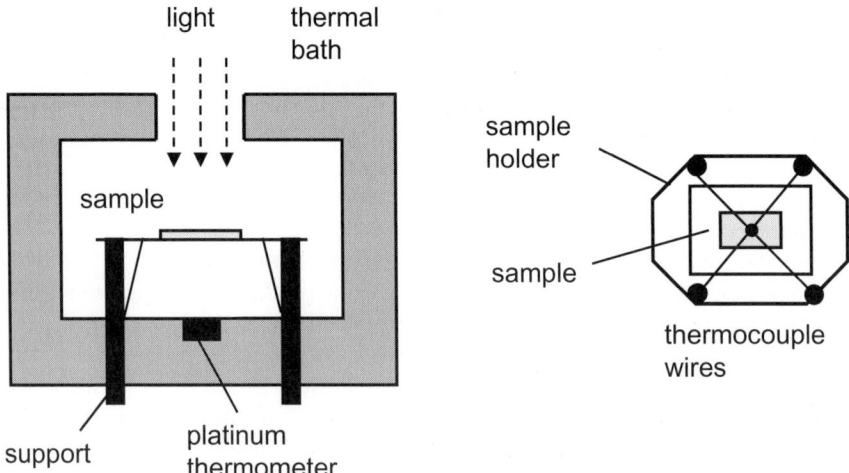

Fig. 5.10. Relaxation technique with light heating (Hatta 1979). The set-up is usable as a modulation calorimeter

Yao et al. (1998) reported on an automated dual-mode calorimeter capable of operation in either modulation or relaxation mode. Instead of the usual step-function input power, a power linearly ramped in time was used for the relaxation measurements. The principal advantage of this technique is its ability to measure small latent heats on small samples. The authors studied phase transitions in some liquid crystals.

5.5 Rapid-Heating Experiments

The rapid-heating technique allows measurements of thermophysical properties to be made over very wide temperature intervals, including the liquid state. Many new results were obtained by this method (for reviews see Lebedev and Savvatimskii 1974; Gathers 1986; Cezairliyan et al. 1990). The primary motivations for trying to conduct thermophysical experiments on subsecond time scales are to extend the measurements of thermophysical and related properties to temperatures far above the limits of steady-state methods or to study systems far removed from the thermodynamic equilibrium.

Very High Heating Rates

Pottlacher et al. (1991, 1993) described experiments employing heating rates above 10^9 K.s^{-1}. Starting at room temperature, the measurements are performed far into the liquid phase of the metal under study, up to 10000 K. In the apparatus developed, the energy is stored in a 5.4-μF capacitor, with the charging voltage up to 8 kV (Fig. 5.11). The wire samples are typically 40 mm long and 0.25 mm thick. Water serves as the ambient medium to avoid peripheral electric discharges. The highest pressure in the vessel is 0.2 GPa. An initial pulse triggers a flashlight for background illumination of the sample, while a three-electrode spark gap triggers the main discharge. The quantities measured during the entire run are as follows: (i) the current through the sample, by means of an induction coil; (ii) the voltage drop across the sample, using a coaxial voltage divider; (iii) the radiance temperature of the sample, by a fast pyrometer; and (iv) the final volume of the sample, by a shadowgraph technique with a 30-ns exposure. The melting temperatures served as the calibration points, and the spectral emissivity of the surface of the sample was assumed to be independent of temperature. Special care should be taken to avoid superheating the samples, which is quite probable in rapid-heating experiments.

From the measurements, the enthalpy, the electrical resistivity, and the thermal expansion of the sample are available. The authors pointed out that static measurements provide more accurate results for the solid phase. However, rapid heating allows measurements far above the melting point of the sample. This technique has the potential to monitor the vacancy equilibration and hence to reveal vacancy contributions to the enthalpy and electrical resistivity of metals at high temperatures.

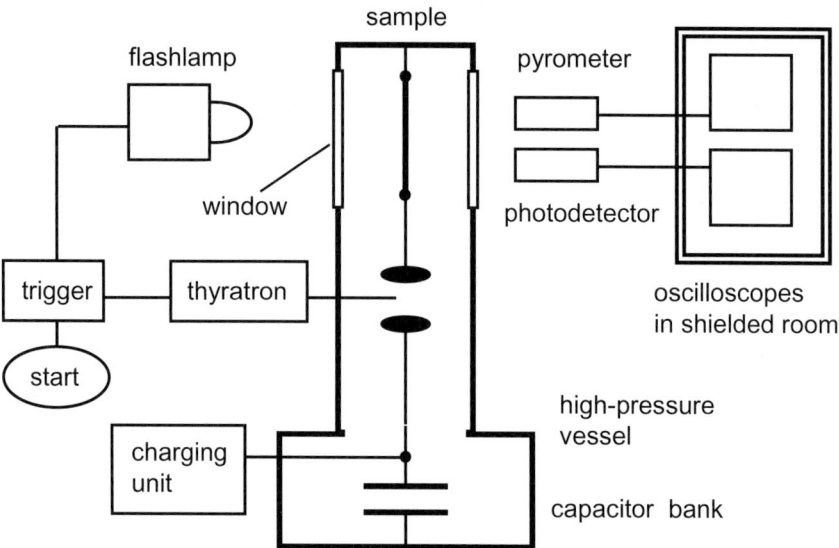

Fig. 5.11. Simplified diagram of apparatus for rapid-heating experiments (Pottlacher et al. 1991, 1993). Measurements are possible far above the melting point of the sample

Set-up of Microsecond Resolution

The same group designed a microsecond-resolution system (Kaschnitz et al. 1992). Lower heating rates, 10^7–10^8 K.s^{-1}, permit more accurate measurements. Wire or tube-shaped samples are resistively heated using an RCL discharge circuit. Energy is stored in a capacitor, 240–500 µF, which may be charged up to 10 kV. Typically, the heating current is about 5000 A and the pulse is 80 µs long. To measure the temperature of the sample, a lens produces its magnified image at the rectangular entrance of an optical fibre. The light passes through the fibre and enters a photodiode detector. The detector is self-calibrated with the plateau of the melting transitions. The thermal expansion of the samples is determined photographically, by means of a Kerr cell. The authors estimated the uncertainty of the data to be 3% for the enthalpy and 3% for the electrical resistivity, without corrections for the thermal expansion of the samples.

High-Speed Scanning Microcalorimetry

In the last decades, many authors used scanning calorimeters capable of high-speed measurements on extremely small samples. Fröchte et al. (1990) developed a high-speed calorimeter for the range 300–2500 K (Fig. 5.12). A conducting sample in the form of a wire or ribbon is heated from room temperature by passing a constant electric current through it. The heating current and the voltage across a central portion of the sample are measured and stored by digital storage oscilloscopes. The heating rate of the calorimeter is between 10^5 and 10^7 K.s^{-1}. At high enough heating rates, the total heat losses become practically negligible, and the measured energy becomes about equal to the increase in the enthalpy of the portion of the sample between the voltage probes. The temperature of the central portion of the sample is measured by means of a fast photodiode. The photodiode signal is amplified and stored. The time resolution of the system is 1 µs, with a storage capacity of 4000 points per channel. The data are then transferred to a computer.

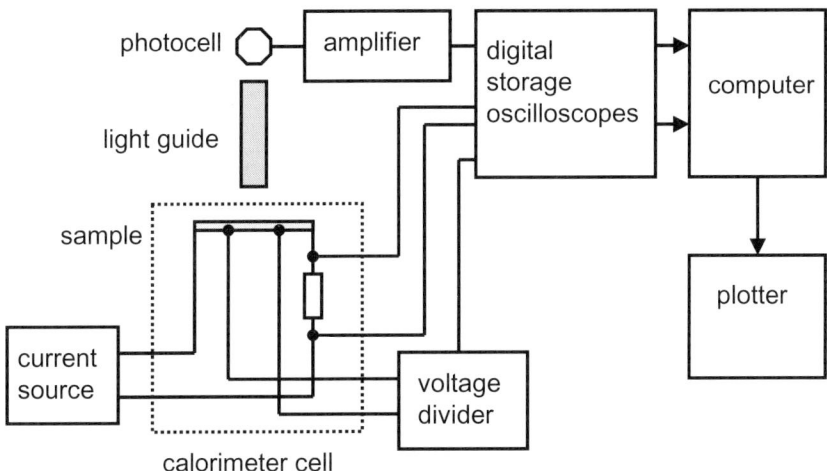

Fig. 5.12. Simplified diagram of high-speed calorimeter for the range 300–2500 K (Fröchte et al. 1990). The heating rate is in the range 10^5 to 10^7 K.s^{-1}

Lai et al. (1995) designed a sensitive scanning nanocalorimeter usable at heating rates up to 10^4 K.s^{-1}. The authors combined the sample, the sample holder, the heater, and the thermometer into a single multiplayer thin-film configuration (Fig. 5.13). The sample holder is a 180-nm silicon nitride membrane supported at the perimeter by a silicon substrate. A 180-nm nickel stripe functions both as a heater and a thermometer. The sample is deposited on the opposite side of the membrane and has approximately the same width as the heater. The thin membrane ensures good thermal conduction between the sample and the heater, while it isolates them electrically. Using this technique, Efremov et al. (2000) observed multiple periodic maxima in the heat capacity of indium clusters in the range of 2–4 nm. Efremov et al. (2003) reported on a scanning nanocalorimetry technique of high sensitivity (10^{-9} J.K^{-1}) and high scan rate (10^4–10^5 K.s^{-1}). Nanocalorimetric measurements were also reported by Zhang et al. (2000, 2002).

Berger et al. (1996) developed a fast and extremely sensitive method for thermal analysis. Their nanocalorimeter is based on the deflection of a bimetallic cantilever. A V-shaped silicon nitride cantilever with a gold layer, 50 nm thick, was used as a micromechanical sensor in the form of a bimetallic strip. Since the gold layer has a larger thermal expansivity than silicon nitride, the cantilever bended towards the silicon nitride side when heated up. The samples with volume of the order of 10^{-9} cm^3 were placed on the silicon nitride side of the cantilever. The mass of the sample was extracted from a resonance frequency shift between an unloaded and an end-loaded cantilever. A proximal resistive heater was used to ramp the temperature of the cantilever and its support. A scanning force microscope served to detect the deflection of the cantilever by measuring the displacement of a laser beam reflected by the cantilever. With no sample mounted, the cantilever showed a deflection proportional to temperature. The response time of the nanocalorimeter is about 1 ms, and it is capable of measuring changes in enthalpy with a resolution of about 10 nJ at room temperatures. Using this technique, Nakagawa et al. (1998) presented a device with a sensitivity of 0.5 nJ and time resolution of 0.5 ms. The mass of the sample was about 10^{-11} g.

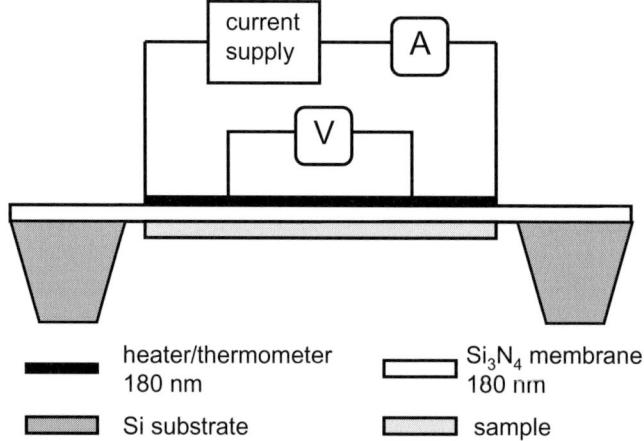

Fig. 5.13. Simplified diagram of calorimeter cell designed by Lai et al. (1995). The measurements are performed at heating rates up to 10^4 K.s^{-1}

Lerchner et al. (1999) discussed the properties of integrated-circuit calorimeters and compared several types of commercially available chips suitable for this purpose (Fig. 5.14). The integrated circuit consists of a thin silicon chip with a rim for the stabilisation, assembled in a massive chip carrier. The sensitive area for the heat detection contains a heater suitable for calibration purposes. Between the sensitive area and the rim, a thermopile is integrated in the membrane. For the application of an integrated circuit as a calorimeter, the heat flow should take place mainly through the membrane. The heat-power sensitivity of the chip is a very important property of such a calorimeter. It depends on the thermal resistance of the membrane, on the Seebeck coefficient of the thermocouple, and on the number of couples in the thermocouple. Lerchner et al. (2002) considered the accuracy achievable with integrated-circuit calorimeters.

Winter and Höhne (2003) explored possibilities of employment commercially available chips suitable for calorimetric measurements and presented some conclusions important for practical use of such chips. Adamovsky et al. (2003) reported on scanning microcalorimetry at very high cooling rates up to 5×10^3 K.s^{-1}. A commercial thin-film sensor (thermal conductivity gauge) served as a calorimeter cell. The cell consisted of a silicon-nitride membrane with a thin-film heater and a thin-film thermopile. A 0.1-µg polyethylene sample was placed at the centre of the membrane. The heat leakage from the sample was calibrated and taken into account.

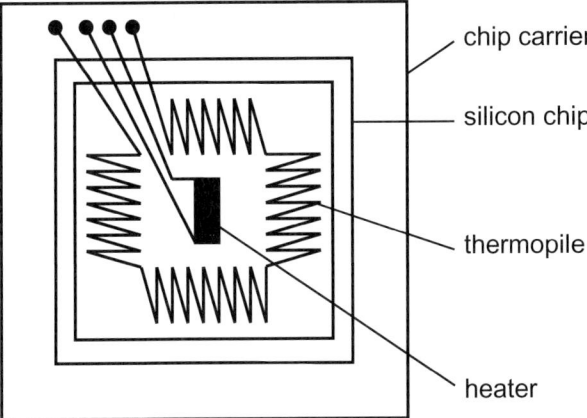

Fig. 5.14. Simplified diagram of a commercially available chip suitable for calorimetric measurements on very small samples. The chip incorporates a thin-film heater and a thin-film thermopile. In many cases, such integrated-circuit calorimeter may be useful

6 Modulation Calorimetry I

This chapter considers basic modulation measurements, including calorimetry under high pressures. For our purpose, it is convenient to divide the whole temperature range of modulation calorimetry into three regions, namely: (i) high temperatures, $T > 1000$ K, (ii) medium temperatures, 300 K $< T < 1000$ K, and (iii) low temperatures, $T < 300$ K. The high-temperature region was the first one, where modulation measurements were carried out.

6.1 High Temperatures

Modulation calorimetry at high temperatures includes measurements on wire and bulk samples of metals and alloys, studies of nonconducting materials, and non-contact calorimetric measurements.

6.1.1 Wire Samples

In the first modulation measurements of specific heat, an AC current passing through wire samples heated them. In this case, it is easy to measure the amplitude of the heating-power oscillations because it equals the mean applied power. The temperature oscillations were determined through oscillations in the resistance of the sample. For such measurements, Corbino developed two techniques, the supplementary-current method and the third-harmonic method.

A very convenient approach turned out the equivalent-impedance technique. It employs a bridge or potentiometer circuit, whose balance is independent of the AC component of the heating current. The heat capacity of the sample is available from the resistance and capacitance corresponding to the balance of the circuit. A disadvantage of this technique is the necessity to know the temperature dependence of the resistance of the sample and of its temperature derivative. This drawback is peculiar to all methods based on determinations of the temperature oscillations through the resistance of the sample. The bridge circuit was used in measurements of the specific heat of tungsten in the range 1500–3600 K (Kraftmakher and Strelkov 1962). Tungsten has a high melting point, and for a long period it served for the fabrication of filaments for light bulbs and cathodes for vacuum tubes. The temperature dependence of its resistivity was carefully studied and now is well known. The measurements provided good opportunity to check the equivalent-impedance technique. In the range 1500–2500 K, the results obtained were in good agreement with existing data. At higher temperatures, a strong non-linear increase in the specific heat was observed and attributed to point-defect formation in the crystal lattice. Later, this strong non-linear increase was confirmed by other calorimetric techniques (Fig. 6.1). By the use of a quasi-adiabatic calorimeter operating in the continuous regime, Schmidt et al. (1970) measured the specific heat of tungsten in the range 300–1900 K. In the range 1500–1900 K, their data are in fair agreement with the results of the modulation measurements: the difference is less than 1%.

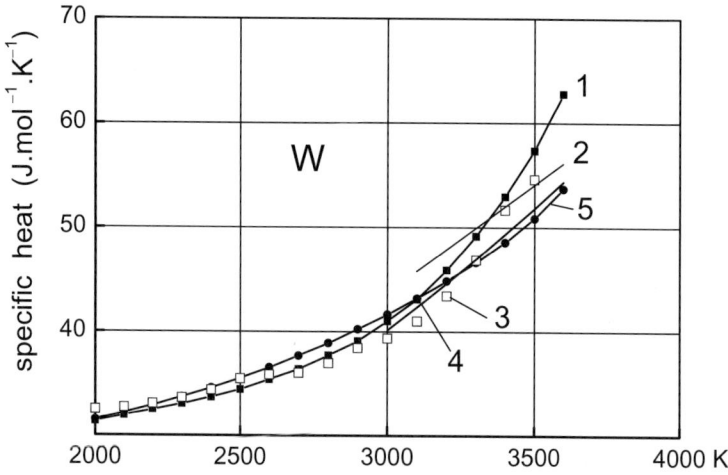

Fig. 6.1. Specific heat of tungsten at high temperatures. 1 – curve representing data from modulation measurements (Kraftmakher and Strelkov 1962), pulse calorimetry (Affortit and Lallement 1968), and relaxation method (Zinov'ev and Lebedev 1976); 2 – drop method (Arpaci and Frohberg 1984); 3, 4 – pulse calorimetry (Yakunkin 1983; Senchenko and Sheindlin 1987); 5 – dynamic calorimetry (Righini et al. 1993)

The equivalent-impedance method was then employed in studies of other high-melting-point metals (Table 6.1): tantalum (Kraftmakher 1963a), niobium (Kraftmakher 1963b), molybdenum (Kraftmakher 1964), and platinum (Kraftmakher and Lanina 1965). The non-linear increase in the specific heat of all the metals was treated as a result of point-defect formation in the crystal lattice (Kraftmakher 1966c).

The results obtainable with the equivalent-impedance technique directly depend on the accepted values of the temperature derivative of the resistance of the samples. For instance, the specific heat of niobium (Kraftmakher 1963b) appeared to be several percent lower than the data from dynamic measurements by Righini et al. (1985) and Maglić et al. (1994). However, when more accurate data on the temperature derivative of resistance (Righini et al. 1985) are used for the calculations, the original specific-heat data shift upward between 4 and 6% (Fig. 6.2).

Table 6.1. Specific heat of high-melting-point metals measured by the equivalent-impedance technique (smoothed values, J.mol^{-1}.K^{-1})

T (K)	W	Ta	Mo	Nb	Pt
1000					29.95
1100					30.5
1200		27.75			31.05
1300		28.05	30.5	27.55	31.65
1400		28.35	31.2	28.05	32.35
1500	28.65	28.65	31.8	28.6	33.15
1600	29.2	28.9	32.5	29.15	34.1
1700	29.75	29.2	33.25	29.7	35.3
1800	30.3	29.5	34.05	30.3	36.8
1900	30.85	29.8	34.95	31.0	38.7
2000	31.4	30.1	36.0	31.75	41.05
2100	32.0	30.4	37.3	32.55	
2200	32.6	30.7	38.9	33.55	
2300	33.2	31.1	40.9	34.65	
2400	33.9	31.5	43.4	35.95	
2500	34.65	32.0	46.5	37.45	
2600	35.55	32.55		39.2	
2700	36.55	33.2		41.15	
2800	37.8	34.05			
2900	39.25	35.0			
3000	41.05				
3100	43.25				
3200	45.9				
3300	49.05				
3400	52.8				
3500	57.25				
3600	62.4				

The equivalent-impedance technique enables one to employ various samples, from thinnest wires or films to 1-mm wires. The heating current increases with the thickness, while the modulation frequency can be reduced. The modulation frequency satisfying the adiabaticity conditions is inversely proportional to the thickness of the sample. With a proper choice of the modulation frequency and a given relation between the AC and DC components of the heating current, the amplitude of the temperature oscillations does not depend on the thickness of the sample. In vacuum, the current necessary to heat up a wire sample to a given temperature is proportional to $d^{3/2}$ (d is the diameter of the wire), while the resistance of the sample and its temperature derivative are proportional to d^{-2}. The AC voltage across the sample caused by the temperature oscillations is therefore proportional to $d^{-1/2}$. The decrease in the signal due to an increase in the thickness of the sample is thus not as strong as one might expect.

The equivalent-impedance technique reduces measurements of the enthalpy to determinations of the resistance of the sample at given temperatures:

$$H_2 - H_1 = \int_{T_1}^{T_2} mc\,dT = \int_{R_1}^{R_2} (2I_0^2/\omega^2 RC)dR, \tag{6.1}$$

where R_1 and R_2 are the resistances at the temperatures T_1 and T_2.

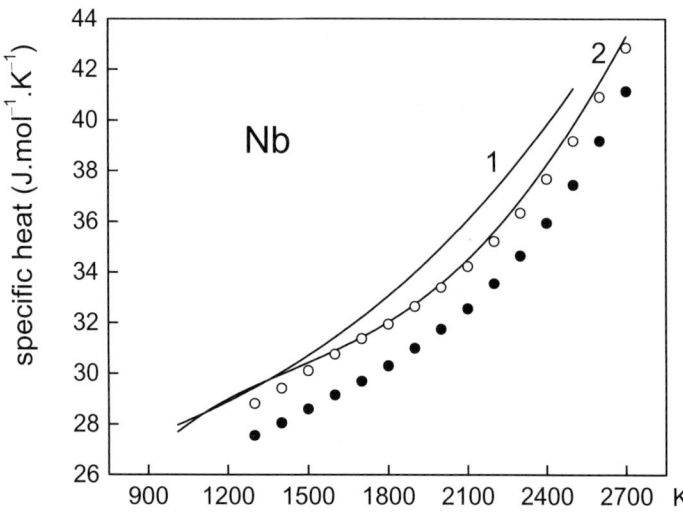

Fig. 6.2. Specific heat of niobium: ● modulation measurements, original data (Kraftmakher 1963b), 1 – Righini et al. (1985), 2 – Maglić et al. (1994), ○ modulation measurements, corrected data

When the thickness of the sample is about 1 mm, the temperature derivative of its resistance can be measured with a thin thermocouple welded to the sample. A DC current and an AC current modulated by an infra low frequency pass through the sample. Temperature oscillations with a period of several seconds and corresponding oscillations of the resistance thus occur in the sample. The voltage across a central portion of the sample contains a component of the infra low frequency. This component is proportional to the amplitude of the temperature oscillations, to the DC current, and to R', the temperature derivative of the resistance of the sample. A two-channel recorder or X-Y plotter records the oscillations of the voltage of the thermocouple and of the voltage across the central portion of the sample. An adiabatic regime is not required in such measurements. The modulation frequency may therefore be reduced to provide significant temperature oscillations and to reduce the influence of the thermal inertia of the thermocouple to a negligible level. The data on R' derived in this way can be then used in measurements of the specific heat at higher modulation frequencies corresponding to an adiabatic regime.

A moderate vacuum suffices for the measurements at medium temperatures to avoid instabilities due to convection. However, a perfect vacuum is needed for high-temperature studies, and the best method to obtain it is the cryopumping. The most difficult case is when an inert gas atmosphere is necessary to keep the samples at high temperatures.

6.1.2 Nonadiabatic Regime

An adiabatic regime of the measurements is achievable by increasing the modulation frequency. However, in some cases such an increase is undesirable (e.g., when the heat is generated at the surface of the sample or when a thermometer used has a long time constant). In a nonadiabatic regime, it is necessary to determine both the amplitude and the phase of the temperature oscillations, and the specific heat is given by Eq. (2.5). It was shown that measurements in a nonadiabatic regime are possible with about the same accuracy as in an adiabatic regime (Varchenko and Kraftmakher 1973). Although the heat losses under nonadiabatic conditions are significant, they are readily taken into account. Moreover, in a nonadiabatic regime one can measure the specific heat using the heat transfer coefficient as a reference quantity. In the case of the radiative heat transfer, this coefficient is governed by the temperature dependence of the hemispherical total emittance of the sample.

When a DC current with a small AC component heats a wire sample, the oscillations of the temperature obey Eqs. (2.10) and (2.11). By combining Eqs. (2.10c) and (2.11c), it is easy to determine any of the quantities c, R' and Q' in terms of one of the other two:

$$mc = 2I_0^2 R'/(\cot\psi - \cot\varphi)\omega, \tag{6.2a}$$

$$mc = 2Q'/(\cot\psi + \cot\varphi)\omega, \tag{6.2b}$$

$$R' = Q'(\cot\psi - \cot\varphi)/I_0^2(\cot\psi + \cot\varphi), \tag{6.2c}$$

where φ is the phase difference between the temperature oscillations and the AC component of the heating current, ψ is the phase difference between the temperature oscillations and the AC component of the voltage across the sample, R' is the temperature derivative of the resistance of the sample, and Q' is the heat transfer coefficient.

The choice of either R' or Q' as a reference quantity depends on their reliability. The heat transfer coefficient Q' may be useful at extremely high temperatures, where equilibrium point defects contribute to the electrical resistivity and especially to its temperature derivative. In the case of the radiative heat exchange,

$$Q = S\varepsilon\sigma T^4, \tag{6.3a}$$

$$Q' = Q(4 + T\varepsilon'/\varepsilon)/T = nQ/T, \tag{6.3b}$$

where S is the surface area of the sample, σ is the Stefan-Boltzmann constant, ε is the hemispherical total emittance, and $\varepsilon' = d\varepsilon/dT$. As a rule, the second term in the brackets is several times smaller than the first one and decreases with increasing temperature. Especially favourable is a case when ε is close to unity and weakly temperature dependent, so that one can take $n = 4$, as for a perfect blackbody. In this case, the measurements reduce to finding $\cot\psi$ and $\cot\varphi$. One of the methods is to measure the phase shift between the signal from a photocell viewing the sample and either the AC current through or the AC voltage across the sample. A photodiode and an oscilloscope served to measure these phase shifts (Fig. 6.3). The signal from the photodiode was fed to the Y-input of the oscilloscope. The voltage across either the sample or the bridge arm in series with the sample proceeded to the X-input through a phase-shifting *RC* circuit. The capacitor *C* was adjusted to transform the Lissajous figure into a straight line.

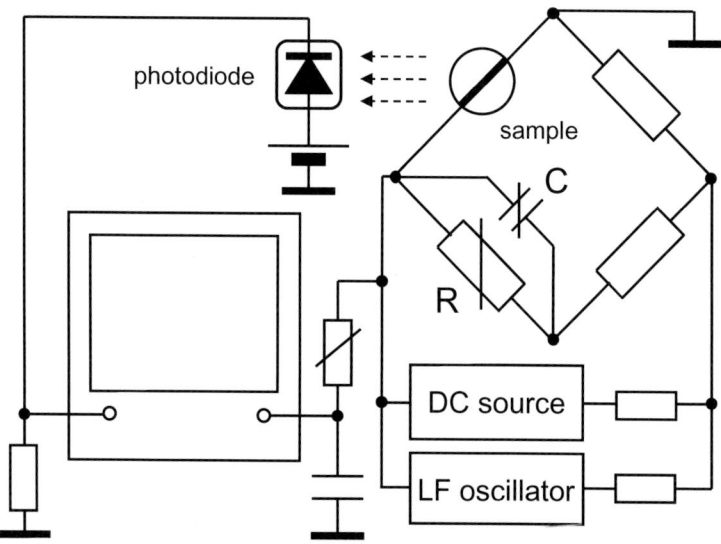

Fig. 6.3. Measurement of specific heat of wire samples in nonadiabatic regime (Varchenko and Kraftmakher 1973). The measurements provide about the same accuracy as in adiabatic regime

The second possibility consists in measuring the electrical impedance of the sample. The sample displays a complex impedance because there exists a phase shift between the AC components of the voltage and the current: $\mathbf{Z} = Z\exp(-i\Phi)$, $\Phi = \varphi - \psi$. Therefore,

$$\cot\Phi = (1 + \cot\psi \cot\varphi)/(\cot\psi - \cot\varphi). \quad (6.4)$$

By equating the amplitudes of the temperature oscillations expressed through the AC current (2.10b) and the voltage (2.11b), the modulus of the impedance of the sample equals

$$Z = U/i = R_0 \sin\varphi/\sin\psi, \quad (6.5)$$

where R_0 is the DC resistance of the sample.

A bridge circuit with one arm consisting of a resistor *R* shunted by a capacitor *C* was used in the measurements. The modulus of the impedance of this arm is $R(1 + \omega^2 R^2 C^2)^{-½}$, and the phase angle is $\tan^{-1}(-\omega RC)$. Under the balance of the bridge,

$$\cot\psi = (R/R_0 - 1)/\omega RC, \qquad (6.6a)$$

$$\cot\varphi = R_0(\cot\psi - \omega RC)/R. \qquad (6.6b)$$

Thus, $\cot\varphi$ and $\cot\psi$ are available from the parameters R, R_0, and C. The results of the measurements by the impedance method and by the photodiode are in good agreement. The quantity mc/R' was measured by the equivalent-impedance method in the 3–60 Hz range. The absence of frequency dependence up to frequencies corresponding to adiabatic conditions confirms the validity of measurements in a nonadiabatic regime (Fig. 6.4). The results show that a nonadiabatic regime is suitable for measuring the specific heat by R', as in an adiabatic case, and also by Q'. Moreover, the temperature derivative of resistance itself can be determined from such measurements. In some cases, a nonadiabatic regime may appear to be advantageous.

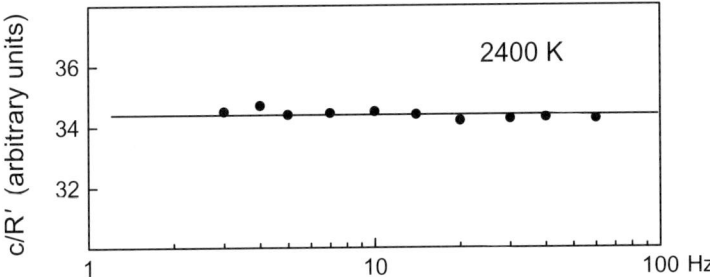

Fig. 6.4. Results of equivalent-impedance measurements at frequencies corresponding to both adiabatic and nonadiabatic conditions do not depend on the frequency (Varchenko and Kraftmakher 1973)

The simplest way to measure the heat capacity by the use of Eq. (6.2b) consists in determining the phase shifts φ and ψ (Fig. 6.5). A sample is heated by a DC current with a small AC component. A photodiode detects the temperature oscillations in the sample. A phase meter measures the phase shifts φ and ψ. One input of the meter is connected to the load resistor of the photodiode, while the second is connected, in turn, to the sample and to a resistor placed in series with the sample.

When the equivalent-impedance technique is used in a nonadiabatic regime, the specific heat obeys a relation similar to (4.12):

$$mc = 2I_0^2 R'/\omega^2 R_0 C[1 + (R/R_0 - 1)^2/\omega^2 R^2 C^2], \qquad (6.7)$$

where the correction factor approaches unity with increasing frequency.

A nonadiabatic regime of the measurements allows one to employ the phase shift between the heating-power oscillations and the temperature oscillations, instead of the amplitude of the temperature oscillations. For instance, one may use a conducting capillary of known heat capacity heated by a DC current with a small AC component. When the mean temperature of the capillary varies, the phase shift ϕ is kept constant, by adjusting the modulation frequency. For instance, the phase shift may be set at 45°. For this shift, $\tan\phi = 1$, and $\omega = Q'/mc$ at every mean temperature.

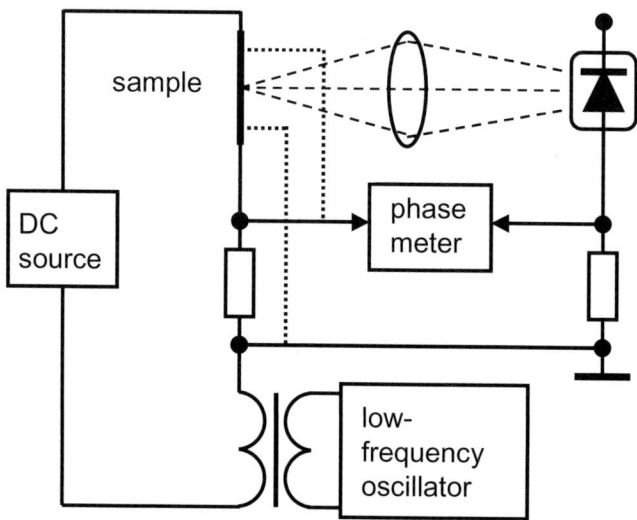

Fig. 6.5. Arrangement for measuring the phase differences and determination of the heat capacity through the heat transfer coefficient

Then the capillary is filled with the sample and the run repeated. The temperature dependence of the heat transfer coefficient Q' remains the same. The new temperature dependence of the frequency ω_1 necessary to maintain the same phase shift depends on the heat capacity of the sample, $m_1 c_1$ (Fig. 6.6). Clearly,

$$\omega/\omega_1 = (mc + m_1 c_1)/mc. \tag{6.8}$$

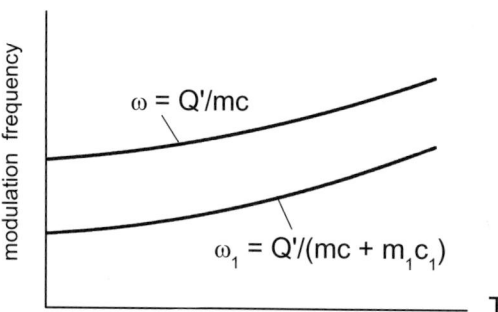

Fig. 6.6. Temperature dependence of the modulation frequency necessary to keep a constant phase shift between the heating-power oscillations and the temperature oscillations, schematically. The two curves correspond to the empty and the filled calorimeters

The capillary serves here as a heater and a reference of specific heat, and there is no need to calibrate the sensor detecting the temperature oscillations. With this method, the heat capacity of the sample is represented by the modulation frequency. Hatta and Katayama (1998) also considered the possibility to determine the heat capacity through measurements of the phase of the temperature oscillations.

6.1.3 Bulk Samples

Electron-bombardment heating is a method suitable for bulk samples or samples of irregular shape. The temperature oscillations in the sample are measured through the radiation from it or by a thermocouple. This method was used in measurements of the specific heat of iron (Varchenko et al. 1978). A blackbody model is necessary to reliably measure temperature oscillations in the sample through the radiation from it.

6.1.4 Molten Metals

To measure the specific heat of molten tin, copper, and gallium, Akhmatova (1965, 1967) filled with them a niobium capillary. The capillary was located in a vacuum chamber and heated by passing an AC current. A photomultiplier detected the temperature oscillations. The method rests on the fact that the temperature dependence of the radiation from the empty and the filled capillary is the same. This allows one to directly compare the temperature oscillations in the two cases at a given mean temperature. The oscillations are inversely proportional to the total heat capacity of the capillary, empty or filled with the melt. The heat capacity of the melt thus is fitted to that of the capillary. The mean temperature is determined by an optical pyrometer or from the electrical resistance of the empty capillary. An advantage of this approach is that one has no need to know absolute values of the temperature oscillations. The capillary serves as a calorimeter and a reference of specific heat. This method seems to be the simplest one and quite adequate.

6.1.5 Active Thermal Shield

When the heat losses are due to the radiation from the sample, the heat transfer coefficient grows rapidly with temperature being proportional to T^3. In fact, the increase is somewhat stronger because of the temperature dependence of the hemispherical total emittance. As a rule, it increases with temperature. Therefore, the modulation frequency required to meet adiabaticity conditions grows with the temperature. Low modulation frequencies can be maintained using a nonadiabatic regime.

The second possibility is the compensation for heat losses from the sample by an active thermal shield (Kraftmakher and Cherepanov 1978). A sample in a vacuum chamber is surrounded by a shield whose temperature oscillates with the same frequency and phase as the temperature of the sample (Fig. 6.7). The sample is heated by an AC current or by a DC current with a small AC component. The amplitude of the temperature oscillations in the shield is adjusted to nullify the AC component of the heat losses from the sample. The criterion of the compensation is a 90° phase shift between the oscillations of the heating power and of the temperature of the sample. The heat-balance equation now takes the form

$$mc\Theta' + (K_1 - bK_2)\Theta = p\sin\omega t. \tag{6.9}$$

Here K_1 and K_2 are the temperature derivatives of the mutual heat transfer coefficients for the sample and the shield, and b is the ratio of the amplitudes of the temperature oscillations in the shield and in the sample. The solution to this equation is

$$mc = (p/\omega\Theta_0)\sin\phi, \tag{6.10a}$$

$$\tan\phi = mc\omega/(K_1 - bK_2). \tag{6.10b}$$

The criterion of adiabaticity, $\tan\phi \gg 1$, thus can be satisfied at low frequencies by adjusting the amplitude of the temperature oscillations in the shield.

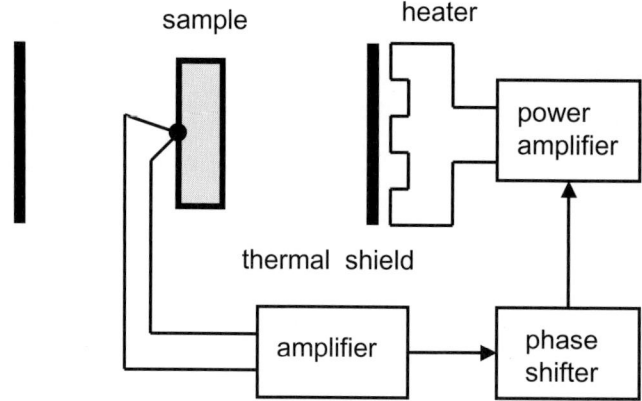

Fig. 6.7. Compensation for heat losses by means of active thermal shield (Kraftmakher and Cherepanov 1978). Temperature oscillations in the shield are adjusted to make temperature oscillations in the sample to be lagged by 90° to the oscillations of the heating power

6.1.6 Nonconducting Materials

To measure specific heat of nonconducting materials, one has to satisfy some additional requirements. In such measurements, separate heaters and thermometers are necessary. The conditions of adiabaticity and of thermal equilibration in the calorimeter pose contradictory requirements for the modulation frequency. At low and medium temperatures, where the heat transfer coefficient is small, it is easy to find a modulation frequency that meets both requirements. Thin samples, of several tenths of a millimetre, ensure internal thermal equilibration. Many modulation measurements on nonconducting samples were carried out using the modulated-light heating. As a rule, no accurate determinations of absolute magnitudes of the heating-power oscillations were made. The imprecision of such measurements in narrow temperature intervals is of the order of 0.01%, but the inaccuracy of the absolute values may amount to 5–10%.

Temperatures up to 1500 K were achieved in measurements on some melts (Derman and Bogorodskii 1970). The liquid under study filled a hollow platinum crucible placed in a furnace. A heater of low thermal inertia passed along the axis of the crucible. The set-up served to create radial temperature waves in the crucible. However, over a definite frequency range the amplitude of the temperature oscillations depended only on the heat capacity of the sample. According to the authors, the uncertainty in the specific-heat data amounted to 4%.

Several methods are known for assembling calorimeter cells for nonconducting materials (Fig. 6.8). For the measurements, it is sufficient to prepare samples of suitable shape, which are compatible with the container and the heater at high temperatures. Probably, modulation calorimetry is applicable to many nonconducting materials.

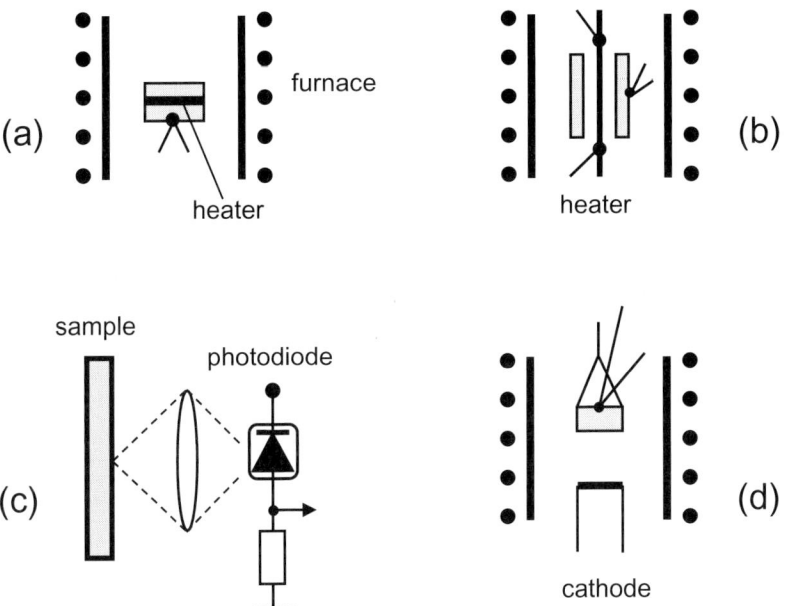

Fig. 6.8. Variants of modulation measurements on nonconducting materials: (a) planar heater is sandwiched between two portions of the sample; (b) heating by radiation or electron bombardment; (c) sample in metal capillary is heated by electric current; (d) sample in crucible is heated by electron bombardment. Thermocouples or photodetectors sense the temperature oscillations in the samples

The equivalent-impedance technique is usable for measuring specific heat of a thin nonconducting layer deposited on a conducting wire or strip. A metal of known specific heat, tungsten or platinum, may serve as the main sample. In this case, one directly compares the heat capacities of the coating and of the main sample. Moreover, there is no need to know the temperature derivative of the resistance of the sample because it is the same for the sample with and without the nonconducting coating. The main sample thus serves as a heater, a thermometer, and a reference of specific heat. Pochapsky (1953) successfully employed this method in pulse-heating measurements of the specific heat of AgBr deposited on a platinum wire.

The nonconducting layer must be of constant thickness, and good adhesion to the main sample must be ensured. A relation for evaluating the specific heat of the layer follows from the basic expression for the equivalent-impedance method. Under adiabatic conditions, Eq. (4.12) is valid for the main sample. For the composite sample, including the coating of the heat capacity $m_1 c_1$, the corresponding relation is

$$mc + m_1 c_1 = 2 I_{01}^2 R' / \omega^2 R C_1. \tag{6.11}$$

At a given temperature, the values of R and R' are the same in both measurements, as well as the modulation frequency. Only the DC current heating the sample and the capacitance corresponding to the equivalent impedance of the sample alter. One thus obtains

$$m_1 c_1 / mc = I_{01}^2 C / I_0^2 C_1 - 1. \tag{6.12}$$

A similar expression can be derived for a nonadiabatic regime.

6.1.7 Noncontact Calorimetry

Noncontact modulation calorimetry involves electromagnetic levitation and optical pyrometry. The heating power is supplied to the sample by means of a laser beam or induction heating. This technique allows measurements of specific heat of small solid particles and supercooled liquids. Monazam et al. (1989) employed an electrodynamic balance for contactless measurements of absorptivities and heat capacities of single particles, 0.05–0.2 mm in diameter, kept in a vacuum chamber. The apparatus included a 50-W CO_2 laser and the optics needed to direct the beam into the chamber. The laser provided 3-ms heating pulses at a 100-Hz repetition rate. A high-speed optical pyrometer monitored the resulting periodic temperature changes in the samples. The authors estimated the absorptivity and the heat capacity of carbon particles in the range 800–1200 K.

Modulation Calorimetry in Space

Fecht and Johnson (1991) proposed a method for an accurate determination of specific heat of metastable undercooled liquid metals and alloys using electromagnetic levitation and induction heating. Measurements on solid niobium samples in a levitation system TEMPUS ("tiegelfreies elektromagnetisches Prozessieren unter Schwerelosigkeit") demonstrated the feasibility of this method (R. Wunderlich and Fecht 1993; R. Wunderlich et al. 1993; Fecht and Wunderlich 1994). A 10-mm solid niobium sphere was suspended in the centre of a heating coil in a vacuum chamber. These preliminary measurements made it possible to determine specific heat of metallic samples in space (R. Wunderlich and Fecht 1996; R. Wunderlich et al. 1997). Modulation of the radio-frequency current in the heating coil was used for determining the heat capacity of metallic samples.

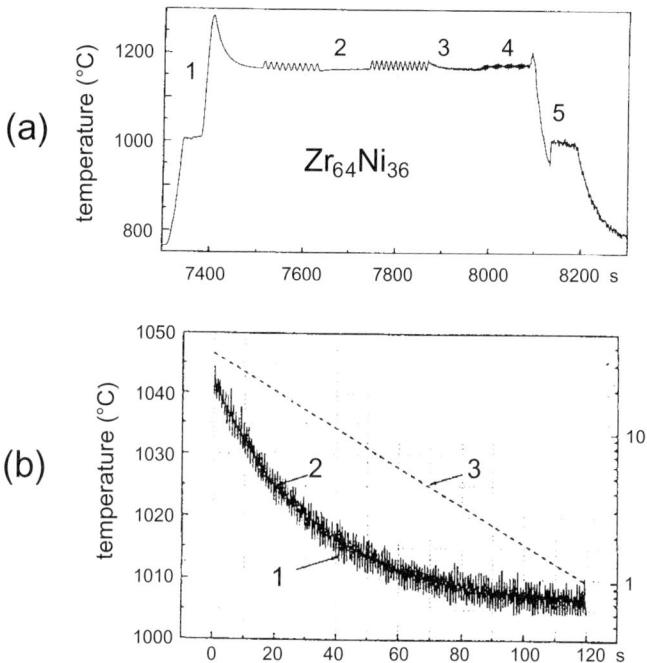

Fig. 6.9. (a) Temperature-time profile of $Zr_{64}Ni_{36}$ sample as processed in microgravity experiment (R.Wunderlich et al. 1997). 1 – isothermal melting, 2 – temperature oscillations due to power modulation, 3 – temperature decay to equilibrium, 4 – variation of modulation frequency for determination of the internal relaxation time, and 5 – recalescence plateau at low undercooling. (b) Part of a temperature profile: 1 – decay to equilibrium, 2 – filtered signal, and 3 – logarithmic plot for determination of the relaxation time

During the IML-2 (International Microgravity Laboratory) mission in July 1994, measurements were performed on ZrNi, ZrCo, and ZrFe samples. The samples had a diameter of 8 mm, corresponding to a mass of typically 1.9 g. The temperature of the samples was measured with a two-channel InAs pyrometer with sensitivity in the ranges of 1.8–2.8 and 3–4 μm. Measurements were performed in the stable and 30 K undercooled melt. Starting from a crystalline sample, five steps are clearly discerned (Fig. 6.9a): (1) the melting plateau, overheating, temperature equilibration in the stable melt; (2) modulation calorimetry at different frequencies; (3) decay to equilibrium temperature; (4) variation of the modulation frequency for determination of the internal relaxation time; and (5) undercooling and recalescence plateau. From the temperature relaxation following a step function change of the heating power (Fig. 6.9b), the relaxation time was determined.

Table 6.2. Modulation measurements of specific heat at high temperatures

Item	Reference
W, 1500–3600 K, EI	Kraftmakher and Strelkov 1962
Ti, 600–1345 K, TH	Holland 1963
Nb, Mo, Ta, W, 1200–2400 K, P	Lowenthal 1963
Ta, 1200–2900 K, EI	Kraftmakher 1963a
Nb, 1300–2700 K, EI	Kraftmakher 1963b
Mo, 1300–2500 K, EI	Kraftmakher 1964
Molten Sn, 1300–2000 K, P	Akhmatova 1965
Molten Cu, 1460–1640 K, P	Akhmatova 1967
Molten Ga, 1380–1620 K, P	Akhmatova 1967
Pt, 1000–2000 K, EI	Kraftmakher and Lanina 1965
Graphite, 1750-2850 K, P	Kraftmakher and Shestopal 1965
Ti, 1400–1900 K, P	Shestopal 1965
Zr, 1300–2000 K, P	Kanel' and Kraftmakher 1966
Au, 700–1300 K, EI	Kraftmakher and Strelkov 1966a
W, 1300–3000 K	Kraev 1967
Cu, 550–1250 K, T	Kraftmakher 1967c
WRe, 1600–2900 K, P	Sukhovei 1967
Mo, 1100–2400 K, P	Makarenko et al. 1970a
Nb, 1100–2400 K, P	Makarenko et al. 1970b
Au, 470–1220 K, TH	Skelskey and Van den Sype 1970
Ir, 1500–2500 K, P	Trukhanova and Filippov 1970
W, dC/dT, SC	Kraftmakher and Tonaevskii 1972
Pt, 1200–1900 K, search for relaxation	Seville 1974
Au, relaxation phenomenon at 1064 K	Skelskey and Van den Sype 1974
W, 800–2500 K, via thermal noise	Kraftmakher and Cherevko 1974
WRe, 1000–2700 K, via thermal noise	Kraftmakher and Cherevko 1975
Pt, active thermal shield, T	Kraftmakher and Cherepanov 1978
W, C_p/C_v via temperature fluctuations, P	Kraftmakher and Krylov 1980ab
W, relaxation in specific heat, P	Kraftmakher 1985
Pt and PtRh, T	Glazkov and Kraftmakher 1986
Ni, 900–1400 K, T	Glazkov 1987
Pt, relaxation in specific heat, P	Kraftmakher 1990
Nb, noncontact measurements, P	R. Wunderlich and Fecht 1993; R. Wunderlich et al. 1993
Noncontact calorimetry in space, P	R. Wunderlich and Fecht 1996; R. Wunderlich et al. 1997, 2001; Egry 2000; Egry et al. 2001

EI – equivalent impedance, TH – third-harmonic method, P – photosensor, SC – supplementary current, T – thermocouple

Egry (2000) reviewed the TEMPUS facility. Specific heat, density, electrical conductivity, viscosity, and surface tension are obtainable as a function of temperature over a wide temperature range, up to 2400°C, including the undercooled state, for a variety of metals and alloys. High electromagnetic fields necessary for the levitation in normal gravity deform the molten samples and induce turbulent currents inside them. These drawbacks are avoided when the levitation is performed in microgravity and the positioning forces are reduced by at least a factor of 100. The TEMPUS facility uses a quadrupole field for positioning, while a dipole field accomplishes heating of the samples. An ultrahigh-vacuum chamber surrounds the levitation coils. The chamber has windows for pyrometers and video cameras. During the MSL-1 (Microgravity Science Laboratory) mission in July 1997, 22 experiments were performed on 18 samples. The specific heat of a number of glass-forming alloys was measured at temperatures up to 1200°C.

Reports on containerless processing in space were also given by Egry et al. (2001) and R. Wunderlich et al. (2001). Table 6.2 lists modulation measurements of specific heat at temperatures above 1000 K.

6.2 Medium Temperatures

During a long time, modulation calorimetry was used only at high temperatures, and the temperature oscillations in a sample were detected through either the electrical resistance of the sample or radiation from it. Modulation measurements at medium temperatures appeared in 1960s. Using the third-harmonic technique, Holland (1963) determined the specific heat of titaniun in a temperature range 600 to 1345 K. The measurents in the vicinity of the phase-transition point (1155 K) appeared to be doubtful because of an anomaly in the resistance of the sample and its temperature derivative near the transition point. Gerlich et al. (1965) determined the specific heat of silicon, germanium, and their alloys in the range 300–1000 K. The authors used the supplementary-current method with the supplementary current of high frequency. Using the third-harmonic technique, Smith (1966) measured the specific heat of germanium in the range 600–900 K. A whisker sample with a mass of about 15 µg was used, and the frequency of the heating current was varied from 5.7 to 92 Hz.

Thermocouples are very convenient tools for measuring temperature oscillations at medium temperatures. First measurements with thermocouples were performed on nickel in a temperature range including its Curie point (Kraftmakher 1966a; Handler et al. 1967). The simplest method to perform such measurements consists in direct electric heating of a rod sample (Fig. 6.10). The sample is placed in a vacuum chamber and heated by an AC current modulated with an infra-low frequency. This modulation defines the frequency of temperature oscillations in the sample. A thin-wire thermocouple measures the mean temperature and the temperature oscillations at the centre of the sample. To avoid end effects, thin potential probes are used to measure the voltage across a central portion of the sample. A power meter measures the oscillations of the electric power dissipated in this portion. The meter has current and voltage inputs, and provides an output voltage proportional to the mean value of the product IU. The oscillating part of this voltage is recorded along with the AC component of the voltage from the thermocouple. Adjusting the heating current changes the temperature of the sample. With this simple set-up, the specific heat of nickel was measured in the vicinity of the Curie point. That time, the behaviour of specific heat at second-order phase transitions was of great interest because of development of a new theory of phase transitions. It is therefore not surprising that the modulated-light heating was invented to measure the specific heat of nickel near the Curie point (Handler et al. 1967). In this experiment, a thermocouple served for measurements of the temperature oscillations in the sample. The sample was placed in a furnace controlling the mean temperature, and the temperature gradients in the sample were greatly reduced. The temperature resolution was therefore significantly better than when using direct electric heating. Another advantage of this technique was due to gradually changing the temperature and continuously recording the results.

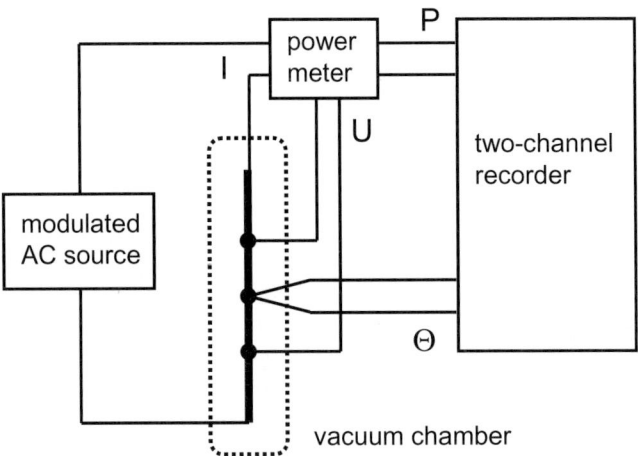

Fig. 6.10. Simplified diagram of arrangement for calorimetric measurements with a thermocouple (Kraftmakher 1966a)

Glass (1968) was the first to apply modulation calorimetry to a nonconducting material. He determined the specific heat of $LiTaO_3$ in the range 300–1000 K. A sample 0.1 mm thick was placed in a furnace and illuminated by chopped light from an incandescent lamp. The modulation frequency was 30 Hz. A thermocouple and a lock-in amplifier measured the temperature oscillations in the sample. A photocell provided the reference voltage. The output voltage of the lock-in amplifier was fed to the Y-input of a plotter, while a thermocouple measuring the mean temperature in the furnace was connected to the X-input. Hence, the set-up was similar to that developed by Handler et al. (1967). Bręczewski et al. (1984) employed a separate heater in measurements on $LiKSO_4$ in the range 400–750 K. A thin nickel film on one side of the sample served as the heater. A thermocouple junction was attached to the backside of the sample.

Using the modulated-light heating, Ohsawa et al. (1978) measured the specific heat of boron monophosphide in the range 300–850 K. For determination of absolute values of the specific heat, the oscillations in the incident power were evaluated through the increase of the mean temperature of the sample.

6.3 Low Temperatures

At low temperatures, oscillations in the power heating the sample are readily measurable when using direct electric heating or a separate heater. Resistance thermometers or thermocouples measure the mean temperature of the sample and the temperature oscillations in it. Platinum thermometers are usable down to about 20 K, and semiconductor thermometers are necessary for lower temperatures. The sensitivity of the latter increases with decreasing temperature. With lock-in detection, very high sensitivity is achievable. For simplicity, many workers employ the modulated-light heating. Usually, the front surface of the sample is painted black with a thin layer of graphite or PbS to enhance and stabilise the absorption of light, which thus becomes independent of temperature.

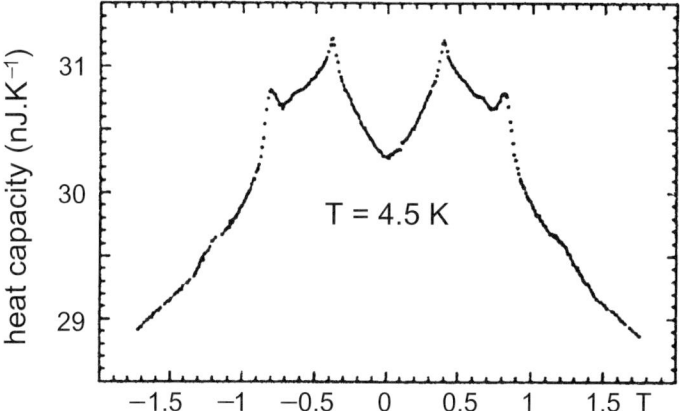

Fig. 6.11. Heat capacity of 1-μg Mn_{12}-acetate crystal as a function of external magnetic field at 4.5 K (Fominaya et al. 1997b). This example demonstrates features of modern low-temperature nanocalorimetry developed by the authors

The advantages of modulation calorimetry are especially significant when specific heat is measured, at a given temperature, as a function of an external parameter such as magnetic field or pressure. If the sensitivity of the thermometer employed does not depend on the magnetic field, one immediately obtains the field dependence of the specific heat. Sullivan and Seidel (1966, 1967, 1968) have shown this important feature in the first modulation measurements at low temperatures. Quantum oscillations in the specific heat were observed in beryllium, as well as the destruction of superconductivity in indium. Changes in the specific heat of beryllium are due to the variation of the density of electronic states at the Fermi level produced by an external magnetic field. A 82-μg sample of beryllium having a heat capacity of approximately 3×10^{-6} J.K^{-1} at 1.5 K was coupled to the bath with a time constant of 1.2 s. The modulation frequency was 10 Hz, and the temperature oscillations were 2.9 mK. At 1 K and 30-s averaging time, the sensitivity to changes in the temperature oscillations was nearly 10^{-7} K. The imprecision of the measurements was better than 0.1%. The heat capacity was measured when scanning the magnetic field near 2.2 T. The results have clearly shown that variations in heat capacity with an external parameter can be recorded directly.

Later, many investigators measured specific heat versus external magnetic field. An example of such data is shown here (Fig. 6.11). Recently, Park et al. (2003) measured the specific heat of a single crystal of the superconductor YNi_2B_2C as a function of the orientation of external magnetic field (see Sect. 1.2).

Table 6.3 lists some modulation measurements at low temperatures. Further examples are given in Chap. 14.

Table 6.3. Modulation calorimetry at low temperatures

Item	Reference
Be, 1.5 K, magnetic field	Sullivan and Seidel 1967, 1968
Pb-Ag-Pb sandwiches, 0.3–1.2 K	Manuel and Veyssié 1972
$Au_{1-x}Ni_x$, amorphous, 2–40 K	Eno et al. 1977
$Pb_{70}Bi_{30}$, 0.5–2 K	Kämpf and Buckel 1977
As, amorphous, 1.8–20 K	Lannin et al. 1978
$FeCl_2$, magnetic field	Shang et al. 1978; Shang and Salamon 1980
$Au_{1-x}Fe_x$, amorphous, 1.5–18 K	Dawes and Coles 1979
Si, amorphous, 2–50 K	Mertig et al. 1984
$SbCl_5$-intercalated graphite	Bittner and Bretz 1985
M_xTiS_2 intercalates (M = Mn, Fe, Co, Ni)	Inoue et al. 1986; Takase et al. 1994
$Al_{80}Mn_{20}$, $Al_{78}Mn_{22}$, 0.5–5 K	Machado et al. 1987ab
$GaAs/Al_xGa_{1-x}As$, 2-D electron gas	Wang et al. 1988, 1992
UBe_{13}, 0.3–1.8 K, magnetic field up to 20 T	Graf et al. 1989
Graphite-ICl intercalate, 2.2–5 K	Tashiro et al. 1990
Ne, Ar, Kr, Xe, Ne-Xe, Ar-Xe films, 0.1–7 K	Menges and v. Löhneysen 1991
Mn_{12} acetate, magnetic field	Fominaya et al. 1997b, 1999b
Fe_8 crystal, 1.5–10 K, magnetic field	Fominaya et al. 1999a
Carbon nanotubes	Yi et al. 1999
Rb_2KScF_6, 4–280 K	Flerov et al. 2003
YNi_2B_2C	Park et al. 2003

Fe_8 = [(triazacyclononane)$_6Fe_8O_2(OH)_{12}$]$^{8+}$

6.4 Measurements Under High Pressures

Chu and Knapp (1973) reported on the first modulation measurements under high pressures, up to 1.2 GPa. They determined the phase diagram of α-U at low temperatures. Nb_3Sn and V_3Si also were investigated (Chu 1974; Chu and Testardi 1974; Chu and Vieland 1974).

Bonilla and Garland (1974) determined the specific heat of chromium along its Néel line, up to 0.3 GPa. The specific heat was measured as a function of pressure for four isotherms between 27.75°C and 37.51°C. The sample was a wire, 125 mm long and 0.125 mm in diameter. One junction of a chromel-constantan thermocouple was in good thermal contact with the sample holder, and the second one was attached to the centre of the sample with a thin coating of varnish (Fig. 6.12). The high pressure was provided by argon atmosphere. The high-pressure cell was completely immersed in a thermostat bath, and its temperature could be measured to within 0.02 K. The pressure could be measured with an absolute accuracy of 0.4 MPa and held constant for a few hours to within 0.03 MPa. The frequency of the heating current was 3.5 Hz. Near the phase transition, 1 to 6 hr were necessary to obtain thermodynamic equilibrium in the sample.

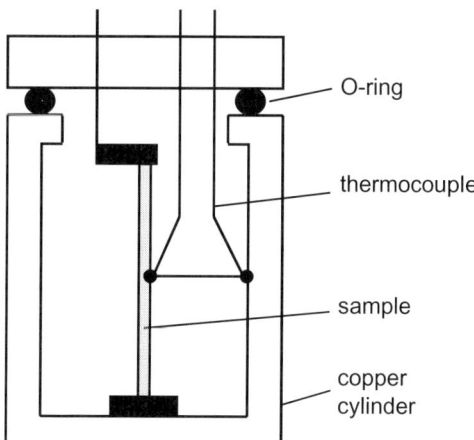

Fig. 6.12. Simplified diagram of high-pressure calorimeter cell used for measurements of specific heat of chromium (Bonilla and Garland 1974)

Baloga and Garland (1977) examined the possibilities of using modulation calorimetry under high pressures with a gas as the pressure-transmitting medium. The sample, a slab of dimension 10×10×0.5 mm, with its addenda, was suspended by leads in a cylindrical copper sample holder (Fig. 6.13). A thin film of resistance epoxy painted directly onto one face of the sample served as the heater. A microbead thermistor was cemented to the opposite face. The main conclusions by the authors were as follows:

• The sample must have large faces to exclude edge effects on the temperature at the centre.

- The sample must be thin enough to avoid significant temperature differences during a period compatible with a reasonable modulation frequency.
- It is desirable to minimise heat leaks from the sample through the heater and thermistor leads.
- The heat capacity of the addenda should be small compared to that of the sample.
- The heater and the thermistor should be electrically and mechanically stable on cycling pressure and temperature.

The authors measured the specific heat of NH_4Cl and ND_4Cl up to 0.3 GPa (Garland and Baloga 1977).

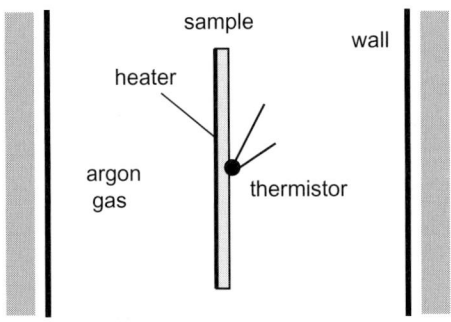

Fig. 6.13. Presentation of calorimeter cell under hydrostatic gas pressure (Baloga and Garland 1977)

Eichler and Gey (1979) developed a modulation calorimeter for measurements under high pressures at low temperatures. Diamond powder with a grain size smaller than 0.5 µm was a pressure-transmitting medium. At low temperatures, it has a low specific heat and a low thermal conductivity. This reduces the contribution of the medium and corrections for heat losses. The sample geometry was chosen to compromise between the requirements of calorimetric and high-pressure techniques. An indium sample consisted of three discs, 3 mm in diameter and 0.5–1.5 mm thick. The heater and thermometer were each mounted between two discs. The heater was made of 30-µm manganin wire, and the thermometer was cut from a standard Allen-Bradley carbon resistor. A circular slice, 1 mm in diameter, was ground and thoroughly polished from both sides to a thickness of about 0.15 mm. The whole assembly consisting of the indium discs and the embodied elements was placed in a hollow cylinder of compressed diamond powder. The upper and lower covers of this cylinder were diamond discs. The surrounding BeCu pressure cylinder served as the temperature bath. The pressure cell was put into a device capable of changing the pressure at any temperature in the range 1.3–300 K. The apparatus contained in a chamber filled with a small amount of helium exchange gas was placed in the bore of a superconducting magnet. The external and internal time constants at 4.2 K, 0.5 s and 0.5 ms, were quite favourable to fulfil the criterion of adiabaticity. The operating frequency, 20 Hz, was near the upper limit of the frequency band, where the quantity $\omega\Theta_0$ remains constant at 4.2 K. The lower limit was nearly 2 Hz. Lock-in amplifiers measured all the AC voltages. A data-acquisition system controlled the measurements. The system measured each quantity repeatedly, 20 to 60 times, and provided the mean values. The total error of the measurements was 3%. There is no way to experimentally determine the heat capacity of the addenda consisting of the heater, the thermometer, and certain proportions of the electric leads and the surrounding diamond powder. In the temperature range 1.3–5 K, this contribution was estimated to be less than 1%.

Chen et al. (1993) performed calorimetric measurements on $Eu_{0.9}Ho_{0.1}Mo_6S_8$ at temperatures 0.15 to 6 K, magnetic fields up to 20 T, and a pressure of 0.8 GPa. This compound is a semiconducting paramagnet at low temperatures and ambient pressure. It becomes metallic and superconducting at high pressures. A sintered sample had thickness of about 0.5 mm, and surface area of 1×2 mm. The sample was sandwiched between a Ni-Cr heater and a RuO_2 thermometer on the larger faces and glued with GE 7031 varnish to allow good thermal coupling and a short relaxation time of the heater-sample-thermometer system (Fig. 6.14).

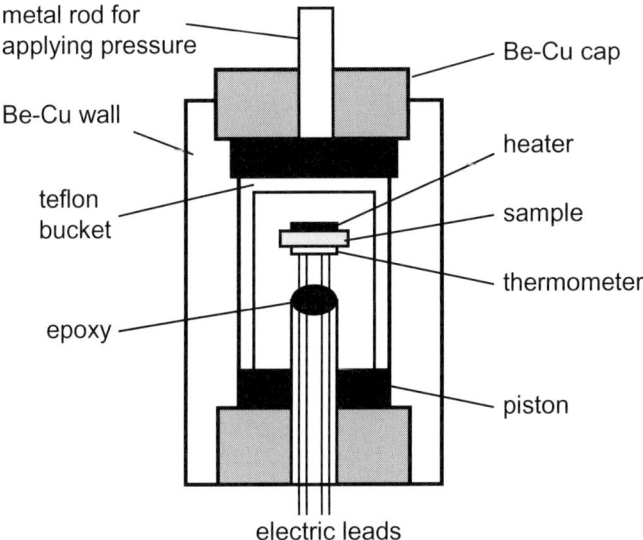

Fig. 6.14. Simplified diagram of high-pressure calorimetric apparatus (Chen et al. 1993). The measurements were carried out in the range 0.15–6 K, under magnetic fields up to 20 T and a pressure of 0.8 GPa

Although a pressure medium surrounds the sample, the various thermal relaxation times can satisfy the prerequisites of modulation calorimetry. The hydrostatic pressure is applied through a liquid, which remains amorphous at low temperatures. Electrical leads are connected to the heater and the thermometer by silver paste. The pressure is locked in by a Be-Cu clamp and remains indefinitely. An Oxford dilution refrigerator provides the low temperature environment. The most effective modulation frequency range was determined by two methods. First, the amplitude of the temperature oscillations was plotted versus $1/f$ to determine a range, over which the criterion of adiabaticity is valid. Second, the related relaxation times were measured directly. An operational frequency of a few hertz satisfies the requirements of modulation calorimetry. At a high pressure and various temperatures, the specific heat of the sample was measured versus the magnetic field. The authors obtained a re-entrant phase diagram H_{c2} versus T at 0.8 GPa by taking the positions of the peaks in the heat capacity as transition points.

Bouquet et al. (2000) developed a calorimeter for low temperatures and high pressures. The authors measured the specific heat of $CeRu_2Ge_2$ in the range 1.5–11 K and under pressures up to 8 GPa. The samples were embedded in steatite, the pressure-transmitting medium. The typical thickness of the sample, thermocouple, and heating wire was 20, 12, and 3 μm, respectively. The modulation frequencies were in the range 500–4000 Hz, while the temperature oscillations in the range 2–20 mK. The phase diagram of the magnetic phases of the sample determined by the authors appeared to be in agreement with previous transport measurements.

Using measurements of specific heat, Demuer et al. (2000) determined the magnetic phase diagram of $CeRu_2Ge_2$ in the temperature range 2–10 K under continuously swept pressures up to 8 GPa. In a diamond anvil cell (Fig. 6.15), solid ^{4}He was used as the pressure-transmitting medium providing the best hydrostatic conditions. The diamond anvil cell had 1-mm diamond culet flats. A mechanical system using bellows allows the pressure to be continuously varied at low temperatures. The pressure was measured by the ruby fluorescence technique with an ionised Ar 50-mW laser as a source of the excitation. In order to check the homogeneity within the pressure chamber, several tiny ruby chips were placed around the sample, a slice of dimensions 0.2×0.2×0.07 mm.

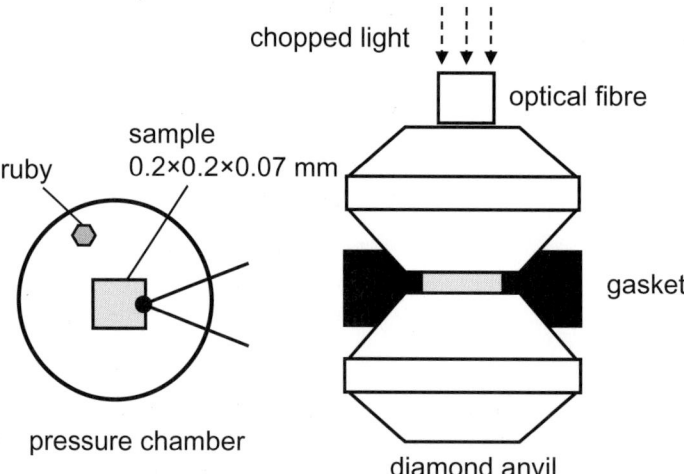

Fig. 6.15. Simplified diagram of diamond anvil apparatus for calorimetric measurements under pressures up to 8 GPa in the range 2–10 K (Demuer et al. 2000)

A chopped laser beam, passing through an optical fibre, provided the AC heating of the sample at frequencies up to 3 kHz. The amplitude of the temperature oscillations, of about 10 mK, was measured by a gold-chromel thermocouple and a lock-in amplifier with a 1:100 input transformer. Under the experiment conditions, the basic requirements of modulation calorimetry could not be met, and the authors limited their analysis of the data obtained to a qualitative description of singularities associated with phase transitions.

Filippov et al. (1976) and Blagonravov et al. (1983, 1984) performed modulation measurements under high pressures and high temperatures.

Table 6.4 lists measurements carried out at high pressures.

Table 6.4. Modulation measurements under high pressures

Item	Reference
α-U, 10–40 K, up to 1.2 GPa	Chu and Knapp 1973
Cr, Néel line, up to 0.3 GPa	Bonilla and Garland 1974
Nb_3Sn, up to 1.8 GPa	Chu 1974
V_3Si, specific heat and dR/dT	Chu and Testardi 1974
Nb_3Sn, specific heat and dR/dT	Chu and Vieland 1974
Molten Cs, up to 1700 K	Filippov et al. 1976
NH_4Cl, ND_4Cl, up to 0.3 GPa	Baloga and Garland 1977; Garland and Baloga 1977
Sn, superconductor, up to 1 GPa	Itskevich et al. 1978
In, superconductor, up to 1 GPa	Eichler and Gey 1979; Eichler et al. 1981
8OCB liquid crystal, 0.05 and 0.1 GPa	Garland et al. 1979; Kasting et al. 1980
Ga, superconductor, up to 3.5 GPa	Eichler et al. 1980
TGSe, up to 1 GPa	Polandov et al. 1981
$KMnF_3$, 2–380 K, up to 0.1 GPa	Stokka and Fossheim 1982b
TGS + TGSe solid solution, up to 1 GPa	Gulish et al. 1983
Molten Cs, up to 2000 K	Blagonravov et al. 1983
Molten Rb, up to 1900 K	Blagonravov et al. 1984
CeSb, 1.7 GPa	Jin et al. 1984
$CeCu_2Si_2$, 0.3–2 K, up to 0.6 GPa	Bleckwedel and Eichler 1985
La, low temperatures, up to 2.5 GPa	Bohn and Eichler 1991
$CrPb_3$, 0.5–2 K, up to 1.3 GPa	Kirsch et al. 1992
$Eu_{0.9}Ho_{0.1}Mo_6S_8$, magnetic field up to 20 T	Chen et al. 1993
o-terphenyl, glass transition, 2–6300 Hz, up to 0.1 GPa	Leyser et al. 1995
$CeRu_2Ge_2$, 1.5–11 K, up to 8 GPa	Bouquet et al. 2000
$CeRu_2Ge_2$, 2–10 K, up to 8 GPa	Demuer et al. 2000
$CePd_2Si_2$, up to 2.5 GPa	Demuer et al. 2001
$CuGeO_3$, up to 9 GPa	Devoille et al. 2002
$CePd_{2.02}Ge_{1.98}$, 0.3–10 K, up to 22 GPa	Wilhelm and Jaccard 2002ab
$CeCu_2Si_2$, up to 6 GPa, 8 T	Holmes et al. 2003

8OCB = octyloxycyanobiphenyl, TGSe = triglycine selenate, TGS = triglycine sulphate

6.5 Measurement of Specific Heat and Thermal Diffusivity

Salamon et al. (1974) employed modulated-light heating for simultaneous measurements of specific heat and thermal diffusivity. The method is a modification of the classical technique of thermal waves known as the Ångström method. An extension of this technique to measure the thermal conductivity and thermal diffusivity simultaneously was developed by Howling et al. (1955). With the method proposed by Salamon et al. (1974), a sample is assumed to be thin compared with the thermal wave length $\lambda = (2D/\omega)^{1/2}$, where D is the thermal diffusivity of the sample, and ω is a frequency of temperature oscillations in it. For sufficiently thin samples, the heat-flow problem becomes essentially one-dimensional and easily solvable.

Modulated light falls onto a thin sample, while a narrow mask closes a portion of the sample (Fig. 6.16). Far away from the mask and at proper modulation frequencies, the temperature oscillations depend only on the heat capacity of the sample. Under the mask, the temperature oscillations depend on the thermal diffusivity of the sample. The thickness of the sample is about 0.1 mm. The frequency of the temperature oscillations is of the order of 10 Hz, and their amplitudes are nearly 1 mK. A fine thermocouple, typically 25-μm chromel-alumel wires flattened to 5 μm, is located on the opposite side of the sample and sufficiently far from the edge of the shadow. This thermocouple provides a signal of an amplitude Θ_0 related to the heat capacity of the sample. A second thermocouple is located behind the centre of the shadow, at the distance a from the lighted region, which equals approximately one thermal wave length. This thermocouple provides a signal of an amplitude $\Theta = \Theta_0 \exp(-a/\lambda)$, which relates to the thermal diffusivity.

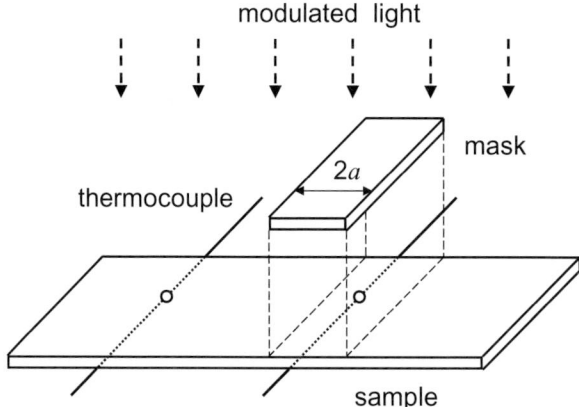

Fig. 6.16. Arrangement for simultaneous measurement of specific heat and thermal diffusivity (Salamon et al. 1974). The left thermocouple provides data for heat capacity, and the right one for thermal diffusivity

The thermal diffusivity thus is obtainable from the ratio of the two signals, Θ and Θ_0:

$$D = a^2 \pi f [\ln(\Theta_0/\Theta)]^{-2}. \tag{6.13}$$

In the measurements, the shadow was provided by a wire about 0.5 mm thick. The phase difference between the two signals may also be used to determine the thermal diffusivity. The method enabled simultaneous measurements of the absolute thermal diffusivity and the relative specific heat. The measurements were performed on CoO, SrTiO$_3$, Cr, and EuO in the range 60–325 K.

Through simultaneous measurements of the specific heat and thermal diffusivity, Salamon et al. (1975) determined the thermal conductivity of tetrathiafulvalene-tetracyanoquinodimethane (TTF-TCNQ) near the metal-insulator transition at 55 K.

Hatta et al. (1985a) improved this technique. In their set-up, a wide mask gradually shifts the closed portion of the sample by means of a fine screw.

Yang et al. (1991) described a fully automated set-up allowing the determination of the heat capacity and the thermal diffusivity on the same piece of a nonconducting thin solid sample, in the 25–300 K temperature range. The sample is heated uniformly at one side, while the temperature oscillations are measured at the opposite side. At low modulation frequencies, the thermal diffusivity is determined from the frequency dependence of the amplitude of the temperature oscillations. The second method involves measurements of the phase of the temperature oscillations in the sample at higher modulation frequencies.

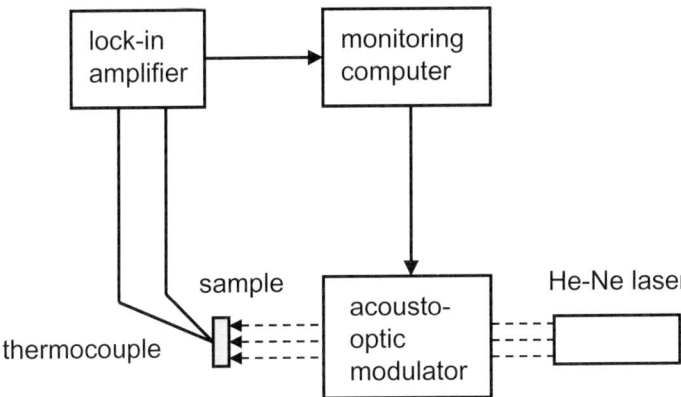

Fig. 6.17. Simplified diagram of apparatus converting thermal diffusivity into modulation frequency (Kato et al. 2001). The negative feedback loop keeps a constant value of the wave number of the temperature wave propagating in the sample

Rudyi (1993) described an elegant method of measuring the thermal diffusivity from auto-oscillations in a system with a thermal feedback. Kato et al. (2001) reported on measurements of thermal diffusivity at low temperatures. The experimental apparatus provided three measurement techniques: (i) a frequency-variation method, (ii) a distance-variation method, and (iii) a constant wave number method. The latter converts the thermal diffusivity D into an equivalent frequency, according to the relation $k = (\omega/2D)^{1/2}$ (k is the wave number). A special feedback loop (Fig. 6.17) continuously checks the deviation of the phase lag from a set value and takes a negative feedback control of the modulation frequency. The method provides relative values of thermal diffusivity and needs a calibration for obtaining absolute data. A method of converting heat capacity into frequency was described in Sect. 6.1.2.

6.6 Sinku-Riko Calorimeter ACC-1

Sinku-Riko, Inc. designed the first commercial modulation calorimeter ACC-1 (AC calorimeter). A sample is set in a furnace or a cryostat, and one its side is heated by chopped light (Sect. 1.2). The temperature oscillations in the sample are measured with a thin thermocouple attached to the other side of the sample. The device includes the following: (i) sample assembly with optical system, (ii) measuring circuit and data-processing unit, (iii) vacuum and atmosphere controller, and (iv) programmable temperature controller. Modifications of the calorimeter encompass the 70–800 K range. The samples are placed in vacuum or a helium atmosphere. The modulation frequency is variable in the range 0.1 to 100 Hz. The sample thickness should be under the thermal wavelength. Samples of dimensions 4×4×(0.1–0.3) mm are necessary for obtaining absolute values of heat capacity. For relative measurements, the dimensions of the sample may be 2×2×0.01 mm. The temperature resolution is 2.5 mK in the range 70–700 K and 25 mK above 700 K. The calorimeter provides a programmable change of the mean temperature. The data are processed by a personal computer and displayed, as specific heat versus temperature, by a monitor or a plotter. The signal-to-noise ratio can be improved by prolonging the measuring interval. When the heating rate is decreased, the measurements become more precise. The inaccuracy of the measurements is 3% but the scatter of data is about 0.01%. Determinations of the thermal diffusivity and the thermal conductivity of the samples are also possible. The calorimeter was commercialised by Sinku-Riko through the technical guidance of A. Ikushima and I. Hatta. The calorimeter has been successfully employed in many studies, including measurements of specific heat of high-temperature superconductors. Many authors studied phase transitions in solids.

With the ACC-1 calorimeter, Okazaki et al. 1990 measured the specific heat of the (Bi, Pb)-Sr-Ca-Cu-O high-temperature superconductor ($T_c \cong 108$ K). A ceramic sample of dimensions 2×2×0.2 mm (about 30 mg) was hung from the thermal bath made of a thick oxygen-free copper block by two pairs of chromel-alumel thermocouple wires 25 µm thick. The thermocouples were attached with a negligible amount of silver paste onto the rear face of the sample. A sample cell including the thermal bath surrounded by a heater was filled with helium gas to optimise the heat exchange between the sample and the bath, and immersed in liquid nitrogen in a cryostat. Through a vertical tube, light from a halogen lamp, chopped at a frequency of 1.8 Hz, irradiated the front face of the sample. The signal from the AC thermocouple was measured with a lock-in amplifier. The relative specific heat of the sample was determined with precision to 0.1%. Each data point represented the averaged specific heat in the time span of 20 s. The temperature dependence of the specific heat was measured, in a continuous heating mode, in the temperature range of 80–250 K. The anomaly related to the phase transition amounted to only about 1% of the total specific heat, but it was separated and presented in the form necessary for the analysis.

With the ACC-1VL modification of the calorimeter employing a helium cryostat, Castro and Burriel (1995ab) performed measurements down to 2 K. A slab-shaped sample 0.3 mm thick was used. The modulation frequency was 2 Hz, and the typical scan rate ranged between 10 and 30 K.h^{-1}. The temperature oscillations were detected with a thermocouple and measured with a digital lock-in amplifier. Absolute values were obtained by scaling the relative values to data from an adiabatic calorimeter.

Studies with the calorimeter ACC-1 are listed in Table 6.5.

Table 6.5. Calorimetric studies with the calorimeter ACC-1

Item	Reference
YBCO, phase transition	Ishikawa et al. 1988
YBCO, 70–300 K	Kishi et al. 1988
AgCrS$_2$, 300–720 K	Kawaji et al. 1989
(CH$_3$NH$_3$)NaSO$_4$.6H$_2$O, (CH$_3$NH$_3$)NaSeO$_4$.6H$_2$O	Miyazaki et al. 1989ab
(Bi, Pb)-Sr-Ca-Cu-O, 80–250 K	Okazaki et al. 1990
Supercooled liquid Te, 220–520°C	Tsuchiya 1991
K$_2$ZnCl$_4$	Gesi 1992
Gd$_{1.85}$Ce$_{0.15}$CuO$_{4-\delta}$, 6–300 K	Hwang et al. 1992
VDF-TrFE copolymers, 30–170°C	Ogura et al. 1992
LaBGeO$_5$, 480–850 K	Onodera et al. 1993
Supercooled liquid Ge$_{15}$Te$_{85}$	Tsuchiya 1993
TlH$_2$AsO$_4$, 80–400 K	Irokawa et al. 1994
Nd$_2$NiO$_4$, 2–80 K	Castro and Burriel 1995a
La$_2$NiO$_4$, Pr$_2$NiO$_4$	Castro and Burriel 1995b
N(CH$_3$)$_4$X (X = MnBr$_3$, NiCl$_3$, NiBr$_3$), 77–470 K	Gesi and Osaka 1995
K$_2$ZnCl$_4$, K$_2$SeO$_4$	Haga et al. 1995a
Rb$_2$ZnCl$_4$	Haga et al. 1995b
Amorphous Hf$_{1-x}$Ta$_x$Fe$_2$ alloys, 4.2–300 K	Murayama et al. 1995
Liquid CdSb, 200–600°C	Tsuchiya 1995
(CH$_3$)$_2$NH$_2$H$_2$PO$_4$, 88–312 K	Hatori et al. 1996
BaTiO$_3$, −170 to 700°C	Tura et al. 1998
SmNiO$_3$, 5–250 K	Pérez et al. 1999
SrTiO$_3$, 4–120 K	Gallardo et al. 2002
Oxide glasses, 300–840 K	Inaba et al. 2002, 2003
Rb$_2$KScF$_6$, 4–280 K	Flerov et al. 2003

7 Modulation Calorimetry II

This chapter considers important applications of modulation calorimetry appeared during the last decades. This list includes modulation microcalorimetry and nanocalorimetry, measurements on organic and biological materials, photoacoustic techniques, specific-heat spectroscopy, and modulated differential scanning calorimetry.

7.1 Modulation Microcalorimetry

Modulation microcalorimetry includes measurements on thin deposited films and freestanding liquid-crystal films containing only several molecular layers, bath modulation, and nanocalorimetry at low temperatures.

7.1.1 Thin Deposited Films

The high resolution peculiar to modulation calorimetry makes it possible to perform measurements on small samples, whose heat capacity is much smaller than that of the addenda. Zally and Mochel (1971, 1972) measured the low-temperature specific heat of thin superconducting BiSb films deposited onto a substrate (Fig. 7.1).

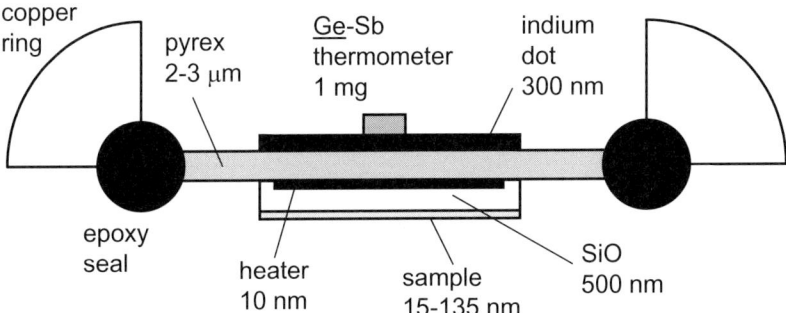

Fig. 7.1. Low-temperature microcalorimeter for thin films (Zally and Mochel 1971, 1972). The heat capacity of the sample is much smaller than that of the addenda

A disc of 2–3 µm Pyrex glass is mounted on a copper ring suspended inside a helium-bath shield. A 10-nm constantan heater is evaporated onto the glass. A 500-nm layer of silicon monoxide evaporated on top of the heater serves as an insulator. A 300-nm dot of indium evaporated on the opposite side of the glass improves thermal diffusion over the region to be occupied by the sample. A thermometer made out of a small chip of Sb-doped germanium is sealed to the indium. At 2.2 K, a typical transition temperature, the heat capacity of the substrate assembly was about 10^{-7} J.K^{-1}. For a 100-nm BiSb film, the heat-capacity discontinuity at the transition point was of the order of 10^{-9} J.K^{-1}. The amplitude of the temperature oscillations was in the range 2–5 mK, while the resolution in the total heat capacity was better than 10^{-3}.

Manuel and Veyssié (1976), Kämpf and Buckel (1977), Kämpf et al. (1981), Rao and Goldman (1981), Suzuki et al. (1982) also studied thin films at low temperatures.

7.1.2 Freestanding Films

Pitchford et al. (1986) found an unusual approach to measuring specific heat of thin freestanding liquid-crystal films (Fig. 7.2). A freestanding film contains 10–1000 smectic layers. The film is kept inside a regulated oven filled with argon gas at a pressure of 50 kPa. Chopped radiation from a 1-mW He-Ne laser at λ = 3.4 µm falls onto the film. A small thermocouple located beneath the film measures the temperature oscillations in it. The separation between the thermocouple junction and the film is about 0.1 mm. At the modulation frequency of 5.6 Hz, this separation is ten times smaller than the thermal diffusion length in the gas. A thermistor located beneath the film but away from the thermocouple junction measures the mean temperature. The measurements have shown that the radiation absorbed by the film is independent of temperature.

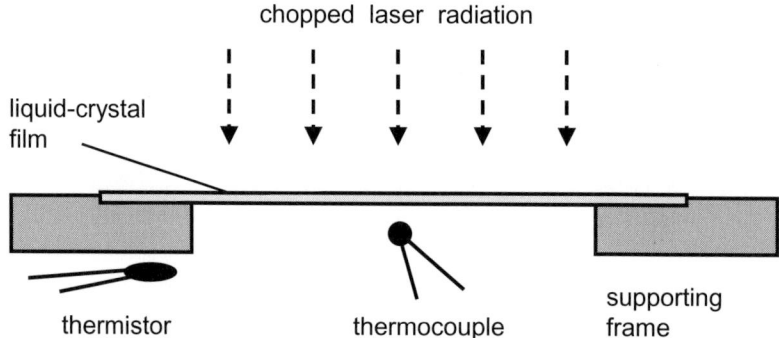

Fig. 7.2. Measurement of specific heat of thin freestanding liquid-crystal films (Pitchford et al. 1986). The thermocouple junction is about 0.1 mm beneath the film. Measurements are possible on films containing only two molecular layers

Later, Geer et al. (1989) reported several improvements in their technique. The heating source was a He-Ne laser of 8-mW output, about ten times greater than used before. The laser was maintained in a temperature-regulated housing to reduce variations in its output power. Much higher modulation frequency was used, 35 Hz. This reduces the effective probing area to approximately 0.25 mm^2, greatly reducing the rounding of the transition from temperature gradients and domain effects. To eliminate the large background signal from direct laser heating at the thermocouple, a second thermocouple was introduced far below the film. This separation, 1 cm, is about twenty times larger than the thermal diffusion length at the frequency employed. These two thermocouples were electrically connected to compensate for the direct laser heating. A clear evolution was observed of the specific-heat anomaly of a liquid-crystal compound near its smectic-A to hexatic-B phase transition as the films were reduced from ten to four molecular layers. This method allowed the authors to measure the specific heat of films ranging from three to a few thousand smectic layers (Geer et al. 1989, 1991ab). Stoebe et al. (1992), Geer et al. (1993), and Jin et al. (1995) studied phase transitions in freestanding films containing only two layers (Fig. 7.3).

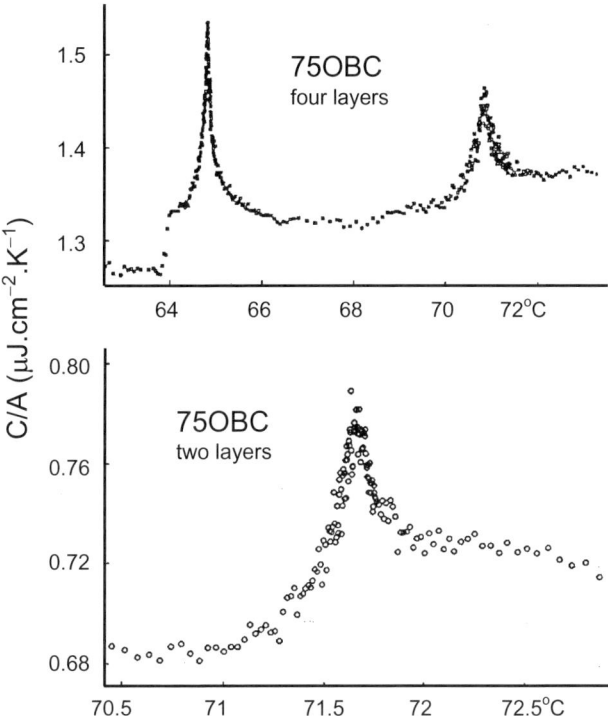

Fig 7.3. Specific heat of freestanding films containing four and two layers in the vicinity of phase transitions (Geer et al. 1993)

7.1.3 Bath Modulation

This method was proposed by Graebner (1989) and employed in studies of very small samples (Fig. 7.4). A small sample is pasted to a thermocouple junction. Short wires of the thermocouple are attached to a light platform, whose temperature is modulated. The thermocouple wires provide the oscillating heat input for the sample. In a certain frequency range, the AC component of the thermocouple response depends on the heat capacity of the sample. The mass of the sample was only 2.9 µg. The method was used in measurements of the specific heat of superconductors in magnetic fields, up to 6 T (Graebner et al. 1989, 1990).

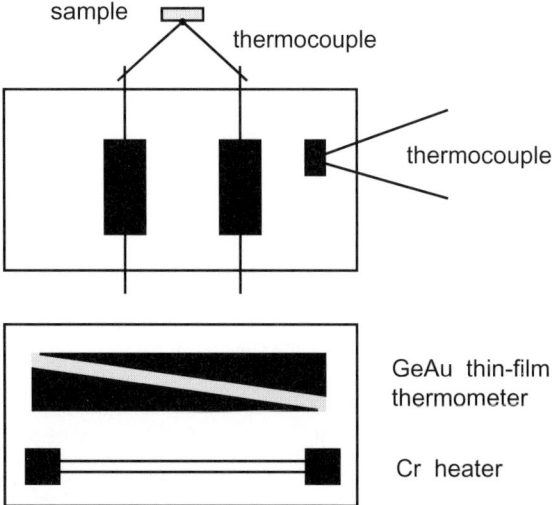

Fig. 7.4. Bath-modulation calorimetry (Graebner 1989). A short-wire thermocouple attached to a modulated platform provides the AC heating for the sample. The face and backside of the platform are shown

7.1.4 Nanocalorimetry at Low Temperatures

During many years, significant efforts have been applied to decrease the size of samples acceptable for modulation calorimetry. Campbell and Bretz (1985) and Kenny and Richards (1990ab) achieved the sensitivity at low temperatures of the order of 10^{-10} J.K^{-1}. Chae and Bretz (1989) improved the sensitivity to 10^{-11} J.K^{-1}. Phelps et al. (1993) and Birmingham et al. (1996) also performed modulation measurements on very small samples. The minimum measurable change in heat capacity was about 10^{-12} J.K^{-1} at 0.4 K and 10^{-10} J.K^{-1} at 2 K.

Riou et al. (1997) designed a high-resolution microcalorimeter optimised for the range 40–160 K (Fig. 7.5). The sample holder is made out of a 5-µm polyphenylquinoxaline (PPQ) membrane suspended on a copper ring. The advantage of PPQ is its high thermal stability, good adhesion on copper, and high elasticity. A thin gold film sputtered in the centre of the membrane defines an isothermal area. The lithographic elements (DC and AC heaters and a copper thermometer) sputtered on the other side of the membrane are connected to the copper ring through narrow metallic pads. The ring is coupled via a calibrated thermal link to a copper plate at liquid helium temperature.

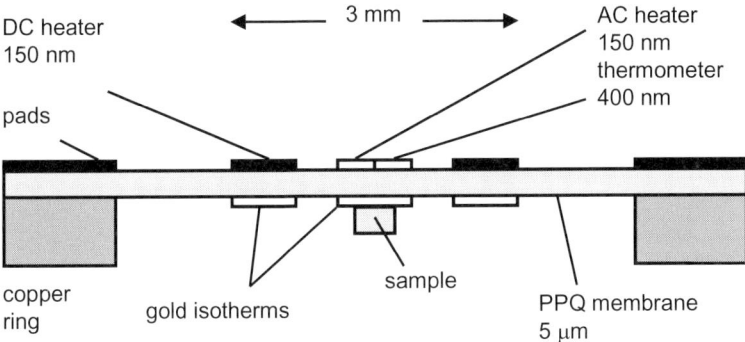

Fig. 7.5. Simplified diagram of modulation microcalorimeter designed by Riou et al. (1997). The mass of the sample, typically 10 μg, is much smaller than that of the addenda

An electric heater and two thermometers, a standard platinum thermometer and a germanium thermometer, are mounted on the copper ring. The temperature of the ring is kept constant at 10 K. The DC heater of the membrane provides the desired mean temperature. An AC power is supplied to the AC heater at the centre of the membrane. The mass of the sample is typically 10 μg. At 100 K, the specific heat of the addendum is 1.5×10^{-6} J.K^{-1}. The temperature oscillations, in the range 5 to 70 mK, are measured with a lock-in amplifier. The averaging time varies from 30 s to 3 min. With this calorimeter, Charalambous et al. (1999) investigated the critical behaviour of an YBCO single crystal.

Fominaya et al. (1997a) further developed modulation microcalorimetry. A low-temperature calorimeter was designed for measurements on thin films and small single crystals. The heat capacity of the addenda was 3×10^{-9} J.K^{-1} at 4 K and 5×10^{-10} J.K^{-1} at 1.5 K. The resolution of the order of 10^{-4} makes it possible to see variations in specific heat, which are smaller than 10^{-12} J.K^{-1}. This achievement offered new possibility of the modulation technique, modulation nanocalorimetry. In the fabrication of the nanocalorimeter, the approach was similar to that taken in designing electronic microchips, and the corresponding technology was employed (Fig. 7.6). A silicon substrate (10×15×0.28 mm) constitutes a base for the calorimeter. A Si$_3$N$_4$ protection layer, 200 nm thick, was removed from a square area of 5×5 mm by means of SF$_6$ plasma reactive ion etching. This area was further etched in KOH to obtain a 5-μm silicon membrane suspended on a thick frame. Then a 100-nm NbTi film and a 30-nm platinum film were deposited by magnetron sputtering to form leads and pads. Optical lithography, ion beam etching, and chemical etching (HF-HNO$_3$) served for this purpose. Further, a sputtered 150-nm CuNi layer formed three heaters, one on the membrane and two on the silicon frame. A

150-nm NbN film was then sputtered and patterned to create thermometers. To improve the thermal isolation of the membrane, holes were etched with SF_6 plasma into the membrane, so that a 3.3×3.3-mm membrane remained, being suspended by twelve 40-μm silicon bridges. The thermal coupling of the membrane to the frame thus depends on the geometry of the bridges and the doping of the silicon.

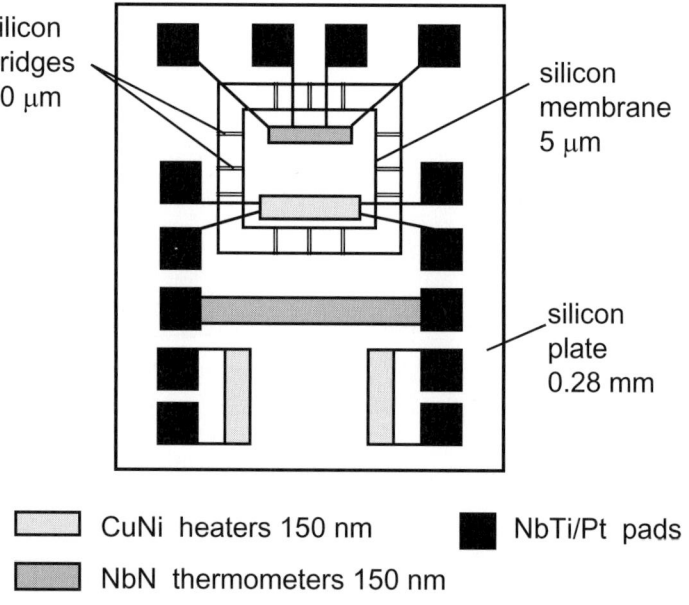

Fig. 7.6. Modulation nanocalorimeter developed by Fominaya et al. (1997a). The calorimeter was fabricated by technology used for manufacturing electronic microchips

The sample was grown on the membrane, between the heater and the thermometer. It is also possible to paste a small sample on the backside of the membrane. The silicon frame is anchored to a copper holder provided with a germanium thermometer and linked to the bath. The whole is installed in a conventional low-temperature calorimeter. A 5-T superconducting coil provides the magnetic field measured by a Hall probe. A low-noise preamplifier and a lock-in amplifier measure the temperature oscillations. A heater governed by the germanium thermometer on the holder or by the NbN thermometer on the frame regulates the temperature.

The feasibility of the nanocalorimeter was confirmed by measurements on a thin lead layer and on single crystals of Mn_{12} acetate. In the 150-nm lead film, a deviation from bulk behaviour was found under external magnetic fields. In a magnetic field, the specific-heat curve near the transition point has a shape similar to that in type II superconductors. The reason is that the thickness of the film is smaller than the coherence length, so that a magnetic flux penetrates the sample. The specific heat of Mn_{12} acetate was measured, at several temperatures, as a function of magnetic field (Fominaya et al. 1997b, 1999b). Fominaya et al. (1999a) carried out a similar study on a Fe_8 [(triazacyclononane)$_6Fe_8O_2(OH)_{12}]^{8+}$ single crystal.

Recently, Merzlyakov (2003) proposed a modulation nanocalorimeter based on a commercially available sensor with a thermopile (Sect. 5.5.3). The thermopile and a thin electric heater are integrated into a thin silicon membrane. The heater drives temperature oscillations in a sample. The frequency response of three types of integrated circuits was checked. The set-up is suitable for heat-capacity measurements over a frequency range 1 mHz–100 Hz. The measurements were performed on aluminium, polystyrene, and polycaprolactone. The resolution of the measurements is as high as 6×10^{-11} J.K^{-1} at room temperature in ambient air. At frequencies above 1 Hz, heat capacity of very small samples is measurable.

7.2 Organic and Biological Materials

Schantz and Johnson (1978), Smaardyk and Mochel (1978), and Tanasijczuk and Oja (1978) were the first to carry out modulation measurements on organic liquids. The main obstacle for applying modulation calorimetry to nonconducting liquids arises from the short thermal diffusion length in the sample. For example, at a 1-Hz modulation the thickness of a water sample should not exceed 0.2 mm.

Organic Liquids and Liquid Crystals

Many organic materials exhibit molecular ordering in intermediate phases between usual liquid and solid, the so-called mesophases (see, e.g., Priestley 1974). The partial ordering of the molecules in a given mesophase may be either translational or rotational, or both. The rotational order has meaning only when the molecules are nonspherical. Two different types of mesophases are known. There are those that retain a 3-dimensional crystal lattice but are characterized by substantial rotational disorder, called plastic crystals. Second, there are mesophases with no lattice, which are therefore fluids but exhibit considerable rotational order, called liquid crystals. In the mesophases, the molecules show some degree of rotational order (and in some cases partial translational order as well), although the crystal lattice is destroyed. In thermotropic liquid crystals, changing temperature most naturally effects phase transitions. Lyotropic liquid crystals are formed by solutions of rod-like entities in normally isotropic solvents. In such cases, a natural parameter governing phase transitions is the solute concentration.

The ordering in liquid crystals may be of three main types: nematic, cholesteric, and smectic. The nematic ordering is characterized by long-range orientational order, i.e., the molecules tend to align parallel to each other. In the cholesteric ordering, the preferred orientation changes gradually from one layer to another. In the smectic ordering, three main phases are A, C and B phases. All three have one degree of translational ordering, resulting in a layered structure. Within the layers of a smectic-A phase, the molecules are aligned perpendicular to the surface, while there is a short-range correlation of the centre of mass positions. The layer thickness is identical to the molecule's length. In the smectic-C phases, there is a uniform tilting of the molecules with respect to the surface. The tilt angle may be temperature dependent. In the smectic-B phases, the molecules exhibit ordering within the layers. This means that the layers are no longer fluids. However, the ordering is not of the sort familiar in solids. The smectic-B phase may be rather a plastic crystal. For materials having nematic and smectic mesophases, the order of their stability with increasing temperature is as follows: solid → smectic-B → smectic-C → smectic-A → nematic → isotropic liquid. For materials having cholesteric and smectic mesophases, the order of the transitions is: solid → smectic-A → cholesteric → isotropic liquid.

Schantz and Johnson (1978) measured the specific heat of 4-n-pentylphenyl-thiol-4'-n-octyloxybenzoate in the nematic, smectic-A, and smectic-C phases. A resistive heater heated the sample, while a thermistor served for temperature measurements (Fig. 7.7). The modulation frequency was $\omega = 0.1$ s^{-1}. A logarithmic divergence of the specific heat was observed for the nematic–smectic-A transition.

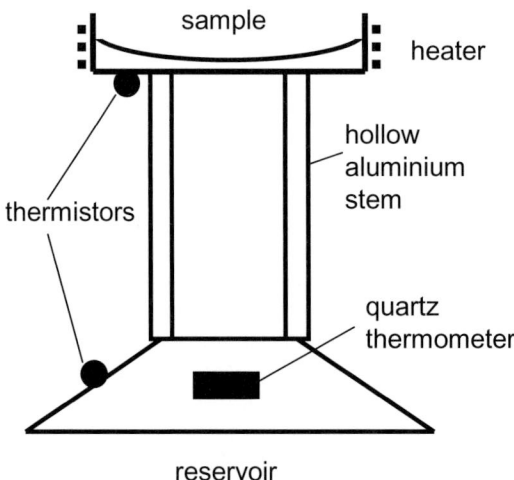

Fig. 7.7. Simplified diagram of calorimeter for organic liquids designed by Schantz and Johnson (1978)

LeGrange and Mochel (1980, 1981) developed a modulation calorimeter for organic liquids, in the range from room temperature to 200°C (Fig. 7.8). The set-up provides a temperature resolution better than 1 mK and detects changes in heat capacity of 0.01%. The sample contains approximately 0.1 mg of liquid material. The liquid is confined between the Pyrex (3 μm) and Mylar (6 μm) sheets. The heater is a thin film of chromel, vacuum evaporated onto the Mylar sheet. The temperature difference between the sample and the heat sink is measured with a thin-film copper-bismuth thermocouple evaporated onto the Pyrex sheet. Measurements with this apparatus were done on the nematic–smectic-A transition in octylcyanobiphenyl (8CB), octyloxycyanobiphenyl (8OCB), cyanobenzylidene-octyloxyaniline (CBOOA) and cyanobenzylidene-nonyloxyaniline (CBNOA).

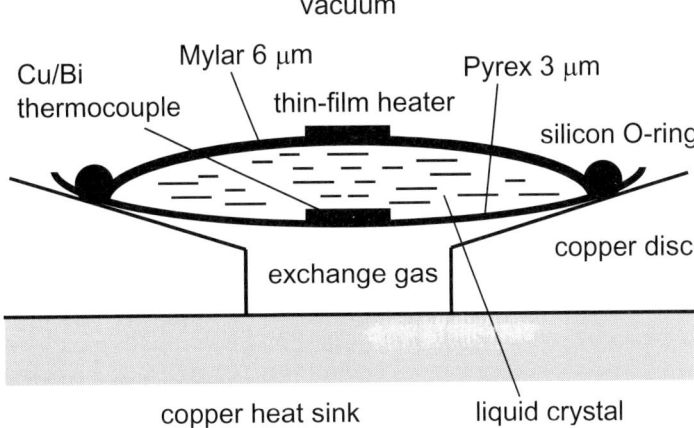

Fig. 7.8. Simplified diagram of calorimeter for organic substances designed by LeGrange and Mochel (1981)

A calorimeter for fluids and liquid crystals was described by Garland (1985). The body of the calorimeter cell is a cup 1 cm in diameter and 0.12 mm deep pressed from 0.25 mm silver sheet (Fig. 7.9). A lid of 0.08-mm silver foil is attached using a cold-welded indium or solder seal. The filled cell weighs about 600 mg and contains 50–100 mg of sample. Three or four short (2 mm) nickel rods serve for magnetic stirring by means of an external rotating permanent magnet (not shown in the figure). In the case of liquid crystals, a long coil of fine gold wire, which greatly reduces temperature gradients inside the cell, replaces these rods. A resistance heater supplies the AC heat input, and a micro bead thermistor measures the resulting temperature oscillations. The thermistor comprises one arm of a sensitive AC bridge driven at 1050 Hz. A fully automated calorimeter was developed by the use of a programmable digital multimeter. This system operates at a 32-s modulation period. Near second-order phase transitions, a very slow scan rate was used, 12 mK.h^{-1}. With the automated calorimeter, binary fluid mixtures, liquid crystals, and micellar solutions were studied.

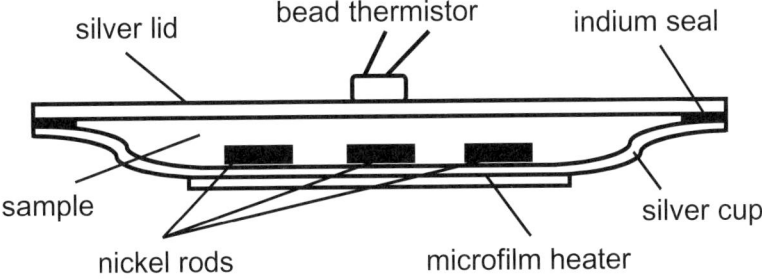

Fig. 7.9. Simplified diagram of calorimeter cell for fluids and liquid crystals described by Garland (1985)

Yao and Hatta (1988) developed a microcalorimeter for liquids employing a thin-wall stainless steel capillary as the calorimeter cell (Fig. 7.10). The capillary is 39 mm in length, and the sample volume is 0.5 µl. Polyethylene tubes of low thermal conductivity are connected to both ends of this capillary, and a liquid is injected through them into the cell by a microsyringe. Light from a halogen lamp modulated at 1 Hz heats the cell, and a 12.5-µm thermocouple measures the temperature oscillations in it. To determine the absolute values of the heat capacity of the sample, the temperature oscillations were also measured for the empty cell and the cell filled with a standard sample, pure water. It was found that the heat capacity of a liquid could be obtained with 1% accuracy. Later, Yao et al. (1999) used a similar microcalorimeter with direct electric heating of the capillary. The authors measured the specific heat of some biological materials. Table 7.1 presents only a minor part of numerous studies of liquid crystals employing modulation calorimetry.

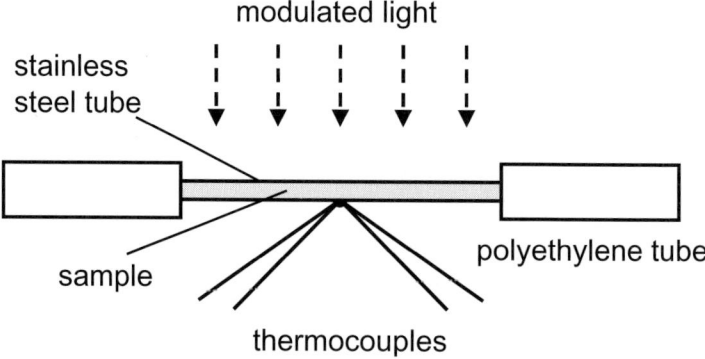

Fig. 7.10. Microcalorimeter cell for liquid samples designed by Yao and Hatta (1988). The volume of the liquid sample is 0.5 µl

Biological Materials

To investigate melting of organic and biological materials, Tanasijczuk and Oja (1978) designed a high-resolution calorimeter. The authors stressed that the calorimeter cell has to meet the following general requirements:

- It must be very thin (0.1 mm) and contain a volume of approximately 10 µl.
- Must be vacuum tight, demountable, and easily fill able with a syringe.
- Must have accurately established and stable dimensions, with heater and thermistor sides accurately parallel.
- The heater and thermistor should not be in electrical contact with the fluid.

The dimensions of the working volume of the calorimeter were highly stable and reproducible to within approximately 10%. The modulation frequencies were in the range 0.3–1 Hz. A microchip resistor element heated the sample, and a micro bead thermistor included in a Wheatstone bridge measured the temperature oscillations in the sample.

Table 7.1. Studies of liquid crystals

Item	Reference
8OCB	Johnson et al. 1978;
	Viner and Huang 1981
8S5	Schantz and Johnson 1978
CBOOA, CBOOA + CBNOA	Smaardyk and Mochel 1978
CBNOA, 8CB, 8OCB, CBOOA, 8S5	LeGrange and Mochel 1980, 1981
4O.8	Lushington et al. 1980a
Liquid-crystal mixtures	Lushington et al. 1980b;
	Garland and Huster 1987;
	Das et al. 1989b;
	Ema et al. 1989ab;
	Garland et al. 1989ab;
	Stine and Garland 1989;
	Boerio-Goates et al. 1990;
	Fontes et al. 1990;
	Wen et al. 1990. 1992;
	Wu et al. 1993;
	Castro and Puértolas 1997
4O.8, 65OBC	Huang et al. 1981
2M4P9OBC	Huang and Viner 1982
4O.7	Bloemen and Garland 1981;
	Garland et al. 1983
AMC-11, 4O.7, 7O.6, 7O.4	Meichle and Garland 1983
65OBC	Viner et al. 1983;
	Nounesis et al. 1986;
	Haga et al. 1997
2M45OBC	Lien et al. 1984
DOBAMBC	Dumrongrattana et al. 1986
Freestanding films	Pitchford et al. 1986;
	Geer et al. 1989, 1991ab, 1993;
	Stoebe et al. 1992;
	Jin et al. 1995
T7, T8	Evans-Lutterodt et al. 1987
A7	Liu et al. 1988
34COOBC, 64COOBC	Mahmood et al. 1988
TBBA	Das et al. 1989a
Critical and tricritical behaviour	Ema et al. 1995, 1996ab, 1997;
	Iannacchione et al. 1995a;
	Ema and Yao 1998;
	Haga and Garland 1998
Chiral liquid crystals	Chan et al. 1995;
	Garland 1996;
	Iannacchione et al. 1998b
8OCB, C_8tolane	Beaubois et al. 1997

The calorimeter cell was placed in an oven, and the temperature scans lasted in the range 3–8 h for a temperature span 20–60°C. The operation of the calorimeter was tested by means of three substances: n-octadecane, tert-butyl alcohol, and di-palmitoyl L-α-lecithin in water. Tert-butyl alcohol was investigated also in the glassy state, after rapidly cooling.

A brief description is given here of a microcalorimeter designed by Imaizumi et al. (1983) after a certain experience in this field was accumulated. Two phosphor-bronze plates, 0.1 mm thick, confine the calorimeter cell (Fig. 7.11). The distance between them could be kept constant by using platinum-wire spacers. Usually, the thickness of the sample is 0.1 mm. A 50-W halogen lamp and a chopper provide modulated-light heating of the cell. The modulation frequency is 0.75 Hz, and the temperature oscillations in the calorimeter are less than 10 mK. Two 25-μm chromel-alumel thermocouples measure the mean temperature and the temperature oscillations in the cell. To confirm the adiabaticity conditions, the phase shift between the input AC power and the temperature oscillations is measured by means of a lock-in amplifier. The measurements are performed using a data-acquisition system and a computer. At every point, the temperature of the cell is kept constant during 60 s, and the temperature increments range from 0.02 to 0.1 K. The calorimeter was used in studies of phosphatidylcholine. The lipid suspensions in water ranged from 0.05 to 0.2 in the weight ratio.

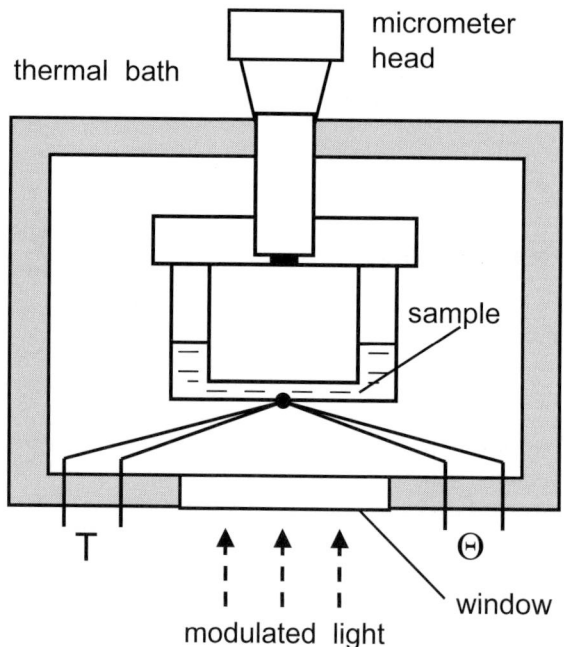

Fig. 7.11. Modulation calorimeter for organic and biological materials with a variable thickness of the sample (Imaizumi et al. 1983)

The authors proposed a method of the determination of absolute values of the heat capacity of the sample under study taking water as a reference. The reciprocal of the amplitude of the temperature oscillations was a linear function of the thickness of the water layer. From this dependence, the amplitude of the power oscillations and the heat capacity of the plates were determined. The same procedure with the lipid suspension allowed absolute determinations of its heat capacity. A sharp maximum in the specific heat was observed at the transition point.

A differential modulation microcalorimeter for organic and biological materials was designed by Dixon et al. (1982).

Modulation studies of biological materials are listed in Table 7.2.

Table 7.2. Modulation calorimetry of biological materials

Item	Reference
Dipalmitoyl L-α-lecithin	Tanasijczuk and Oja 1978
Lysozyme crystal, denaturation point	Imaizumi et al. 1979
Protein aqueous solutions	Imaizumi et al. 1981
DPPC	Hatta et al. 1983, 1984; Imaizumi et al. 1983; Imaizumi and Garland 1987
DMPC, DPPC (+ cholesterol)	Imaizumi and Hatta 1984
CsPFO micelles	Imaizumi and Garland 1989
DMPC, unilamellar vesicles	Nagano et al. 1995ab
Phosphatidylcholines, unilamellar vesicles	Uchida et al. 1997

DPPC = dipalmitoylphosphatidylcholine, DMPC = dimyristoylphosphatidylcholine, CsPFO = caesium perfluoro-octanoate

7.3 Photoacoustic Techniques

The photoacoustic phenomenon was known for about a century before it was applied to modulation measurements of specific heat and other thermophysical properties of solids. Rosencwaig and Gersho (1976) developed a theory of such measurements. The authors claimed that the main source of the acoustic signal in a photoacoustic cell arises from the periodic heat flow from the solid to the surrounding gas when the solid is periodically heated. Only a relatively thin layer of air adjacent to the surface of the solid, about 2 mm under a 100-Hz modulation, participates in the production of the acoustic signal. The general solution for the thermal-diffusion equations contains many parameters related to the solid, the backing, and the gas. However, for special cases readily achievable, the solution becomes much simpler.

Pichon et al. (1979) described a photoacoustic cell operating between 5 and 300 K in a gas-flow cryostat. The authors studied the specific-heat anomaly of $CrCl_3$ and MnF_2 at magnetic phase transitions. Bechtold et al. (1980) studied phase transitions in metal-hydrogen interstitial alloys NbH_x, TaH_x, and VH_x. The authors designed a photoacoustic cell suitable for measurements at temperatures up to 1050 K (Fig. 7.12). A He-Ne laser provides the AC heating of the sample. Its power is sufficiently low to avoid local heating of the surface of the sample. The modulation frequency is 40 Hz. The photoacoustic cell was built up as a Helmholtz resonator to keep the inner gas volume smaller than 1 cm^3. At 400 K, the temperature in the cell can be stabilised within 25 mK. A condenser microphone in conjunction with a sensitive preamplifier and a lock-in amplifier acts as the sound sensor. To protect the microphone from the high temperature, it is placed outside the furnace.

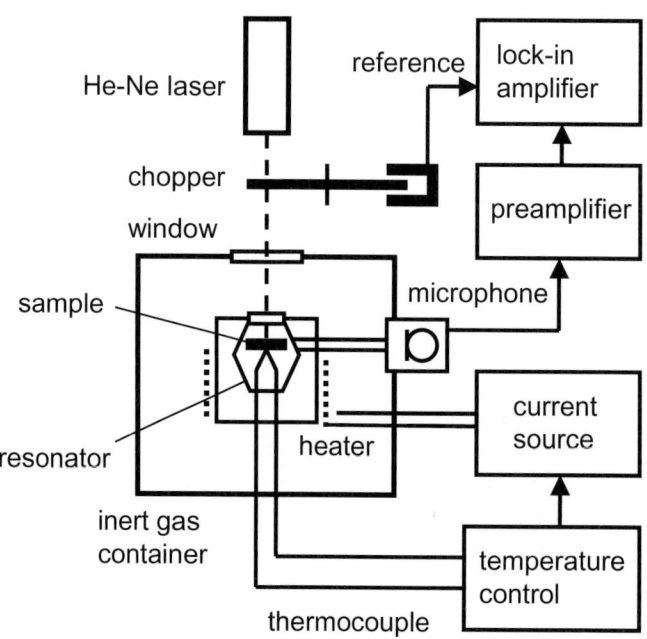

Fig. 7.12. Simplified diagram of photoacoustic cell for temperatures up to 1050 K (Bechtold et al. 1980). The microphone is placed outside the furnace

Bibi and Jenkins (1983) studied a phase transition in CoSiF$_6$·6H$_2$O. Zammit et al. (1988) described an apparatus for the photoacoustic calorimetry. With this technique, Zammit et al. (1990), Marinelli et al. (1992, 1998), Schoubs et al. (1994) studied phase transitions in liquid crystals. Using the photoacoustic and photopyroelectric techniques, Glorieux et al. (1994) investigated antiferromagnetic phase transitions in CoO and Cr$_2$O$_3$. At low modulation frequencies, the thermal diffusivity, $D = \kappa/\rho c$, and the thermal effusivity, $e = (\rho c \kappa)^{1/2}$, govern the amplitude and the phase of the signal (κ is the thermal conductivity, and ρ is the density of the sample). At high frequencies, the signal becomes inversely proportional to the thermal effusivity of the sample. By combining the data, the specific heat and the thermal conductivity are available: $\rho c = e/D^{1/2}$, $\kappa = eD^{1/2}$.

Glorieux and Thoen (1994) and Glorieux et al. (1995) used the photoacoustic technique for simultaneously determining the specific heat and the thermal conductivity of gadolinium near its Curie point. The mean temperature of the photoacoustic cell was automatically controlled within 0.01 K, while the heating rate was 0.02 K.min^{-1}. The sample, in the form of a thin slab, was kept in contact with air on the top and bottom sides. A modulated beam from a 10-mW He-Ne laser illuminated it. A microphone followed by a lock-in amplifier detected the photoacoustic signal. The magnitude and the phase of the temperature oscillations were obtained from a calibration by means of a reference sample with known optical and thermal properties, carbon-coated copper. Through iterative calculations, the specific heat and the thermal conductivity were determined. In the range 10 to 240 Hz, both quantities were frequency independent. Fitting the data to reference values at 40°C yielded absolute values. The measurements clarified the behaviour of gadolinium samples of different quality in zero and nonzero magnetic fields. It turned out that employment of high-quality single-crystal samples avoids rounding effects. In addition, the temperature oscillations in the samples were created also by an AC magnetic field.

Vassilev et al. (1995) built a variable-temperature photoacoustic cell on a conventional differential scanning calorimeter. The system operates in the range 300–800 K.

7.4 Specific-Heat Spectroscopy

The specific-heat spectroscopy is one of important achievements of modulation calorimetry during the last two decades. It presents an elegant technique for studying relaxation phenomena in specific heat.

7.4.1 Measurement of Thermal Effusivity

The technique developed by Birge and Nagel (Birge and Nagel 1985, 1987; Birge 1986) relies on the heat diffusion from a thin planar heater immersed in a liquid to be studied. The heat flow from such a heater is described by the one-dimensional heat-diffusion equation. The heater functions also as a thermometer. The complex amplitude of the temperature oscillations $\Theta(x, \omega)$ meets the equation

$$i\omega\rho c\Theta(x, \omega) = \kappa \partial^2 \Theta / \partial x^2, \quad (7.1)$$

where ρ, c and κ are the density, the specific heat, and the thermal conductivity of the medium surrounding the heater.

The appropriate boundary condition is $\Theta(x, \omega) \to 0$ as $x \to \infty$. The solution to Eq. (7.1) consistent with this requirement is

$$\Theta(x, \omega) = \Theta(0, \omega)\exp(-k|x|), \quad (7.2)$$

where $\Theta(0, \omega)$ is a complex amplitude of the temperature oscillations at the heater, and the complex thermal wave vector k equals

$$k = (\omega\rho c/\kappa)^{1/2}\exp(-i\pi/4) = (\omega\rho c/2\kappa)^{1/2}(1 - i). \qquad (7.3)$$

There is an additional boundary condition at the heater, which relates the heat flux j to the temperature gradient at the heater:

$$\kappa\partial\Theta/\partial x\big|_{x\to 0_-} - \kappa\partial\Theta/\partial x\big|_{x\to 0_+} = j. \qquad (7.4)$$

The two derivatives are calculated on opposite sides of the heater. From Eqs. (7.3) and (7.4), one obtains

$$\Theta(0, \omega) = j/2\kappa k = (j/2e\omega^{1/2})\exp(-i\pi/4), \qquad (7.5)$$

where the quantity $e = (\rho c\kappa)^{1/2}$ is called the thermal effusivity.

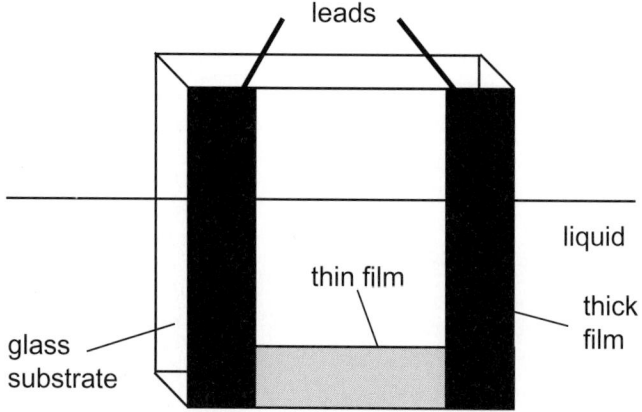

Fig. 7.13. The heater-thermometer probe for specific-heat spectroscopy developed by Birge and Nagel (1985, 1987). The temperature oscillations in the probe depend on the thermal effusivity of the surrounding medium

In a normal liquid, the temperature oscillations at the heater lag in phase behind the heat oscillations by $\phi = \pi/4$, and their amplitude is proportional to $\omega^{-1/2}$. However, when the thermal effusivity of the liquid becomes complex, an additional contribution appears to the phase and to the frequency dependence of the amplitude.

The heater-thermometer probe is a thin nickel film evaporated onto a substrate of window glass (Fig. 7.13). Nickel was chosen because of its large temperature coefficient of resistance and ease of evaporation. Thick silver contacts at the edges of the substrate were evaporated on top of nickel. A typical probe had a length of 2 cm, a width of 0.3–0.6 cm, a thickness of about 0.1 µm, and a total resistance of 10–20 Ω. The wider heaters are needed for low-frequency operations, since the lateral dimensions of the heater must be long compared to the thermal wave length. The probe is included into a Wheatstone bridge and driven with an AC current at frequency ω, so that the heat flux is produced at frequency 2ω. The temperature oscillations are measured with the third-harmonic technique (Sect. 4.1.2). The bridge is balanced at the fundamental frequency, while the 3ω component is measured with a lock-in amplifier (Fig. 7.14). The reference voltage for the amplifier is taken from a frequency tripler. For frequencies below 0.2 Hz, the authors designed a digital lock-in amplifier.

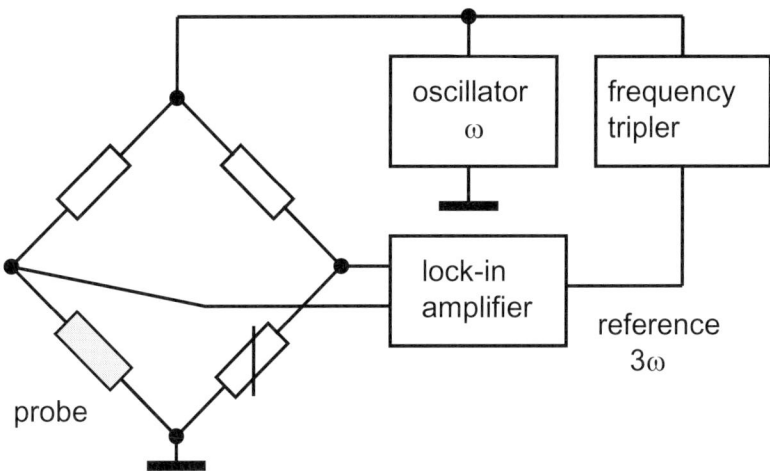

Fig. 7.14. Simplified diagram of bridge circuit for measuring thermal effusivity (Birge and Nagel 1986). The bridge is balanced at frequency ω, and a lock-in amplifier measures 3ω signal generated by the probe. The small quadrature component of the fundamental frequency remains unbalanced, but the lock-in amplifier does not sense it

The prime consideration in the construction of the heater is that its thickness is much smaller than the thermal wavelength, $\lambda = |k|^{-1}$. For operating frequencies f less than 10 kHz, $\lambda > 1$ μm. One can satisfy the above criterion by evaporating a thin-film heater onto a substrate, whose thickness is much greater than the longest thermal wavelength encountered. For $f = 0.01$ Hz, $\lambda \approx 1$ mm. The thermal waves are exponentially damped, so that a thickness of 6 mm is ample. The solution (7.5) must be modified taking into account different thermal parameters of the sample and of the substrate. The result is

$$\Theta(0, \omega) = j\exp(-i\pi/4)/\omega^{1/2} (e + e_{sub}), \qquad (7.6)$$

where e_{sub} relates to the substrate.

A measurement with an empty sample cell gives the term due to the substrate alone, which must be known to deduce the term related to the sample. It is important that e_{sub} not to be too large compared to e. From the temperature oscillations in the probe, one obtains the thermal effusivity of the sample, $e = (\rho c \kappa)^{1/2}$. The quantity $c\kappa$ becomes complex and frequency dependent when the sample approaches the glass-transition region.

7.4.2 Frequency-Dependent Effusivity

A specific-heat spectrometer developed by Birge and Nagel (1985, 1987) served for measurements in the range 0.01-3000 Hz. It was used to study organic liquids near the glass transition, where the thermal effusivity becomes frequency-dependent. It turned out that this frequency dependence should be attributed to the specific heat. This dependence varies along with the temperature of the liquid because of changes in the relaxation time (Fig. 7.15). Glycerol and propylene glycol were the first liquids studied by the authors.

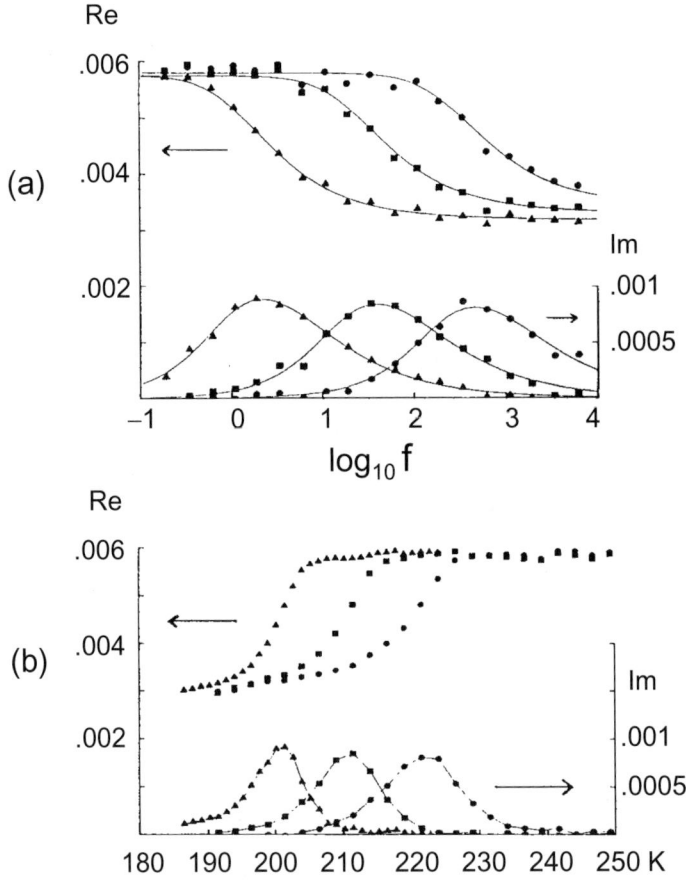

Fig. 7.15. (a) The real and imaginary parts of $\rho c \kappa$ in $J^2 \cdot cm^{-4} \cdot s^{-1} \cdot K^{-2}$ for glycerol as a function of frequency. The temperatures are: ▲ 203.9 K, ■ 211.4 K, and ● 219.0 K. **(b)** The real and imaginary parts of $\rho c \kappa$ as a function of temperature. The frequencies are: ▲ 0.62 Hz, ■ 34 Hz, and ● 1100 Hz (Birge and Nagel 1985). The authors founded a new branch of calorimetry, the specific-heat spectroscopy

Later, Birge et al. (1997) formulated the main requirements for the probe used in the specific-heat spectroscopy as follows:

- The substrate must be thick compared to the characteristic thermal wavelength λ in the liquid under study. For the lowest frequency to be used, the substrate thickness must be greater than 2λ for obtaining errors less than 2% of the temperature at the heater.
- The surface of the heater should be of optical quality.
- The thermal effusivity of the substrate material should be as small as possible.
- It might be possible to construct planar heaters on a thin film instead of a thick substrate. This would optimise the signal by allowing heat diffusion into the liquid on both sides of the film.
- The operating frequency bandwidth of a heater is determined by the thermal properties of the material under study and the heater dimensions. The upper frequency limit is determined by the minimum measurable 3ω signal. Typical values of this minimum are of the order of 1 µV.
- For given electrical properties of the heater and a minimum measurable signal, the upper frequency limit is proportional to the ratio L^2/W^6, where L is the length of the heater, and W is its width. Minimising the width of the heater is critical for maximising the upper frequency limit.
- The lower frequency limit arises when the thermal wavelength λ becomes a significant fraction of the heater width W. For a given tolerance on errors in phase shift measurements, this limit is proportional to W^{-2}.

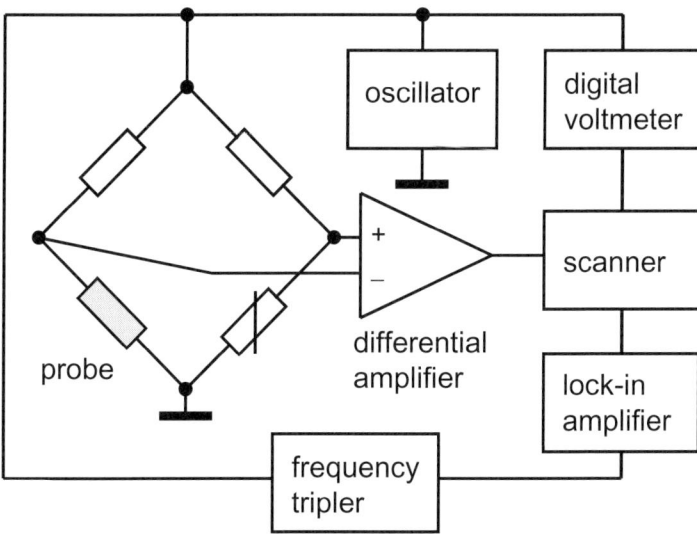

Fig. 7.16. Automated specific-heat spectrometer employing the third-harmonic technique (Jung et al. 1992; Moon et al. 1996). The spectrometer operates in the range 10 mHz to 10 kHz

Measurements employing the third-harmonic technique are possible in a wide frequency range. Jung et al. (1992) designed an automated calorimeter for the range 10 mHz–10 kHz (Fig. 7.16). At its early stage, a drawback of the specific-heat spectroscopy was that it could not separate the specific heat and the thermal conductivity. Since then, the technique has been improved (Dixon and Nagel 1988; Inada et al. 1990; Menon 1996). A reliable separation of the specific heat and thermal conductivity became possible by using probes of different widths evaporated onto a single substrate and employed in different frequency bands. In the probe described by Menon (1996), the widest probe is used between 2 mHz and 1 Hz, the medium from 0.02 to 16 Hz, and the narrowest from 1 to 4000 Hz (Fig. 7.17).

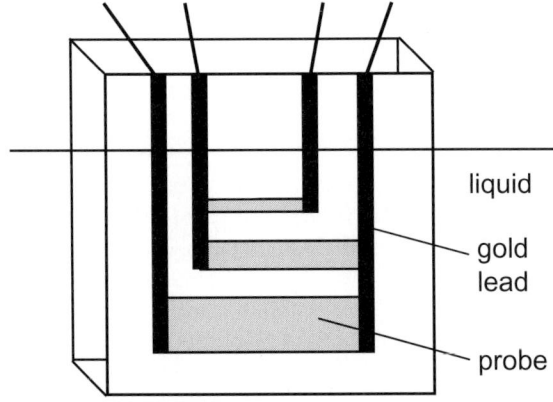

Fig. 7.17. Sensor with three probes of different width are used for measurements in various frequency ranges and separation of specific heat and thermal conductivity (Menon 1996)

Jonsson and Andersson (1998) performed a theoretical analysis and measurements of temperature oscillations in planar 3ω sensors used in the specific-heat spectroscopy. For low frequencies, the response of the sensor is far from the prediction that does not account for the lateral heat flow. To obtain results within 1% of the theory, the ratio between the width of the sensor and the quantity $(D/\omega)^{1/2}$ must be greater than 30. At high frequencies, the temperature response of the sensor moves from that of the sample toward that of the material of the sensor. For a thickness of 0.1 μm, the deviation from the model reaches 5% at 5 kHz. The numerical results relate to a nickel-film sensor evaporated onto a glass sample. The phase of the temperature oscillations in the sensor also deviates from the ideal value. The gradual change in the response can be accounted for by using a two-layer model considering properties of the sensor and the sample. An isolating layer between the sensor and the sample also causes deviations from the ideal situation.

Recently, Jung et al. (2003) developed a differential calorimeter for liquid samples. The calorimeter based on the third-harmonic technique includes two identical probes, so that the imbalance of the Wheatstone bridge, without a sample, is zero. Once a sample is placed on one probe, a third-harmonic signal is produced. The frequency range of the calorimeter was extended up to 30 kHz.

7.4.3 How to Extend the Frequency Band – a Proposal

In a review of the specific-heat spectroscopy, Birge et al. (1997) pointed out two main avenues for improvement of this technique: (i) extending the bandwidth to higher frequencies, and (ii) enhancing the sensitivity. The authors described the situation as follows: "Unfortunately, at frequencies greater than 10 kHz an additional third-harmonic signal generated by a different mechanism becomes the dominated signal. Although it has not been proved, it appears that the mechanism is due to thermally induced acoustic resonances of the heater substrate... We have not yet devised a scheme to defeat this spurious signal." Menon (1996) stated that when the third-harmonic signal is to be measured within 1% accuracy, stray third-harmonic voltages must be held below about 20 nV. He claimed: "in principle, the high-frequency limit in the specific-heat spectroscopy is set only by the frequency scale, at which the thermal diffusion length becomes comparable to the thickness of the heater film. This limit is at least two decades above the highest frequency used in this experiment, which is only 4 kHz. Above this frequency, the parasitic third-harmonic signal becomes too large to be reliably treated as a calibrated background."

As a rule, a bridge balances the fundamental-frequency voltage across the sensor to extract the third-harmonic signal generated by the probe. However, instead of directly measuring the output voltage of the bridge, voltages across the sensor and across another resistor of the bridge are fed to two inputs of a differential amplifier. Relatively large voltages are thus fed to the inputs of the amplifier. Jeong (1997) pointed out that the upper limit in the modulation frequency is posed by an insufficient common-mode rejection by the differential amplifier. However, this ratio governs appearance of an output voltage of the fundamental frequency. Much more important in this case is that, due to the non-linearity, the amplifier produces also a third-harmonic voltage.

This conjecture was checked by means of a lock-in amplifier with differential input, PAR model 124A. A sine-wave voltage from a multifunction synthesizer, HP model 8904A, was fed to its inputs. At 3 kHz, the amplitude of this voltage was varied in the range 0.1–3 V. The lock-in amplifier employed incorporates a selective amplifier. It was in turn tuned to the fundamental and the third-harmonic frequency. The amplified voltage was observed by an oscilloscope and measured by the lock-in amplifier. The second channel of the synthesizer provided a 3ω reference when measuring the third-harmonic signal created by the amplifier. The results obtained (Fig. 7.18) show that the output signal at the fundamental frequency is proportional to the input signal, and the common-mode-rejection ratio amounts to 114 dB. More important, the differential amplifier generates also a third-harmonic signal. This signal non-linearly depends on the input voltage, increasing from about 10 nV for 0.3 V at the input to about 300 nV for 3 V at the input. This third-harmonic signal should not depend on frequency, in agreement with the observations by Menon (1996). The above conjecture is thus quite realistic. A very simple measurement is possible to check whether the differential amplifier itself generates a 3ω signal: it is enough to replace the probe by a usual resistor of equal resistance and to monitor the output voltage.

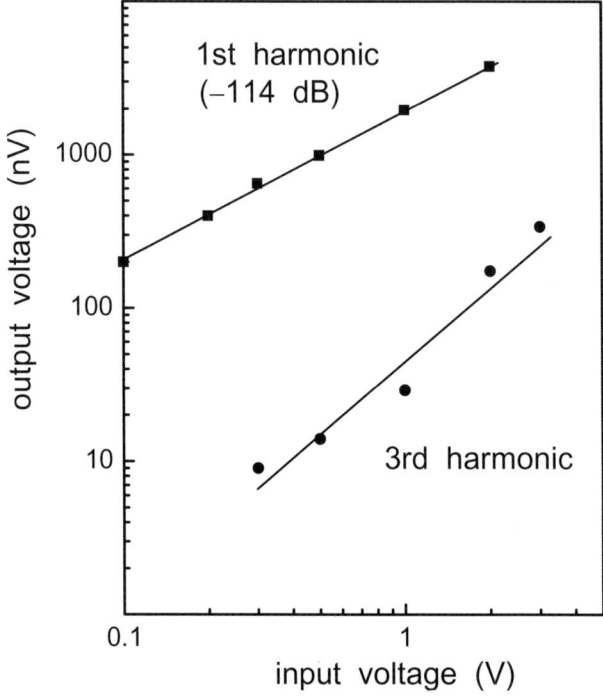

Fig. 7.18. Output voltages of fundamental and of third-harmonic frequencies versus input voltage at inputs of differential amplifier. The results confirm that the non-linearity of the differential amplifier generates a third-harmonic signal

Probably, the unwanted effect is avoidable by directly measuring the output voltage of the bridge, which can be made several orders of magnitude smaller than the voltage across the sensor. An isolating transformer is sufficient in this case. However, such a transformer may introduce additional phase shifts, which are difficult to calibrate for (Menon 1996). To circumvent this difficulty, it is probably possible to completely balance the third-harmonic signal generated by the probe. For this purpose, it is sufficient to introduce a 3ω signal of adjustable amplitude and phase into the arm of the bridge containing the probe (Fig. 7.19). A multifunction synthesizer, HP 8904A, may serve for producing the necessary compensating voltage. A selective amplifier measures the output voltage of the bridge directly by the use of an input transformer. Therefore, the amplifier itself would not generate a 3ω signal.

An extension of the frequency band seems to be achievable also by using, instead of the third-harmonic technique, the frequency-conversion method. With this method, the signal to be measured has a frequency equal to the difference between the frequency of the temperature oscillations and that of the supplementary current. The difference frequency can be selected in a favourable range. The supplementary current must be several times smaller than the heating current. The difference-frequency signal is therefore weaker than the third-harmonic signal, but it can be measured more reliably. This was shown in determinations of the temperature derivative of specific heat (Kraftmakher and Tonaevskii 1972), where the second-harmonic component in the temperature oscillations was measured in the presence of a much stronger fundamental signal. The frequency-conversion method is considered in Sects. 4.1.1 and 13.2.

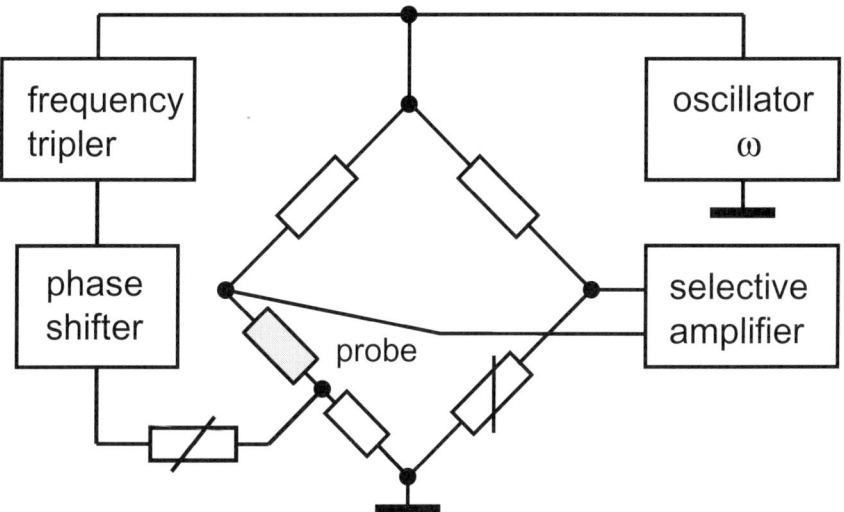

Fig. 7.19. Circuit for balancing the third-harmonic voltage generated by the probe. The voltage across a resistor included in series with the probe is adjusted to balance the third-harmonic voltage. The output voltage of the bridge is measured directly with a selective amplifier with an input transformer. The signal of the fundamental frequency is close to zero, and the amplifier itself does not generate 3ω voltage

7.4.4 What is Worth Remembering

It is worth remembering that the basic components of the specific-heat spectroscopy were found and explored long before. First, the third-harmonic method was invented by Corbino (1911) and further developed by Filippov (1960), Rosenthal (1961, 1965), and Holland (1963). Second, Filippov (1960) was the first to employ a heater-thermometer probe for modulation measurements of thermal effusivity of liquids. Rosenthal (1965) also used this method to sense the medium, in which the probe was immersed. Third, Jackson and Koehler (1960) and Korostoff (1962) considered frequency-dependent specific heat in relation to point-defect equilibration. Van den Sype (1970) further elaborated this approach. Meissner and Spitzmann (1981), Loponen et al. (1982), and Berret et al. (1992) considered time-dependent specific heat.

7.5 Modulated Differential Scanning Calorimetry

A new calorimetric technique was recently proposed, namely, modulated differential scanning calorimetry, MDSC (Gill et al. 1993; Reading et al. 1993). MDSC technique offered new possibilities for the traditional thermal analysis.

7.5.1 Principle of Modulated Differential Scanning Calorimetry

In MDSC, the temperature of the sample is sinusoidally varied about a constant ramp, so that

$$T = T_0 + qt + A\sin\omega t, \qquad (7.7)$$

where q is the underlying constant heating rate, and T_0 the initial isotherm at the beginning of the scanning experiment. The resulting heating rate dT/dt thus oscillates about the underlying heating rate q. This approach allows one to separately measure a reversible heat flow depending on the heat capacity of the sample, and an irreversible heat flow related to the latent heat of a first-order phase transition. When the temperature is modulated at the block temperature T_b, i.e.,

$$T_b = T_0 + qt + \Theta_s\sin\omega t, \qquad (7.8)$$

temperatures of the sample and of the reference, T_s and T_r, at steady state are

$$T_s = T_0 + qt - qC_s/K + \Theta_{0s}\sin(\omega t - \varepsilon), \qquad (7.9)$$

$$T_r = T_0 + qt - qC_r/K + \Theta_{0r}\sin(\omega t - \phi), \qquad (7.10)$$

where K is the Newton's law constant, assumed to be the same for the sample and for the reference (Boller et al. 1995).

The calorimeters usually consist of close-to-identical aluminium pans with and without a sample. In MDSC, two time scales are considered, the underlying heating rate and the period of modulation. The reversible heat capacity is extracted from the modulation amplitudes and is given, for the case of equal mass of the empty sample and reference pans, by

$$mc = (\Theta_{0\Delta}/\Theta_{0s})[(K/\omega)^2 + C_1^2]^{1/2}, \qquad (7.11)$$

where $\Theta_{0\Delta}$ represents the amplitude of the temperature difference $T_r - T_s$, and C_1 is the pan heat capacity.

The data obtained can be presented as a graph of the AC heat flow versus the temperature. The approach to the steady state and the drift in the heat flow is clearly visible from this plot. A calibration is necessary for evaluating the heat capacity of the sample. Usually, a standard sample serves for the calibration. The advantage of the use of the modulation regime lies in the elimination of any slow drifts of the calorimeter. The total heat flow is evaluated by taking sliding averages over full cycles of modulation. This type of averaging gives data similar to standard DSC. The irreversible part of the heat capacity is taken as the difference between the total and the reversible heat capacities.

Hensel and Schick (1997) reported on calibration of MDSC set-ups. Donth et al. (1997) compared results for several polymers from the specific-heat spectroscopy with those from DSC. Saruyama (1998) described a light-modulated differential scanning calorimeter. Optical fibres were used to separate the light source from the furnace, and light intensities for the sample and reference sides were controlled independently from each other. Mesquida et al. (1999) reported on a new cooling system for MDSC apparatus, which operates down to about 60 K. To illustrate the capabilities of the modified cooling system, the authors presented results of specific-heat measurements for betaine borate, betaine phosphate, and a standard silicon rubber, SilGel 604. B.Wunderlich et al. (2000) proposed to carry out MDSC measurements with multiple frequencies. The method is based on modulation with a complex saw tooth. B.Wunderlich et al. (2001) discussed instrumentation and interpretation of MDSC.

Huth et al. (2001) formulated main advantages of MDSC measurements, as compared with the third-harmonic technique, as follows:

- The absolute heat capacity values are more precise.
- No problems arise with the thermal conductivity of the samples apart from the high frequency limit.
- The available frequency range can be significantly extended towards lower frequencies.
- MDSC has no problem with stationary temperature fields, and precise temperature calibration procedures are applicable.

Merzlyakov and Schick (2001ab) developed methods of simultaneous multifrequency heat-capacity measurements by MDSC. Carpenter et al. (2002) compared results obtained from MDSC measurements on three glass formers (glycerol, propylene glycol, and salol) with those from the specific-heat spectroscopy. Danley (2003) reported on a variant of MDSC technique with independent sample and reference calorimeters. Menczel and Judovits (1998) presented a literature summary on MDSC. Table 7.3 presents a part of works employing the MDSC technique. Some calorimetric set-ups described in detail are listed in Table 7.4.

7.5.2 What is Worth Remembering

It is worth remembering that the idea of the modulation differential scanning calorimetry was proposed long ago by Gobrecht et al. (1971). The authors employed temperature modulation of a Perkin-Elmer DSC-1B differential scanning calorimeter in the frequency range 0.01–1 Hz and considered the complex specific heat. The real and imaginary parts of the thermal admittance of glassy selenium were found to be frequency dependent. The authors stressed the advantage of this approach as follows: "A combination of linear and periodic heating offers the advantage of the good temperature resolution of slow scan speeds and the higher output signal due to the faster oscillations."

Table 7.3. Modulated differential scanning calorimetry (MDSC)

Item	Reference
Presentation of MDSC technique	Gill et al. 1993; Reading et al. 1993, 1994; Boller et al. 1994
Compatibility of DSC and AC calorimetry	Hatta 1994, 1996
Mathematical description of MDSC	B.Wunderlich et al. 1994; Lacey et al. 1997; B.Wunderlich 1997
Glass transitions	Boller et al. 1995; Hensel et al. 1996; B.Wunderlich et al. 1996b; Wagner and Kasap 1996; Weyer et al. 1997b; B.Wunderlich and Okazaki 1997
$NaNO_2$	Hatta et al. 1995
Modulated-light heating	Nishikawa and Saruyama 1995
Interpretation of MDSC measurements	Schawe 1995, 1996, 1997; Jones et al. 1997; Reading 1997; Ozawa and Kanari 2000; B.Wunderlich et al. 2001
Experimental test of MDSC	Hatta and Muramatsu 1996
Cyclic MDSC	O'Reilly and Cantor 1996; B.Wunderlich et al. 1998
Complex specific heat	B.Wunderlich et al. 1996b; Simon and McKenna 1997; Baur and Wunderlich 1998; Merzlyakov and Schick 1999ab, 2000
Lead germanate, betaine arsenate, TGS	Bohn et al. 1997
Interpretation of C''	Höhne 1997a
Employment of lock-in technique	Höhne 1997b
Separation from transition effects	B.Wunderlich et al. 1997
$BaTiO_3$, $NaNO_2$, 8OCB	Hatta and Nakayama 1998
$SrTiO_3$, KH_2PO_4, PVA, DFTCE	Krüger et al. 1998
Pb nanoparticles in Al matrix	Li et al. 1998
Phase transitions of first order	Boller et al. 1998; B.Wunderlich et al. 1999
Phase transitions, theoretical considerations	Höhne 1999ab
$Pd_{40}Ni_{10}Cu_{30}P_{20}$, glass transition	Hu et al. 1999
Polymers	Luyt et al. 1998; Ribeiro and Grolier 1999; Castro and Puértolas 2003
$KHSO_4$, $LiNH_4SO_4$	Diosa et al. 2000ac
Calibration for MDSC	Moon et al. 2000a; Höhne et al. 2002; Merzlyakov et al. 2002
Crystallisation of polymers	Schick et al. 2000b, 2001
Multifrequency measurements	Androsch and Wunderlich 1999 B.Wunderlich et al. 2000; Merzlyakov and Schick 2001ab
NH_4HSO_4	Diosa et al. 2001
Time dependence of heat capacity	Saruyama 2001
Square modulation, $KMnF_3$, DKDP	del Cerro et al. 2003

TGS = triglycine sulphate, 8OCB = octyloxycyanobiphenyl, PVA = polyvinyl acetate, DFTCE = 1,2-difluoro-1,1,2,2-tetrachlorethane, DKDP = deuterated potassium dihydrogen phosphate

Table 7.4. Calorimetric set-ups

Item	Reference
Modulated-light heating	Connelly et al. 1971
Thin films, 1.3–8 K	Greene et al. 1972
Calorimeter with ^{3}He cryostat, 0.3–3 K	Manuel et al. 1972
High pressures, low temperatures	Eichler and Gey 1979
Improvement of light-heating method	Ikeda and Ishikawa 1979
Differential calorimeter, biological materials	Dixon et al. 1982
Automated calorimeter, 2–380 K	Stokka and Fossheim 1982a
Thin films, 0.4–3.5 K	Suzuki et al. 1982
Biological materials, 270–400 K	Imaizumi et al. 1983
Automated high-resolution calorimeter	Garland 1985
Specific-heat spectrometer, 0.01–3000 Hz	Birge and Nagel 1987
Small samples, magnetic fields up to 20 T	Schmiedeshoff et al. 1987
Varied frequency, variable-depth calorimeter	Wang and Campbell 1988
Microcalorimeter, low temperatures	Chae and Bretz 1989
Modulated-bath calorimeter	Graebner 1989
Magnetic fields up to 8 T, 4.2–200 K	Calzona et al. 1990
Specific-heat spectrometer, 10–10000 Hz	Inada et al. 1990
Microcalorimeter, adsorbed gases	Kenny and Richards 1990a
Freestanding thin films	Geer et al. 1991b
Quench-condensed films, 0.1–7 K	Menges and v. Löhneysen 1991
Automated calorimeter, 25–300 K	Yang et al. 1991
High-sensitivity calorimeter	Bednarz et al. 1992
Automated specific-heat spectrometer, 0.01–10000 Hz	Jung et al. 1992
Modulation and relaxation modes	Ema et al. 1993; Yao et al. 1998
Microcalorimeter, low temperatures	Minakov and Ershov 1994
Quench-condensed films, 0.4–3 K	Birmingham et al. 1996
Low-temperature differential microcalorimeter	Carrington et al. 1996, 1997
Contactless calorimetry	R. Wunderlich and Fecht 1996
Specific-heat spectrometer	Birge et al. 1997
Automated multifrequency calorimeter, 0.5–50 mHz	Ema and Yao 1997
Calorimeters for liquids in porous media	Finotello et al. 1997
Nanocalorimeter, low temperatures	Fominaya et al. 1997a
Specific-heat spectrometer, 0.2–2000 Hz	Korus et al. 1997
Microcalorimeter, 5–300 K	Marone and Payne 1997
Microcalorimeter of high resolution, 40–160 K	Riou et al. 1997
Calorimeter with optical fibre light guide	Garfield et al. 1998
Modulated DSC set-up, down to 60 K	Krüger et al. 1998
Differential calorimeter	Maesono and Tye 1998
Microcalorimeter for liquids	Yao et al. 1999
Microcalorimeter, Peltier heating	Moon et al. 2000b; Jung et al. 2002
Calibration of modulation calorimeters with 8OCB	Schick et al. 2000a
Calorimeter for 4.2–400 K range	Fraile-Rodríguez et al. 2001
Pressures up to 22 GPa, low temperatures	Wilhelm and Jaccard 2002ab
Differential specific-heat spectrometer, up to 30 kHz	Jung et al. 2003

8 Modulation Dilatometry

At present, methods for measuring the dilatation of solids provide sensitivity of the order of 10^{-10} m and even better. However, difficulties of high-temperature dilatometry are caused by the plastic deformation of the samples rather than by the lack of sensitivity. This is why one had to accept data on thermal expansivity averaged over wide temperature intervals. Important improvements in this field have been made in the last decades. Along with a significant progress in traditional dilatometry, two new techniques appeared, modulation dilatometry and dynamic technique. Modulation dilatometry allows determinations of the linear thermal expansivity with small temperature oscillations and thus greatly improves the resolution of dilatometric measurements. This technique seems to be very useful because it ignores irregular external disturbances and senses only what is necessary for determining the thermal expansivity. Modulation dilatometry is applicable to conducting and nonconducting materials and in wide temperature ranges, and very sensitive methods exist for measuring extremely small periodic displacements.

8.1 Brief Review of Dilatometric Techniques

Most sensitive dilatometers employ optical interferometers and capacitance sensors. Interferometric measurements became much easier by using lasers, owing to the high temporal and spatial coherence of the laser beam. Recently, James et al. (2001) reviewed dilatometric techniques at elevated temperatures.

According to Bennett (1977), the majority of traditional dilatometers fall into one of six categories:

- Dilatometers, in which the expansion of a rod sample is measured relative to a fused silica tube.
- Comparators using some type of micrometer microscope.
- Dilatometers, in which the expansion of a sample causes a mirror to tilt, either directly or via a rotating rod.
- Capacitance methods, where the thermal expansion produces a change of the capacitance in a parallel-plate capacitor.
- X-ray diffraction methods, which measure changes in the spacing of set of atomic planes in crystalline materials.
- Optical interferometry, particularly with Fizeau and Fabry-Perot interferometers.

8.1.1 Optical Methods

A sensitive laser dilatometer was designed by Feder and Charbnau (1966). Changes in the length of the sample are measured with a Fizeau-type interferometer (Fig. 8.1). The light beam is reflected from the bottom optical flat, upon which the sample rests, and from the top flat. The reflected beams produce visible interference fringes. The fringe shifts thus depend only upon the changes in the length of the sample. A He-Ne laser served as the light source, and a special system was designed for automatic fringe counting. The assembly consists of a prism, two adjustable slits, and two phototubes. It is first rotated to arrange the fringes perpendicular to the apex of the prism. Each fringe is then split into two parts with each phototube seeing one half of the fringes. By connecting the phototubes to a two-pen chart recorder, the fringe shifts are recorded as a series of sine waves. This double-recording system senses changes in the direction of the fringe motion. Temperature control is achieved by immersing the tube containing the sample in a regulated oil bath. The accuracy of the measurements depends only on the precision, with which the wavelength of the light is known, and the shift in the fringe pattern is determined.

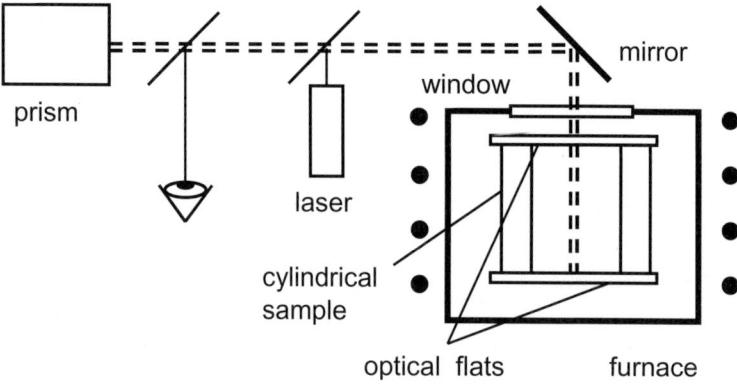

Fig. 8.1. Simplified diagram of interferometric dilatometer designed by Feder and Charbnau (1966). The dilatometer was used for determinations of vacancy contributions to thermal expansivity of some low-melting-point metals

Jacobs et al. (1970) developed a very sensitive dilatometer based on the dependence of Fabry-Perot resonances on the mirror separation (Fig. 8.2). The sample governs a space between the mirrors. The light spectrum consists of variable frequency sidebands obtained by modulation of a stabilised He-Ne laser. A change in the temperature of the sample causes a change in the resonance frequencies. The modulation frequency is adjusted until one of the sidebands coincides with a Fabry-Perot resonance. When the temperature changes, a new modulation frequency is necessary for the maximum transmittance. The change in this frequency shows the relative change in the length of the sample. The sample has the form of a hollow cylinder 10 cm long. Its ends are polished flat and approximately parallel.

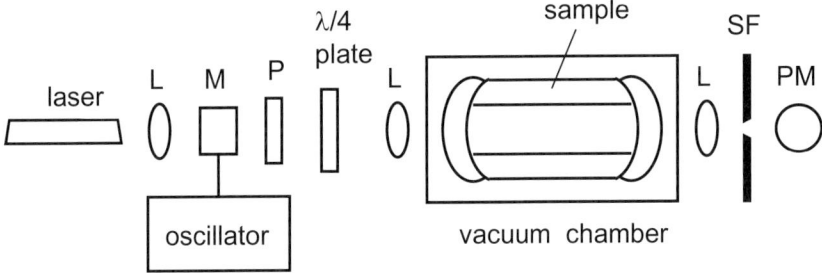

Fig. 8.2. Simplified diagram of interferometric dilatometer designed by Jacobs et al. (1970). Modulation of a He-Ne laser provides sidebands adjustable for achieving a resonance in the Fabry-Perot interferometer according to changes in the length of the sample. *L* – lens, *M* – modulator, *P* – polarizer, *SF* – spatial filter, *PM* – photomultiplier

The laser beam collimated by a lens passes through a modulator. A polarizer and a quarter-wave plate serve as an isolator against reflections back. The polarizer also suppresses the carrier frequency, so that only the sidebands are transmitted. The broadband tuning in the range 10–480 MHz is accomplished by means of an oscillator. The precision of the dilatometer is limited only by the laser instability, which is of the order of 10^{-9}. This technique is a good tool for the traditional dilatometry.

Bennett (1977) described an absolute interferometric dilatometer for the range 0–500°C developed at the NPL (National Physical Laboratory, Teddington). The dilatometer provides high resolution and accuracy. Edsinger and Schooley (1991) reported on an interferometric dilatometer for temperatures –20°C to 700°C. Suska and Tschirnich (1999) designed a sensitive dilatometer based on a Fizeau-type interferometer with two corner-cube reflectors.

The displacement measurements are also possible by the speckle pattern interferometry (Leendertz 1970; Løkberg et al. 1985; Kim et al. 1997). Gangopadhyay and Henderson (1999) reviewed vibration sensing based on this technique.

8.1.2 Capacitance Dilatometers

Capacitance dilatometers also provide very high sensitivity. Johansen et al. (1986) designed a sensitive capacitance dilatometer integrated in an oven. The dilatometer is of the parallel-plate capacitor type. A Peltier element provides fine regulation of the oven. In the range –60°C to 150°C, the oven can be stabilised to better than 10^{-4} K. For large dilatations, a HP 4192 low-frequency impedance analyser measures the capacitance. For a capacitance near 15 pF, the sensitivity is 10^{-3} pF. A computer repeatedly reads and averages the data from the analyser. For the sub micron range, the detection system is based on a manual capacitance bridge and a lock-in amplifier. The output signal of the amplifier is calibrated versus the deviation in capacitance between a reference capacitor and that of the dilatometer. In this case, the sensitivity is 7×10^{-12} m. A quartz-crystal thermometer measures the temperature of the dilatometric cell. Johansen (1987) modified this dilatometer to operate in a modulation regime (Sect. 8.2.5).

Rotter et al. (1998) developed a miniature capacitance dilatometer suitable for measuring thermal expansion and magnetostriction of small samples and samples of irregular shape. The active length of the sample may be smaller than 1 mm. An AC bridge measures the capacitance. The absolute resolution is about 10^{-10} m. The dilatometer was tested in the range 0.3–200 K and in magnetic fields up to 15 T.

8.1.3 Dynamic Techniques

Miiller and Cezairliyan (1982) at NBS (National Bureau of Standards, presently the National Institute of Standards and Technology) developed a subsecond dilatometric technique. A similar dilatometric apparatus was built at the IMGC (Instituto di Metrologia "Gustavo Colonnetti", Torino).

Miiller-Cezairliyan Approach

The dynamic dilatometric technique involves resistively heating the sample from room temperatures to above 1500 K in less than 1 s. The sample is mounted in a chamber providing measurements either in vacuum or in a gas atmosphere. The interferometer is a modified Michelson interferometer. The light source is a plane-polarized He-Ne laser. A laser interferometer measures the expansion of the sample. The sample has the form of a tube with parallel optical flats on opposite sides. The distance between the flats, 6 mm, represents the length of the sample. The sample acts as a double reflector in the path of the laser beam. The interferometer is thus insensitive to translational motion of the sample (Fig. 8.3).

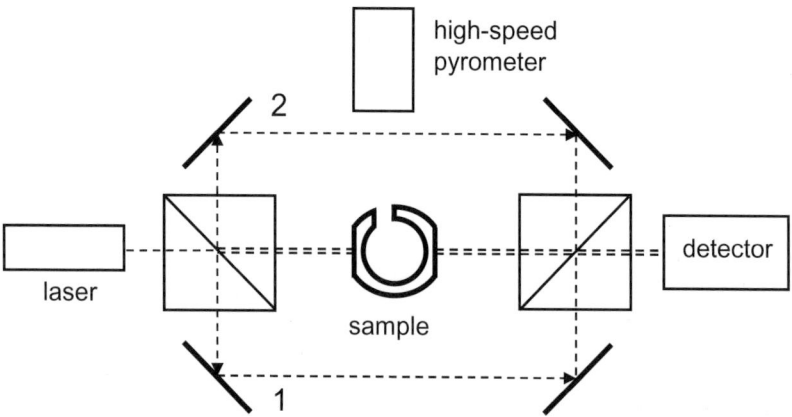

Fig. 8.3. Principle of dynamic dilatometry developed by Miiller and Cezairliyan (1982). The sample acts as a double reflector in the path of one of two beams, and a blackbody model provides determination of true temperature

The rotational stability is monitored by reflecting the beam of an auxiliary laser from a third optical flat on the sample. A small rectangular hole in the wall of the sample, 0.5×1 mm, serves as a blackbody model for the temperature measurements. During the pulse heating, a dual-beam oscilloscope displays the traces of the radiance from the sample and of the corresponding shifts in the fringe pattern.

The power-pulsing circuit (Fig. 8.4) consists of the sample in series with a battery bank (up to 26 V), an adjustable resistor (0 to 30 mΩ), and a fast-acting switch. The sample is rapidly heated in a vacuum environment from room temperature to a desired temperature. Electronic logic circuits and a series of time delay units govern the timing of various events, such as closing and opening of the switch and triggering of electronic equipment. The temperature of the sample is measured with a high-speed photoelectric pyrometer capable of 1200 evaluations per second. A rapidly rotating chopper disc alternatively passes precisely timed portions of radiance from the sample and from a tungsten-filament reference lamp through an interference filter to a photomultiplier. The effective wavelength of the interference filter is 653 nm, with a bandwidth of 10 nm. The circular area viewed by the pyrometer is 0.2 mm in diameter. The sighting hole is fabricated 0.8 mm off the axis of the tube to improve the blackbody quality. In order to compensate for the cross-sectional nonuniformity due to the hole, a portion of the sample is removed by grinding a flat along the length of the tube, excluding the 1-mm length of the hole. Dividing the temperature interval of the measurement into several overlapping temperature ranges optimises the signal resolution of the pyrometer.

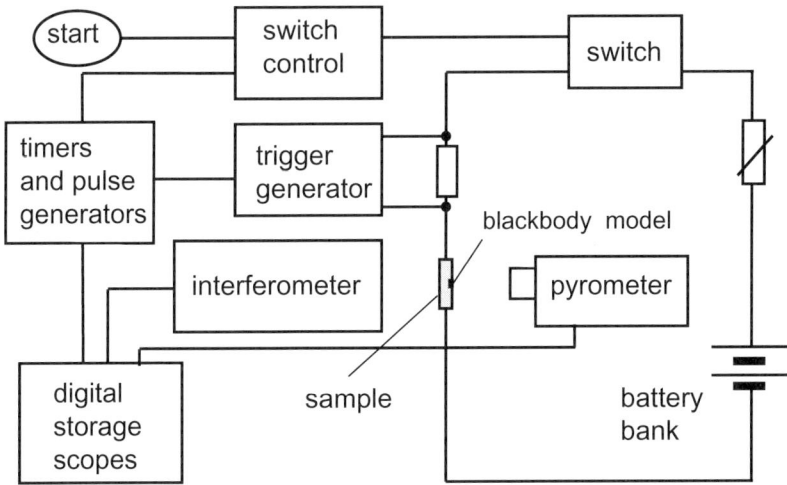

Fig. 8.4. Simplified diagram of the dynamic dilatometer developed by Miiller and Cezairliyan (1982). This technique was successfully employed in studies of refractory metals

During each pulse-heating experiment, the signals from the pyrometer and the interferometric system are recorded by means of two digital storage oscilloscopes, each capable of storing 4000 data points with a full-scale signal resolution of about

1 part in 4000. The system developed by Miiller and Cezairliyan has two very important advantages: (i) the measurements relate to the blackbody temperature, and (ii) only a central portion of the sample is involved in the measurements, so that there is no need to take into account the temperature distribution along the sample. Using this technique, the authors investigated the thermal expansion of Ta, Mo, Nb, and W (Miiller and Cezairliyan 1982, 1985, 1988, 1990, 1991).

Miiller and Cezairliyan (1982) estimated the sources and magnitudes of errors in the measurements of linear thermal expansion as follows: (i) temperature: 0.3% at 2000 K, and 0.5% at 3000 K; (ii) fringe count: 0.3%; and (iii) length at 20°C: 0.2% at 2000 K, and 0.3% at 3000 K. The authors explained the observed decrease in the length of the sample at 20°C as the result of a partial annealing of stresses in the sample induced during fabrication. This observation confirms the unavoidable changes in the length of a sample under high temperatures, which are the main problem of the high-temperature dilatometry. Modulation dilatometry is the only method that does not sense such changes.

IMGC Apparatus

Righini et al. (1986a) at IMGC designed a dynamic dilatometer, which correlates the thermal expansion of the sample to its temperature profile. An interferometer measures the longitudinal expansion of a sample, while a scanning optical pyrometer with a rotating mirror determines its temperature profile (Fig. 8.5). A typical spacing between two consecutive measurements is 0.35 mm with a pyrometer viewing area of 0.8 mm in diameter. Two massive brass clamps maintain the ends of the sample close to room temperature and provide steep temperature gradients towards the ends. Two thermocouples spot-welded at the ends of the sample measure its temperature in the regions, where pyrometric measurements are impossible.

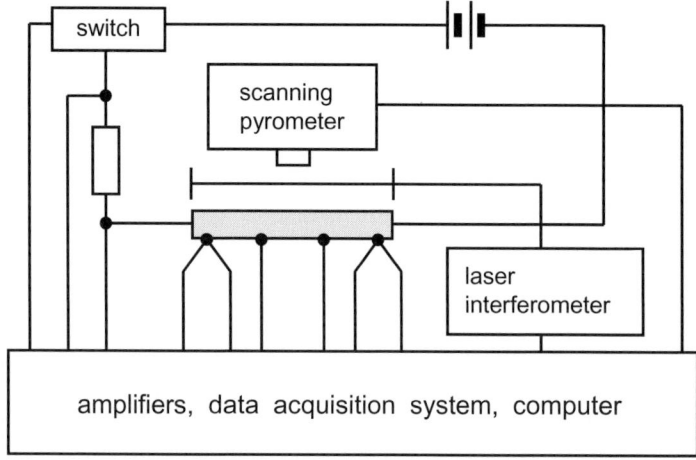

Fig. 8.5. Simplified diagram of dynamic dilatometer developed by Righini et al. (1986a). The total expansion of the sample is correlated to its temperature profile

A corner cube retroreflector is attached to the lower (moving) clamp, while the beam bender to the upper clamp. The resolution of the interferometer is about 0.15 µm, and 2000 measurements per second are feasible. All the results proceed to a data-acquisition system. Radiance temperatures measured by the pyrometer are transformed into true temperatures by either the resistivity of the sample or the normal spectral emittance.

The measurements carried out on a niobium sample (Righini et al. 1986b) lasted from 0.3 s (fast) to 2.2 s (slow). For the profile measurements, fast experiments are preferable because the portion of the profile not known from the scanning pyrometer is below 6%. For slow experiments, this figure increases to 20%. On the other hand, in fast measurements the thermal-expansion polynomial is defined by few data points limited by the speed of rotation of the mirror. A compromise must be found between a better knowledge of the temperature profile and a better definition of the thermal-expansion polynomial. The authors stressed that an ideal experiment with this technique would be either to bring the entire sample to the high temperature or to limit the measurements to a central portion of the sample.

8.2 Principle of Modulation Dilatometry

Measurements of the 'true' thermal expansivity, the thermal expansion coefficient within a narrow temperature interval, became possible with a modulation technique. Modulation dilatometry involves oscillating the temperature of the sample around a mean value and measuring corresponding oscillations in the length of the sample. Under such conditions, the linear expansivity is measured directly. This approach is very attractive because it ignores the plastic deformation of the sample and irregular external disturbances. In modulation dilatometry, only those changes in the length of the sample are measured, which reversibly follow the temperature oscillations. This seems the best way to circumvent the main problems peculiar to the traditional dilatometry at high temperatures. Modulation dilatometry senses only what is necessary, while all unwanted disturbances appear to be beyond the measurements. The measurements are possible with temperature oscillations of the order of 0.1 K and less. At present, even a resolution of 1 K at high temperatures is a great improvement over the traditional methods. To measure the oscillations in the length of the sample, the most sensitive techniques are applicable.

8.2.1 Wire Samples

Modulation dilatometry was proposed long ago (Kraftmakher and Cheremisina 1965). It has been applied to the study of thermal expansion of metals and alloys at high temperatures (Kraftmakher 1967b, 1972; Glazkov 1985, 1987, 1988; Glazkov and Kraftmakher 1986). This technique was reviewed in several papers (Kraftmakher 1973b, 1984, 1992). However, it gained no recognition. After 20 years, modulation dilatometry was reinvented and applied to nonconducting materials by Johansen (1987).

In the first modulation measurements, an AC current or a DC current with a small AC component heated a wire sample. The upper end of the sample is fixed, and the lower end is pulled by a load or a spring and is projected onto the entrance slit of a photomultiplier. The AC voltage at the output of the photomultiplier is proportional to the amplitude of the oscillations in the length of the sample. The temperature oscillations are determined from either the electrical resistance of the sample or the radiation from it. They can be evaluated if the specific heat of the sample is known. When a DC current I_0 with a small AC component heats the sample, the linear expansivity $\alpha = (1/l)dl/dT$ equals

$$\alpha = mc\omega V/2lKI_0U, \qquad (8.1)$$

where m, c, and l are the mass, specific heat, and length of the sample, ω is the modulation frequency, U is the AC voltage across the sample, V is the AC component at the output of the photomultiplier, and K is the sensitivity of the photomultiplier to the elongation of the sample. The sensitivity to the elongation is available from static measurements: $K = dV_0/dl$, where V_0 is the DC voltage at the output of the photomultiplier. Equation (8.1) is valid for an adiabatic regime of the measurements. Determinations of the phase shift between the heating-power oscillations and the temperature oscillations are necessary for measurements in a nonadiabatic regime.

It is easy to assemble a circuit, whose balance is independent of the AC component of the heating current (Fig. 8.6). In this case, the AC signal at the output of the photomultiplier is balanced by a variable mutual inductance M with the heating current passing through its primary winding.

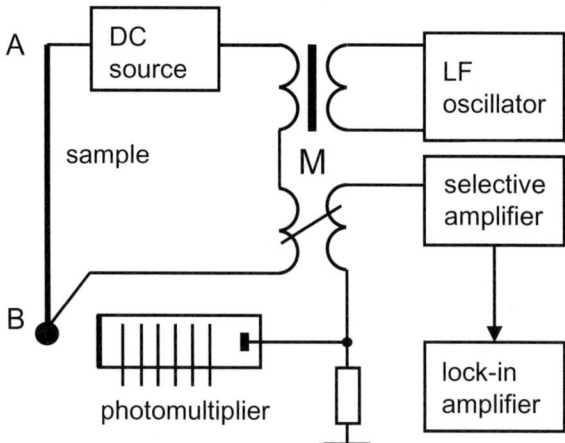

Fig. 8.6. Modulation measurements of thermal expansivity. Only oscillations of the length, whose frequency equals the frequency of the temperature oscillations in the sample, *AB*, are measured (Kraftmakher and Cheremisina 1965). The mutual inductance balances the AC output voltage of the photomultiplier

The voltage generated by the secondary winding becomes equal to the AC voltage at the output of the photomultiplier when

$$\alpha = mc\omega^2 M/2I_0 l R_0 K, \quad (8.2)$$

where R_0 is the resistance of the sample. The measuring system incorporates a selective amplifier tuned to the modulation frequency and/or a lock-in amplifier. The dilatometer thus becomes insensitive to irregular mechanical perturbations or to oscillations of other frequencies. When an AC current heats the sample, the expansivity is given by

$$\alpha = 2mc\omega V/lPK, \quad (8.3)$$

where P is the mean electric power supplied to the sample.

Usually, the modulation frequency in the range 10–100 Hz is sufficiently high to satisfy the criterion of adiabaticity. Otherwise, the temperature oscillations obey formulas for a nonadiabatic regime.

8.2.2 Differential Method

With the differential method (Kraftmakher 1967d), a wire sample consists of two portions joined together: a sample under study AB and a reference sample BC of known linear expansivity (Fig. 8.7). The two portions are heated by DC currents from separate sources and by AC currents from a common oscillator.

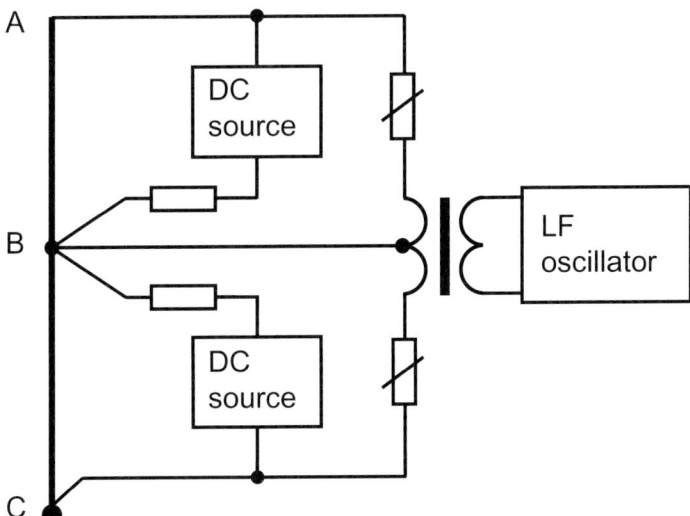

Fig. 8.7. Direct comparison of thermal expansivity of two wire samples. Oscillations in the length of the sample under study, AB, and of the reference sample, BC, balance each other (Kraftmakher 1967d). The reference sample may be the same wire but kept at a constant mean temperature

The temperature oscillations in the two portions are of opposite phase. By adjusting the AC components of the heating currents, the oscillations in the length of the reference portion completely balance those of the portion under study. Now the photomultiplier acts only as a null indicator, and any variations in its sensitivity, as well as in the intensity of the light source, etc., do not contribute. To calculate the expansivity, one has to determine the temperature oscillations in the two portions. If the corresponding specific heats are known, a simple relation holds when the oscillations balance each other:

$$\alpha_1 I_{01} U_1 l_1 / m_1 c_1 = \alpha_2 I_{02} U_2 l_2 / m_2 c_2, \tag{8.4}$$

where subscripts 1 and 2 refer to the main and the reference portions of the sample. The reference sample is kept at a constant mean temperature, and all the related quantities are constant. Hence,

$$\alpha = B U_2 c_1 / I_{01} U_1, \tag{8.5}$$

where B is a coefficient of proportionality.

The measurements are thus reduced to nullifying the oscillations in the length of the composite sample and measuring the DC current I_{01} in the main portion and the AC voltages U_1 and U_2 across the two portions. This technique was employed in studies of the thermal expansion of platinum in the range 1200–1900 K and of tungsten in the range 2000–2900 K (Fig. 8.8). The samples were 0.05 mm thick. Their mean temperature was determined from the electrical resistance, and the temperature oscillations, of about 1 K, were calculated from the specific heat. Similar wires kept at constant mean temperatures served as the reference samples. The sensitivity of the set-up was about 1 nm. In both cases, the non-linear increase of the thermal expansivity was attributed to vacancy formation in the crystal lattice (Kraftmakher 1967b, 1972).

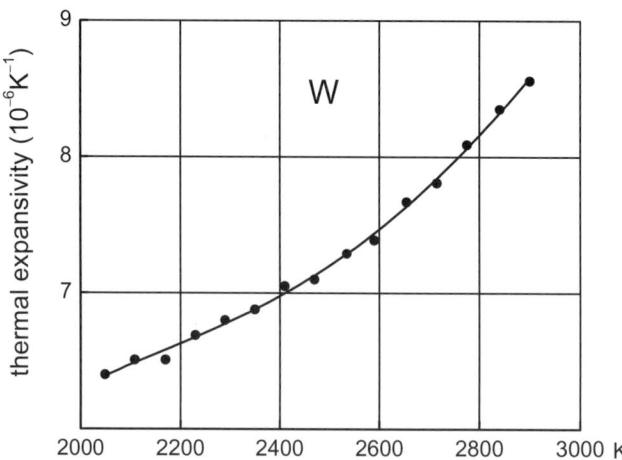

Fig. 8.8. Thermal expansivity of tungsten (Kraftmakher 1972). The non-linear increase in the expansivity was attributed to vacancy formation in crystal lattice

8.2.3 Bulk Samples

Another version of modulation dilatometry was developed for comparatively bulk samples, such as rods and strips (Kraftmakher and Nezhentsev 1971). To eliminate cold-end effects, the temperature oscillations occur only in a central portion of the sample (Fig. 8.9a). A mains current heats the sample. A small current of the same frequency, but with its phase linearly varying with time, is added to the mains current in the central portion. The superposition of the two currents causes oscillations of the power dissipated in the central portion, and its temperature oscillates around a mean value. The modulation period is in the range 1–10 s. A thin thermocouple measures the mean temperature and the temperature oscillations in the central portion of the sample. A photodiode senses the oscillations in the length of the sample, and an electromechanical transducer, a small earphone, balances these oscillations. The transducer is attached to the sample and is controlled by a photodiode sensing the length of the sample. A blade glued to the membrane of the earphone is illuminated, and its image is projected onto the photodiode. The output voltage of the photodetector is partly compensated for and then fed to an amplifier. The transducer is connected to the output of the amplifier to balance the changes in the length of the sample. Owing to the high gain of the amplifier, the system provides almost complete compensation, while the sensitivity is about 10 nm. The changes of the current in the transducer are proportional to the changes in the length of the sample. These oscillations are recorded along with the temperature oscillations in the sample.

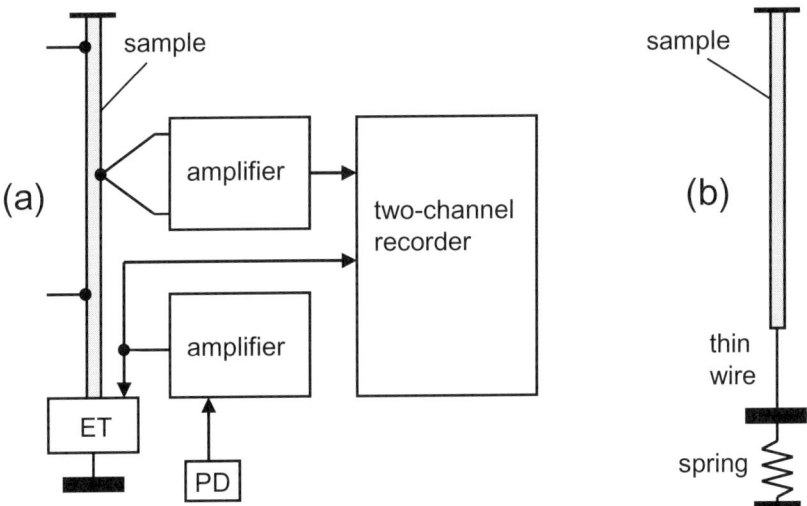

Fig. 8.9. (a) Modulation dilatometer for bulk samples (Kraftmakher and Nezhentsev 1971). The temperature oscillations occur only in a central portion of the sample. Electromechanical transducer *ET* balances oscillations in the length of the sample. *PD* is a photodiode. (b) Thin wire electrically heated provides the compensation

A thin wire heated by an electric current also may provide the necessary compensation (Fig. 8.9b). One of its ends is attached to the sample and the other is pulled by a spring. A blade mounted at the upper end of the spring is projected onto a photodiode. Since the temperature oscillations in the sample are small, only small changes in the temperature of the wire are necessary for the compensation. The mean temperature of the wire is constant, and the changes in its length are proportional to the changes in the heating current. The time response and linearity of a 50-μm tungsten wire at 1500 K appeared to be sufficient. To determine absolute values of thermal expansivity, the compensator requires calibration. Samples of known thermal expansivity are also usable. In many cases, it is sufficient to know the temperature dependence of the expansivity, so that relative values are acceptable. Mean values of the expansivity in wide temperature ranges, which are available from traditional measurements, can be used to normalise data from modulation measurements.

To verify this technique, measurements on a nickel sample were performed in the range 700–1400 K. The length of the sample was 200 mm, and the temperature oscillations were generated in its central portion 40 mm long. The modulation period was 10 s, and the temperature oscillations amount to about 5 K. A thin tungsten wire served as the compensator.

8.2.4 Interferometric Modulation Dilatometer

Oscillations in the length of the sample are measurable by various methods, including those of highest sensitivity. The interferometric method is one of such techniques. In the interferometric modulation dilatometer (Glazkov and Kraftmakher 1983), samples are in the form of a thick wire or a rod (Fig. 8.10). The upper end of the sample is fixed, while a small flat mirror $M1$ is attached to its lower end. A DC current from a stabilised source heats the sample. A small AC current is fed to a central portion of the sample through thin wires. The temperature oscillations thus occur only in this portion. To prevent an offshoot of the AC current to the upper and lower portions of the sample, a coil of high AC impedance is placed in series with the sample (not shown in the figure).

The beam of a He-Ne laser passes through a beam splitter BS and falls onto the mirror $M1$ and a second mirror $M2$ attached to a piezoelectric transducer PT. A photodiode senses the intensity of the interference pattern. Its output voltage, after amplification, is applied to the transducer with such a polarity that the oscillations of both mirrors are in phase. Owing to the high gain of the amplifier, the displacements of the mirror $M2$ follow those of the mirror $M1$. The AC voltage applied to the transducer is therefore strictly proportional to the oscillations in the length of the sample. A selective amplifier tuned to the modulation frequency amplifies this voltage, and a lock-in detector measures it. The reference voltage for the detector is taken from the oscillator supplying the AC current to the central portion of the sample. The DC output voltage of the detector is proportional to the amplitude of the oscillations in the length of the sample.

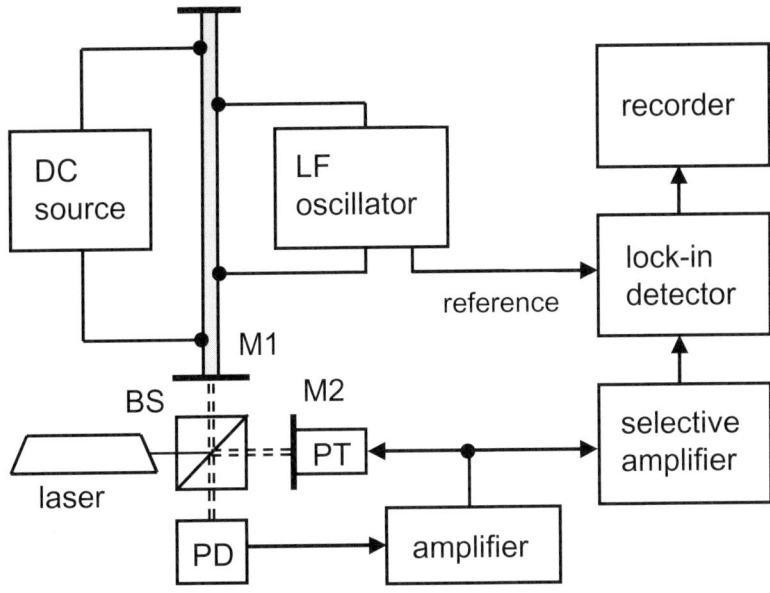

Fig. 8.10. Diagram of interferometric modulation dilatometer (Glazkov and Kraftmakher 1983). Temperature oscillations are created only in a central portion of the sample. Piezoelectric transducer *PT* balances oscillations in the length of the sample. *BS* – beam splitter, *PD* – photodiode, *M*1 and *M*2 – mirrors

A 0.3-mm platinum wire was used to check the dilatometer. The length of the sample was 150 mm, and the temperature oscillations occurred in its central portion 40 mm long. The data showed good linearity and possibility of performing measurements with temperature oscillations of the order of 0.1 K. The results were almost independent of the intensity of the laser beam. The wires confining the central portion were 50 μm in diameter. Clearly, they disturb the temperature around the points where they are welded to the sample. Adjusting the current passing through the wires reduces this effect. Using an additional current to heat the wires, the mean temperature of the wires can be made close to that of the sample. A somewhat more difficult task is to produce, in addition, equal temperature oscillations in the sample and in the wires.

A micropyrometer and a photodiode were used to compare the mean temperatures and the temperature oscillations along the sample. In the central portion, variations of the mean temperature, 1500 K, did not exceed 2 K when an additional heating of the wires was employed. Without such a heating, the decrease of the temperature at the points of welding the wires amounted to 30 K. The amplitude of the temperature oscillations inside the central portion was constant within 1%, while no temperature oscillations were observed beyond it. Radial temperature differences in the sample must be negligible because of its small thickness and high thermal conductivity. With the radiation heat exchange, the calculated temperature differences do not exceed 0.2 K at 2000 K and decrease with decreasing temperature. The relations between the amplitudes and the phases of the temperature oscillations inside the sample and on its surface are also favourable.

To check the validity of a modulation dilatometer, it is probably sufficient to verify that the results obtained do not depend on the modulation frequency (Fig. 8.11). The uncertainty in the oscillations of the length of the sample is about 1%. The errors in the linear thermal expansivity depend mainly on the uncertainty in measuring the temperature oscillations.

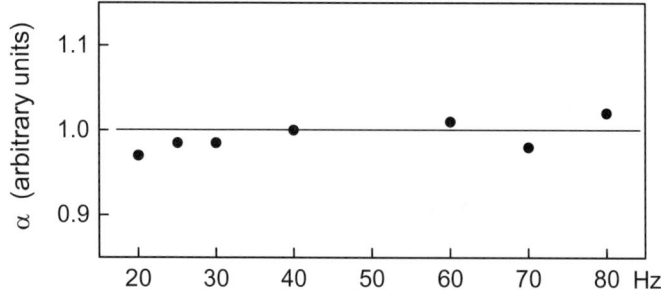

Fig. 8.11. Results of the measurements are almost independent of modulation frequency (Glazkov and Kraftmakher 1983). This may be a general criterion of the validity of a modulation dilatometer

8.2.5 Nonconducting Materials

Johansen (1987) employed a computerized capacitance dilatometer (Johansen et al. 1986) in a regime of periodic changes of the temperature of the sample. A Peltier module provides AC heating of a copper block containing the sample (Fig. 8.12). The dilatometer cell is designed for plane parallel samples of nearly cylindrical shape, 4–5 mm in diameter, and an unspecified height up to 5 mm. The sample is positioned in the central hole of a low-expansivity plane-parallel fused quartz ring supporting one evaporated silver electrode. Placing a disc with the other electrode directly on top of the sample forms an air capacitor. This disc is clamped by a downward force from elastic bellows (not shown in the figure). The vertical dilatation of the sample is derived from the observed variation in the capacitance. Capacitance measurements are made with precision bridge equipment and a lock-in analyser serving as the imbalance detector. The output voltage of the analyser was calibrated to represent variations in the dilatometer capacitance. A 10^{-11} m resolution is achieved over a large dynamic range, of about 10^{-6} m. Unlike Joule heating, the Peltier effect introduces no DC temperature increment. The temperature of the copper block could be controlled at a fixed temperature point or in a linear sweep mode, in the range –60°C to 150°C. To check the dilatometer, the linear thermal expansivity of Rochelle salt, $NaKC_4H_2O_6.4H_2O$, was measured near 24°C. The salt undergoes an orthorombic-to-monoclinic phase transition at this temperature. The modulation frequency was 2.5 mHz, and the amplitude of the temperature oscillations was about 20 mK.

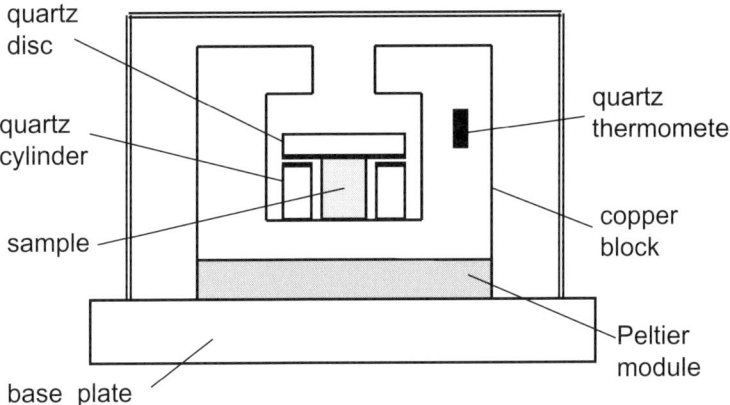

Fig. 8.12. Simplified diagram of capacitance modulation dilatometer for nonconducting samples (Johansen 1987). The temperature range is –60°C to 150°C

Johansen (1987) formulated the advantages of modulation techniques, including the modulation dilatometry, as follows: "Generally, steady-state AC detection methods have several advantages over classical measurements. Periodic excitation permits advanced noise-reducing analog and digital signal processing (phase-sensitive detection) to be applied. In this way the disturbance produced by irrelevant monotonic effects can be eliminated. In measurements of thermal expansion such an effect can be plastic deformation in the specimen, e.g., due to an applied hydrostatic pressure, combined uniaxial stresses, or simply the minimum mounting force acting on a soft sample. Further, in studying second-order or near-second-order phase transition the steady-state nature of the AC method enables a constant DC temperature to be maintained long enough to establish thermodynamic equilibrium, even when relaxation times become very long."

8.3 Extremely Small Periodic Displacements

Several methods are known for measuring extremely small periodic displacements. The periodic nature of the displacements allows employment of selective amplifiers and/or lock-in detectors. This enhances the sensitivity of the measurements far above the sensitivity to usual (not periodic) displacements. The most convenient methods to measure periodic displacements use laser interferometry. The high sensitivity obtained in such cases may seem to be unrealistic, but it is indeed achievable.

Bruins et al. (1975) developed a very sensitive technique for such measurements (Fig. 8.13). A spherical Fabry-Perot interferometer was used to determine small piezoelectric constants, but the authors pointed out that the method could be adapted to other measurements. The interferometer has several advantages over other interferometric arrangements. First, it is inherently much more sensitive than the Michelson or other two-beam interferometer, because of the high sharpness of the multiple-beam interference fringes. Second, it is a compact instrument suitable for a temperature-controlled environment. A piezoelectric transducer served as a reference, and the vibrations of about 4×10^{-14} m were measured with 5% accuracy. The theoretical sensitivity in a well-isolated environment was estimated to be 10^{-15} m. The method is also usable as a null technique.

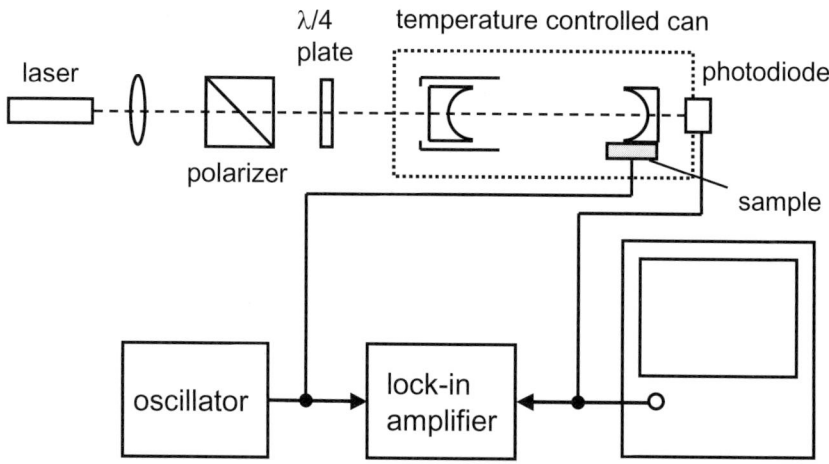

Fig. 8.13. Simplified diagram of set-up for measurement of small periodic displacements by Fabry-Perot interferometer (Bruins et al. 1975). Vibrations of about 4×10^{-14} m were measured with 5% accuracy

Fanton and Kino (1987) designed a simple apparatus for measurements of small periodic displacements (Fig. 8.14). The laser beam passes through a polarizing beam splitter and a birefringent Wollaston prism. The prism splits the beam into equal-amplitude beams angularly separated by 0.5° and focused to two spots on the sample. The reflected beams recombine in the Wollaston prism, pass through the splitter, and interfere on a photodetector. The beam focused onto an undisturbed portion of the sample serves as a reference. At the focus of the other beam, the sample is heated periodically. The resulting thermal expansion of the sample causes phase modulation of the beam, which is converted to intensity modulation when the two beams interfere. Both beams pass through the same optical components making the system resistant to vibrations and drift. The sensitivity of the system was checked by measuring vibrations of a piezoelectric transducer split into halves, one stationary and one active. The background optical noise of the system was 3.4×10^{-14} m. The system was also used to obtain a thermal image of a flawed conducting strip, a nickel film on a silicon wafer. The strip was notched and then heated by passing an 8-kHz current through it. The actual temperature oscillations in the strip were of about 10 mK. The authors pointed out that it is possible to modulate the laser at a frequency close to that of the vibrations and to obtain a signal of the difference frequency.

The sensitivity achievable by the above techniques is much better than necessary for modulation dilatometry. Thermal expansivity becomes measurable even on very short samples. In particular, it can be determined along the thickness of the sample, so that the problem of temperature gradients in the sample would be completely avoided (Tong et al. 1991).

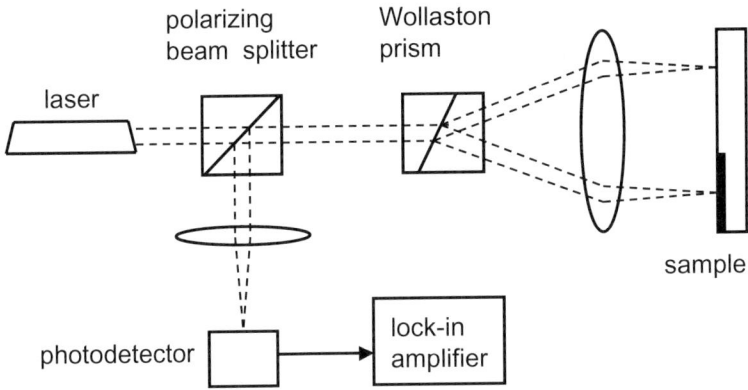

Fig. 8.14. Simplified diagram of apparatus for measuring extremely small periodic displacements (Fanton and Kino 1987). The background noise of the system was 3.4×10^{-14} m

9 Other Modulation Techniques

This chapter describes modulation techniques for measuring the temperature derivative of electrical resistance, the thermopower, and the spectral absorptance. The use of temperature modulation provides important advantages; the main of these is the possibility for direct determinations of differential characteristics, which in many cases are more informative.

9.1 Temperature Derivative of Resistance

The modulation technique allows one to directly measure the temperature derivative of electrical resistance. The method consists in oscillating the sample temperature around a mean value and measuring the oscillations in the resistance of the sample along with the temperature oscillations.

Medium and High Temperatures

Direct measurements of the temperature derivative of resistance were proposed for studying the anomaly of the resistivity of nickel near its Curie point (Kraftmakher 1967a).

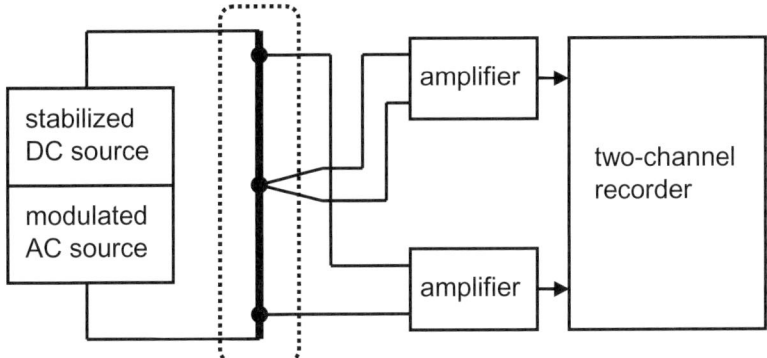

Fig. 9.1. Arrangement for direct measurements of the temperature derivative of resistance used for studying the anomaly of resistance of nickel in its Curie point (Kraftmakher 1967a). The temperature oscillations and corresponding variations in the resistance of the sample are recorded simultaneously

A nickel sample was heated by a DC current from a stabilised source and by an AC current amplitude-modulated with a period of a few seconds (Fig. 9.1). A thermocouple measured the temperature oscillations in the sample. Simultaneously, measurements were made of the voltage oscillations, which appear due to the DC current passing through the oscillating resistance of the sample. To eliminate cold-end effects, the measurements were carried out on a central portion of the sample. A two-channel recorder recorded the AC components of the voltage from the thermocouple and of the voltage across the central portion of the sample. Data on the temperature derivative of resistance clarified the singularity of the resistivity at the Curie point. Later, the method was used for the determination of vacancy contributions to the electrical resistance of aluminium and platinum (Kraftmakher and Sushakova 1972, 1974). The non-linear increase of the temperature coefficient of resistance was attributed to vacancy formation in crystal lattice (Fig. 9.2).

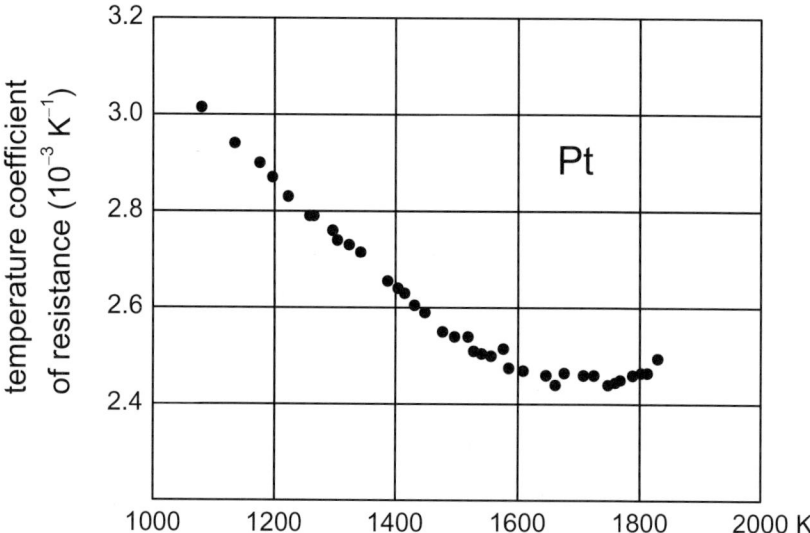

Fig. 9.2. Temperature coefficient of resistance of platinum measured with modulation technique (Kraftmakher and Sushakova 1974). The non-linear contribution at high temperatures was attributed to vacancy formation in crystal lattice (Sect. 13.1)

Salamon et al. (1969) used a similar method in studies of the resistivity of chromium near its Néel point, by means of the modulated-light heating. Lederman et al. (1974) described this technique in more detail. The specific heat and the temperature derivative of resistance of iron near its Curie point were measured simultaneously (Fig. 9.3). The temperature inside the furnace was linearly swept at a rate of 1 K.min^{-1} or less. The modulation frequency was 8 Hz. Lock-in amplifiers measured the AC voltages corresponding to the temperature and the resistance oscillations.

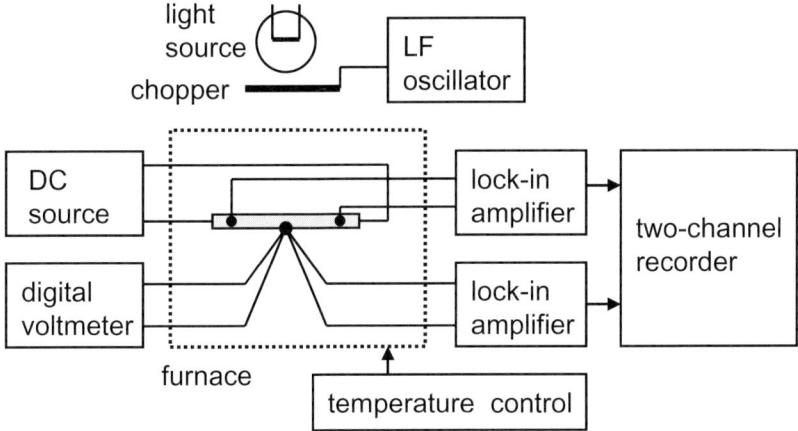

Fig. 9.3. Measurement of temperature derivative of resistance by the use of modulated-light heating. Oscillations in the temperature and in the resistance of the sample are recorded simultaneously (Lederman et al. 1974)

Low Temperatures

Chaussy et al. (1992) reinvented this technique (Fig. 9.4) and employed it to determine the anomaly in the resistivity of erbium near the Néel point (Terki et al. 1992). A single crystal of erbium was in the form of a disc with a diameter of about 3 mm, a thickness of 0.12 mm, and a mass of 4.2 mg. It was glued on a sapphire rod of dimensions 10×8×0.2 mm. The measurements of the temperature derivative of the resistance were made along the c axis of the crystal. An electrical heater provided the AC heating at a frequency 4 Hz. This frequency was a compromise between the low frequency necessary to have sufficient temperature oscillations and the operating frequency band of lock-in amplifiers measuring the AC voltages. The amplitude of the temperature oscillations in the sample was in the range 5–10 mK. Stainless-steel heaters and NbN thermometers were evaporated onto a sample holder in order to improve the thermal coupling. The AC component of the voltage across the sample is proportional to the temperature derivative of the resistance of the sample. The first stage of the measuring system, constituted by a transformer and a field-effect transistor, was immersed in a helium bath to reduce the thermal noise. The superconducting primary winding of the transformer contained only 5 turns, while the secondary 25000 turns. A capacitor was added to the secondary winding for achieving resonance conditions. A compensation scheme with a standard resistor served for avoiding a DC current through the primary winding of the transformer, which could modify its characteristics. With $R = 3$ mΩ, biased current of 100 mA, and averaging over 100 s, the set-up provided a resolution in $\Delta R/R$ of the order of 10^{-9}.

Modulation measurements of the temperature derivative of resistance are listed in Table 9.1.

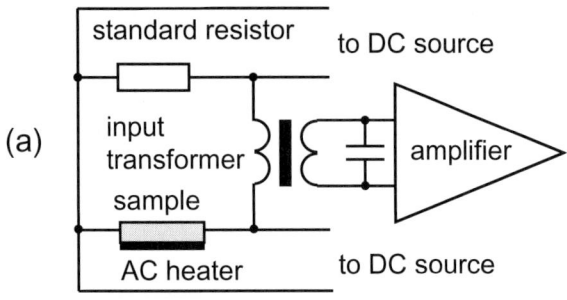

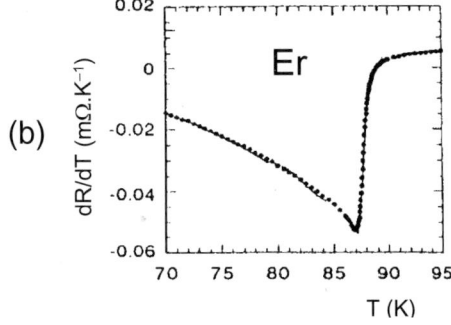

Fig. 9.4. (a) Simplified diagram of the first stage of set-up for measuring temperature derivative of electrical resistance (Chaussy et al. 1992). The input transformer greatly improves the sensitivity. (b) Temperature derivative of the resistance of erbium near the Néel point (Terki et al. 1992)

Table 9.1. Modulation measurements of temperature derivative of electrical resistance

Item	References
Ni, Curie point	Kraftmakher 1967a
Cr, Néel point	Salamon et al. 1969
β-brass, phase transition	Simons and Salamon 1971
Al, vacancy formation	Kraftmakher and Sushakova 1972
V_3Si, high pressures, low temperatures	Chu and Testardi 1974
Fe, Curie point	Lederman et al. 1974; Shacklette 1974; Kraftmakher and Pinegina 1974
Pt, vacancy formation	Kraftmakher and Sushakova 1974
Solid electrolytes	Vargas et al. 1976, 1977
Sm, phase transitions	Kraftmakher and Pinegina 1978
PtRh, vacancy formation	Glazkov 1985
Ni, vacancy formation	Glazkov 1987
Rh, vacancy formation	Glazkov 1988
Er, Néel point	Chaussy et al. 1992; Terki et al. 1992

9.2 Direct Measurement of Thermopower

Among all the modulation techniques, measurements of thermopower (the Seebeck coefficient) are the simplest. The method consists in measuring the same periodic temperature oscillations by a thermocouple under study and by a reference one. Three groups independently proposed this technique (Freeman and Bass 1970; Hellenthal and Ostholt 1970; Kraftmakher and Pinegina 1970). The method was several times reinvented (Korn and Mürer 1977; Kawai et al. 1978; Papp 1984; Howson et al. 1989, 1990).

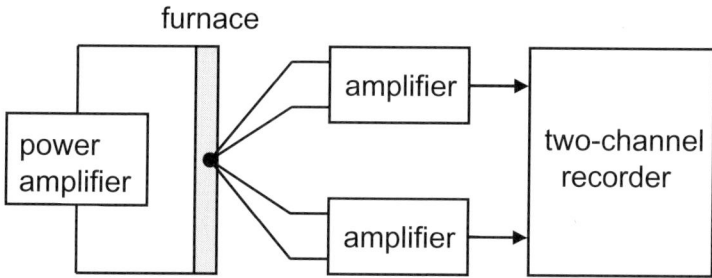

Fig. 9.5. Measurements of thermopower with a low-inertia furnace (Kraftmakher and Pinegina 1970, 1971, 1978; Kraftmakher 1971a). The signals from two thermocouples are recorded simultaneously

In measurements of thermopower at high temperatures, a small low-inertia furnace created temperature oscillations measured by two thermocouples (Fig. 9.5). The modulation period was several seconds. Freeman and Bass (1970) and Hellenthal and Ostholt (1970) described very similar methods for measuring the Seebeck coefficient with the modulated-light heating (Fig. 9.6). A lock-in amplifier measures the signal from the thermocouples. It is possible to compensate the signal at the input of the amplifier or to continuously record it. The high sensitivity of the method allows studying subtle effects, e.g., changes in the thermopower resulting from deformation (Hellenthal and Ostholt 1970). Howson et al. (1989, 1990) measured the thermopower of $YBa_2Cu_3O_{7-\delta}$ single crystals from the transition temperature to 250 K and in magnetic fields up to 2 T. The crystals were typically 1×1×0.05 mm in size. The thermopower was measured relative to Pb reference leads (Fig. 9.7). One half of the surface of the sample was exposed to chopped light from a tungsten lamp, while the other half was masked and thermally anchored. The crystal was mounted on a strip of Mylar 75 μm thick, one half of which was covered with 200 nm of evaporated Pb. The other Pb contact was made to a 1.5-mm thick piece of Pb supported by the Mylar strip. A 25-μm chromel–constantan thermocouple measured the temperature oscillations. The absolute temperature of the sample was measured with a platinum thermometer placed in a copper heat sink. One side of the crystal was thermally anchored through the 1.5-mm Pb piece to the heat sink. The other end of the sample was in thermal contact to the heat sink through the 200-nm film of Pb on the Mylar strip.

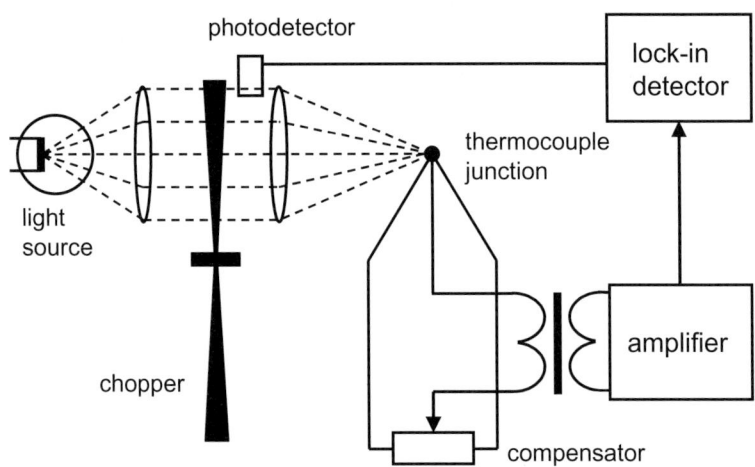

Fig. 9.6. Measurement of thermopower with modulated-light heating (Hellenthal and Ostholt 1970). Thermopowers of two thermocouples are balanced by means of a compensator

Modulation measurements of thermopower are listed in Table 9.2.

Table 9.2. Modulation measurements of thermopower

Item	Reference
Au-Al thermocouple, 30–150°C	Freeman and Bass 1970
Samples after deformation	Hellenthal and Ostholt 1970
Fe, Curie point	Kraftmakher and Pinegina 1970
Co, Curie point	Kraftmakher and Pinegina 1971
Cu, vacancy formation	Kraftmakher 1971a
TCNQ compounds, 1.2–400 K	Chaikin and Kwak 1975
SnCu alloys	Korn and Mürer 1977
Ge, TiO_2	Kawai et al. 1978, 1984
Sm, phase transitions	Kraftmakher and Pinegina 1978
$Fe_xCo_{80-x}B_{20}$, 4–300 K	Kettler et al. 1982
$Fe_xNi_{80-x}B_{19}Si$, 4–300 K	Kettler et al. 1984, 1986
Ni and Ni alloys, Curie point	Papp 1984
YBCO, $La_{1.85}Sr_{0.15}CuO_{4-\delta}$	Yu et al. 1988
YBCO	Howson et al. 1989, 1990; Lowe et al. 1991; Oussena et al. 1992
$Nd_{1.85}Ce_{0.15}CuO_{4+y}$	Mangelschots et al. 1992
$Nd_{2-x}Ce_xCuO_4$	Xu et al. 1992
$Tl_2Ba_2CuO_6$	Lin et al. 1993
$LaNiO_3$, $PrNiO_3$, 1.4–300 K	Xu et al. 1993
$ErCo_2$, $LaRu_2$, 3–300 K	Resel et al. 1996
SiGe, PbTe, $FeSi_2$, SiB_{14}, 300–1100 K	Goto et al. 1997; Kato et al. 1999

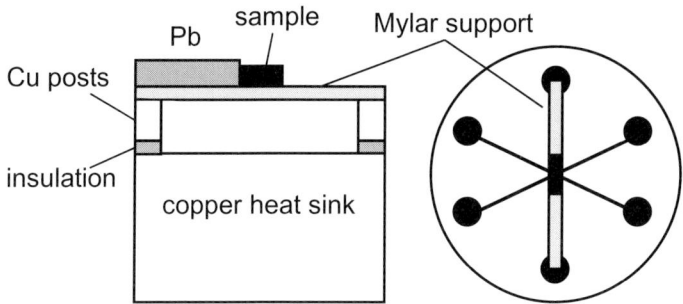

Fig. 9.7. Arrangement for modulation measurements of thermopower at low temperatures (Howson et al. 1990)

9.3 Spectral Absorptance

The radiation properties of materials are emittance, absorptance, reflectance, and transmittance. They control the rate of heat transfer by radiation between noncontacting bodies, and between a body and its surroundings (for a review see Richmond 1984). Spectral emittance and absorptance are equal only under thermodynamic equilibrium. No reliable data on these characteristics are available for many important materials. The number of such materials grows rapidly, and such data are necessary in a wide range of temperatures and wavelengths.

Spectral absorptance is measurable even at low temperatures, where the thermal radiation is still negligible. The absorptance may be related to total or partial radiation from thermal sources of various temperatures or from any other source of radiation. In particular, absorptance of solar radiation is very important for many applications. For samples of very high reflectance, measurements of spectral absorptance provide an important advantage. For instance, it is very difficult to distinguish between samples with reflectance 0.98 and 0.99, but the difference between absorptances 0.01 and 0.02 is easy to measure.

9.3.1 Calorimetric Techniques

Two calorimetric techniques are known for the determination of spectral absorptance. In steady-state measurements, the power of the absorbed radiation is calculated from the increment in the temperature of the sample and the heat transfer coefficient (Biondi 1954, 1956). The pulse technique involves measurements of temperature changes after exposing a sample to radiation. In modulation measurements, the surface of the sample is exposed to periodic pulses of radiation. The absorption of the radiation leads to an increase in the mean temperature of the sample and to periodic temperature oscillations. For short modulation periods, the amplitude of the temperature oscillations is proportional to the period and to the absorbed power and inversely proportional to the heat capacity of the sample. When the heat capacity is known, the basic relation of modulation calorimetry is applicable in straightforward manner to determine the absorbed radiation.

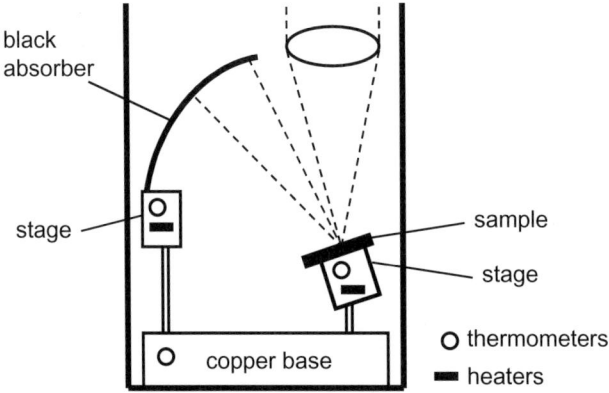

Fig. 9.8. Arrangement for measurements of spectral absorptance at 4.2 K (Biondi 1954, 1956). The absorbed radiation causes an increment in the temperature of the sample

Biondi (1954, 1956) measured the spectral absorptance of gold and copper at 4.2 K (Fig. 9.8). The energy absorbed on the surface of a metal sample is determined from the increment in the temperature of the sample. The source of radiation is a 1000-W projection lamp. The desired wavelength is obtained by the use either of filters or of a monochromator. The light is transmitted through an optical system and focused on the polished surface of the target at an angle of incidence of 15°. The radiation is then reflected onto a gold black absorbing surface, whose absorptivity is greater than 99% over the wavelength range of the experiment. The target and the black absorber are threaded into stages made of copper. Embedded in each stage are a carbon resistance thermometer and a manganin heater. A thin-walled tubing connected to the base supports each stage. Heat leaks between the stages and the base provide the proper thermal time constant of about 10 s for each stage. The measurements were made at 4.2 K over the wavelength range 0.3–3.3 μm. The power incident on the target was in the range 0.1–1 mW. After the steady state is reached, the temperatures of the target and the absorber stages are measured. With the source turned off, the temperatures are measured again. Power is then applied to the target heater to reproduce the temperature increment.

Pulse Technique

Bimberg and Bubenzer (1981) introduced a sensitive pulse technique for the absorption spectroscopy. The samples are suspended in the evacuated chamber of an optical cryostat and kept at 1.5 K (Fig. 9.9).

Commercial Allen-Bradley 110-Ω carbon resistors serve as temperature sensors. Their resistance and the temperature coefficient of resistance strongly increase with decreasing temperature. At 1.5 K, a temperature rise of 1 mK leads to a resistance change of about 100 Ω. The temperature sensor is thermally connected to the sample with thin copper wires. The change of the voltage drop across the sensor is monitored with a recorder. An argon-pumped dye laser or a 200-W high-pressure mercury lamp and a grating monochromator provide the incident radiation. At low temperatures, where the specific heat of the sample decreases drastically, the absorbed energy of the order of 10^{-12} J becomes detectable.

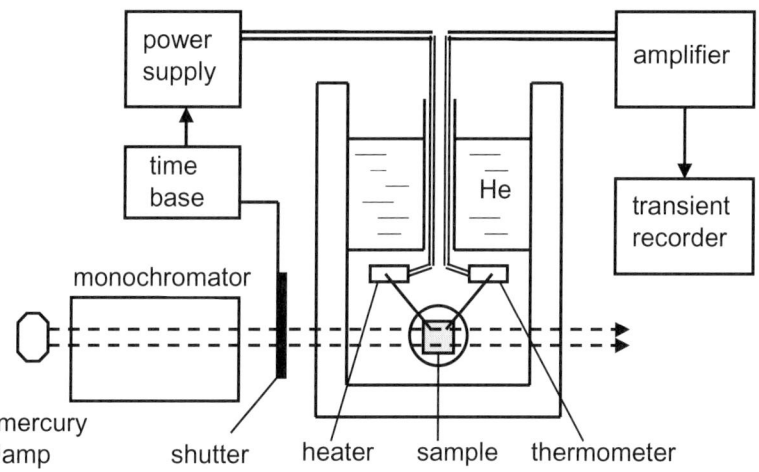

Fig. 9.9. Simplified diagram of apparatus for calorimetric absorption spectroscopy (Bimberg and Bubenzer 1981)

Modulation Measurements

Geraghty et al. (1984) developed a modulation technique for recording spectra of weakly absorbing films on surfaces at low temperatures. The sample substrate is mounted within a vacuum chamber (Fig. 9.10). The samples are prepared in the form of plates 0.1 mm thick. A germanium thermometer and a thin-film nichrom heater are located on the mounting side of the substrate. The heater serves for the calibration of the calorimeter. Thus, while many researchers claimed that it is difficult to determine the absolute value of the absorbed radiation, the authors demonstrated a simple method to solve the problem. The incident radiation provided by a 900-W xenon arc lamp and a monochromator is modulated at 33 Hz. The amplitude of the temperature oscillations in the sample, measured by a lock-in amplifier, is plotted versus the wavelength of the incident light. The photocalorimetric spectroscopy appeared to be a sensitive technique for studies of surface adsorption.

With the above calorimetric techniques, the spectral absorptance is available from observations of the heating rate of the sample when a constant light flux falls onto its surface or of the temperature oscillations in the sample when employing modulated light. In both cases, the heat capacity of the sample must be known for the determination of the spectral absorptance.

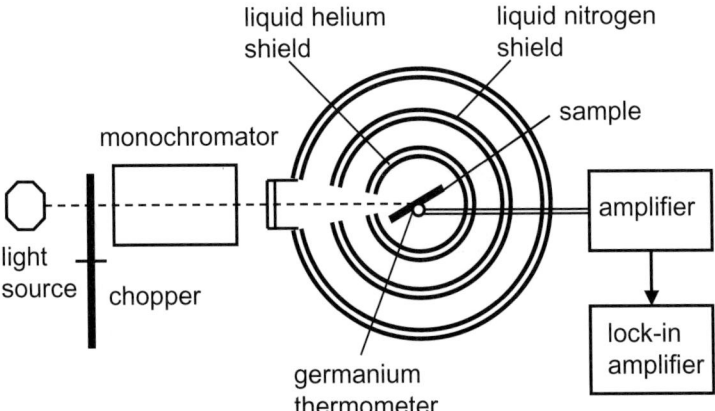

Fig. 9.10. Simplified diagram of arrangement for photocalorimetric spectroscopy (Geraghty et al. 1984)

9.3.2 Compensation Method

A compensation technique, which requires no data on the heat capacity of the sample, was proposed (Kraftmakher and Tarasenko 1987). Pulses of electric current are passed through the sample, being complementary to the pulses of the absorbed radiation (Fig. 9.11).

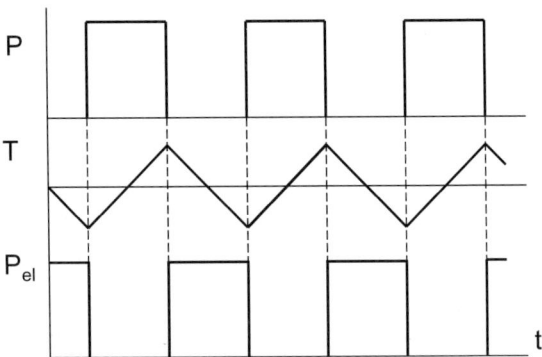

Fig. 9.11. Principle of measuring spectral absorptance by the use of compensation (Kraftmakher and Tarasenko 1987). P is the power of absorbed radiation, T is the temperature of the sample, and P_{el} is the power of the complementary electric pulses. By adjusting P_{el}, the temperature oscillations are reduced to zero

When the power of the electric pulses P_{el} is equal to the power of the absorbed radiation P, the total power applied to the sample becomes constant, and the temperature oscillations in the sample vanish. A device with a perfect blackbody measures the power of the incident radiation.

The advantages of this approach are evident:

- High sensitivity because very small temperature oscillations are measurable.
- Wide temperature range.
- Enhancement of the accuracy by the compensation.

Restrictions also exist because thin conducting samples are necessary for the measurements. A special set-up was assembled to check the method (Fig. 9.12). A 500-W halogen lamp is a source of the radiation. A monochromator provides monochromatic light falling onto the sample. An electrical relay driven by a pulse generator chops the radiation and switches on and off the current passing through the sample. The samples, metal strips 40×2×(0.05–0.1) mm in size, are placed in a vacuum chamber. Thin potential probes confine their central portion, 20 mm long. A thermocouple junction is pasted to the backside of the sample, without electric contact. The signal of the thermocouple is amplified and recorded. The modulation period is 10 s, and the temperature oscillations caused by the radiation amount to 0.1 K.

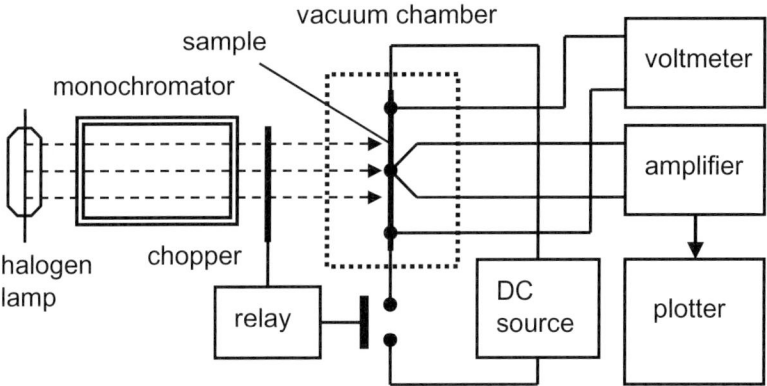

Fig. 9.12. Arrangement for measuring spectral absorptance (Kraftmakher and Tarasenko 1987). Complementary electric pulses reduce the temperature oscillations due to the pulses of absorbed radiation

The measurements were performed by two methods. First, the amplitude of the electric pulses was increased gradually causing a decrease of the temperature oscillations in the sample. When the electric power becomes larger than the power of the absorbed radiation, the phase of the temperature oscillations changes by 180°. A plot of the amplitude of the temperature oscillations versus the power of the electric pulses allows determination of the absorbed power (Fig. 9.13). The power of the incident radiation was measured using a black painted sample, for which the spectral absorptance was considered to be unity.

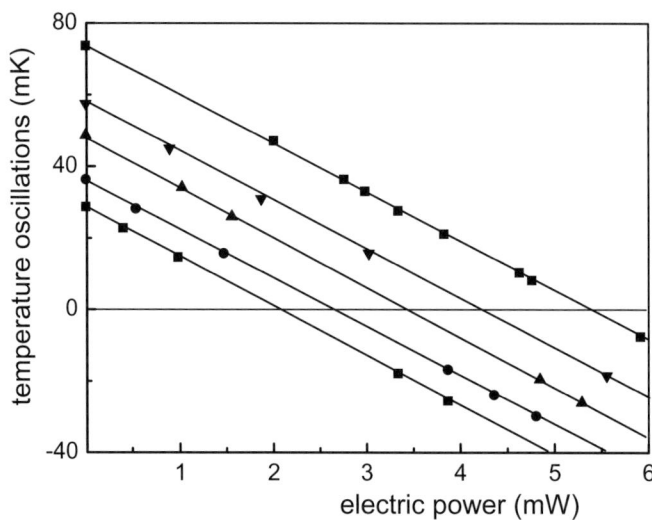

Fig. 9.13. Amplitude of temperature oscillations in the sample versus power of complementary electric pulses, for different blackness of the surface of the sample (Kraftmakher and Tarasenko 1987). Negative values mean that the complementary electric power exceeds the power of absorbed radiation

Second, one surface of the strip under study was painted black. The temperature oscillations in it were measured when the radiation fell, in turn, on both surfaces. If the absorptance of the black surface is unity, the ratio of the amplitudes of the temperature oscillations equals the spectral absorptance. With the absorbed power greater than 1 mW.cm^{-2}, the scatter of the data is less than 5%. To enhance the sensitivity and extend the wavelength range, narrow-band filters should be used instead of a monochromator. The filters provide a higher radiation power necessary for samples of low absorptance. At higher temperatures, the temperature oscillations were detected through the resistance of the samples. A DC current heated the samples, while a modulated AC current generated pulses of complementary electric power.

The modulation technique has significant advantages over traditional methods of measuring spectral absorptance. The measurements are possible on strips of conducting materials (metals, alloys, semiconductors), thin conducting or insulating coatings on such strips, or conducting coatings on insulating substrates of any thickness.

10 Noise Thermometry of Wire Samples

Thin wire samples heated by an electric current passing through them are well suited for modulation measurements at high temperatures. However, an accurate determination of their temperature poses a serious problem. This problem is solvable by measuring the thermal noise of the samples.

The generation of thermal noise by electrical conductors is a fundamental physical phenomenon. Thermal noise is therefore unavoidable and obeys strict quantitative relations. Johnson (1928) was the first to measure this noise and to show that it is independent of the composition of the resistor and the nature of the charge carriers. For this reason, thermal noise is often called Johnson noise. The noise signals are extremely small, and some ingenuity is required to make accurate measurements of thermal noise. Reviews of noise thermometry were given by Actis et al. (1972), Kamper (1972), and by White et al. (1996). Soulen et al. (1992) reported on noise thermometry for low temperatures developed at the National Institute of Standards and Technology (Gaithersburg).

10.1 Brief History of Noise Thermometry

As was theoretically shown by Nyquist (1928), the mean squared thermal-noise voltage generated by a resistor in a narrow frequency band Δf equals

$$<\Delta V^2> = 4k_B RT\Delta f, \tag{10.1}$$

where R is the resistance. This equation including Boltzmann's constant k_B and the absolute temperature T thus confirms the fundamental origin of thermal noise. More generally, Eq. (10.1) should be written as follows:

$$<\Delta V^2> = 4hfR\Delta f/[\exp(hf/k_B T) - 1], \tag{10.2}$$

where h is Planck's constant, and f is the frequency. Usually, $hf \ll k_B T$, and Eq. (10.1) remains valid in all cases except very low temperatures. After amplification by an amplifier, the noise voltage equals

$$<V^2> = 4k_B RT \int_0^\infty K^2 df, \tag{10.3}$$

where K is the gain of the amplifier depending on frequency.

The above equations offer a method of measuring the absolute temperature. An additional expression is important for estimating the attainable precision of noise thermometry. The noise voltage is a fluctuating quantity and must be time-averaged. The statistical uncertainty of such temperature measurements depends on the averaging time τ:

$$\Delta T/T = (2\pi\tau\Delta f)^{-\frac{1}{2}}. \tag{10.4}$$

Noise thermometers are in practical use since the pioneering work by Garrison and Lawson (1949). The authors used two resistors at different temperatures, the resistance ratio of which was adjustable to balance the noise voltages generated by the two resistors. The two noise voltages were measured with an amplifier in turn switched to the resistors R_1 and R_2 (Fig. 10.1). Under the balance,

$$R_1 T_1 = R_2 T_2, \tag{10.5}$$

provided both noise voltages are measured in the same frequency bandwidth. When the frequency band is governed by the time constant of the input circuit of the amplifier, the last requirement makes it necessary to adjust the capacitances across the resistors to cause $R_1 C_1 = R_2 C_2$.

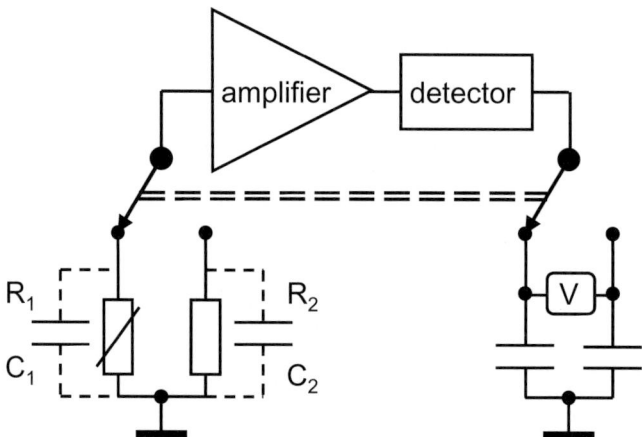

Fig. 10.1. Principle of operation of noise thermometer developed by Garrison and Lawson (1949). The noise-generating resistors and the capacitors charged by the amplified and rectified noise voltages are switched simultaneously. One of the resistors is adjustable, and a voltmeter serves as null indicator

Unfortunately, this simple and straightforward method of noise thermometry is not accurate because the inherent noise of an amplifier depends on the impedance of a source at its input. Several solutions were proposed to circumvent this difficulty. Pursey and Pyatt (1959) used two equal resistors at different temperatures, and the balance was achieved by attenuating the thermal noise from the resistor at the higher temperature. Actis et al. (1972) at the IMGC (Instituto di Metrologia "Gustavo Colonnetti", Torino) used this method for measuring temperatures in the range 400–1000°C with 0.03% accuracy. Another possibility is to add a calibrated noise voltage from a noise generator to the thermal noise from the resistor at the lower temperature.

Borkowski and Blalock (1974) proposed to measure the available thermal-noise power, i.e., the product of the open-circuit noise voltage and the short-circuit noise current generated by a resistor. The noise power is a linear function of the absolute temperature. In the temperature range 725–1275 K, the deviations of the experimental data from a straight-line calibration through the absolute zero were not more than 0.1%. For their device, the authors designed amplifiers whose inherent noise was equal to thermal noise of a 10-Ω resistor at room temperature (the so-called equivalent noise resistance).

When a noise-generating resistor is included into a tuned RLC circuit (Fig. 10.2), the mean square noise voltage across the capacitor C can be made independent of the resistance by integrating the noise signal over all frequencies (Pepper and Brown 1979). If the bandwidth of the amplifying and detecting circuitry is much wider than that of the tuned circuit, then

$$<V^2> = k_B T/C. \tag{10.6}$$

This relation was the basis of a noise thermometer developed by the authors.

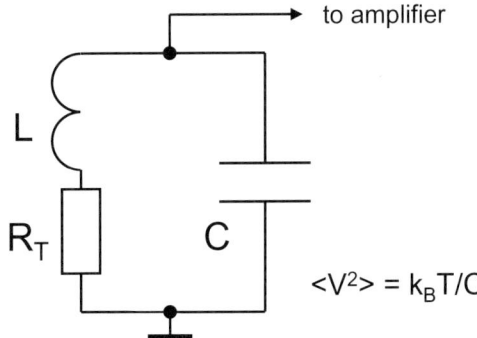

Fig. 10.2. Tuned circuit employed for measuring thermal noise generated by the resistor R_T (Pepper and Brown 1979). The noise voltage across the capacitor does not depend on the resistance

Pepper and Brown (1979) designed a noise thermometer with a direct measurement of the difference between the absolute temperatures of the reference and sensor resistors even without knowledge of the capacitance of the tuned circuit (Fig. 10.3). A vacuum diode acts as a calibrated source of noise, which can be added to the thermal noise of the resistor R_{ref} at a reference temperature. The latter is lower than the temperature to be measured. This makes it possible to obtain the same noise voltage as from the sensor resistor R_T. Both resistors, as well as the resistances introduced by the electronic switch and the inductance coil, and the current through the diode should be known for calculating the temperature.

A vacuum diode is a source of fluctuating current due to the so-called shot effect. The thermionic current through the diode comprises of pulses introduced by individual electrons. Superposition of the pulses leads to fluctuations of the current around a mean value. In a saturation regime of the diode, these pulses are statistically independent.

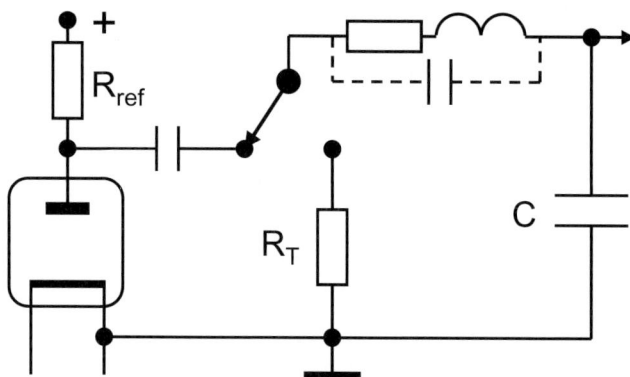

Fig. 10.3. Simplified diagram of the first stage of noise thermometer developed by Pepper and Brown (1979). A vacuum diode serves as a calibrated source of noise, which can be added to the thermal noise of the resistor R_{ref} at a reference temperature

It was shown by Schottky (1918) that the mean square of the fluctuations of the current, in a saturation regime, equals

$$<\Delta I^2> = 2eI\Delta f, \tag{10.7}$$

where I is the mean value of the current, e is the electron charge, and Δf is the frequency band, in which the fluctuations are measured. It is seen from Eq. (10.7) that the shot effect also represents a fundamental physical phenomenon, and it provided good opportunity for determining the electron charge. When a load of the diode has an impedance Z (Fig. 10.4), the mean squared noise voltage across the load is

$$<\Delta V^2> = 2eIZ^2\Delta f, \tag{10.8}$$

where Z is the modulus of the impedance.

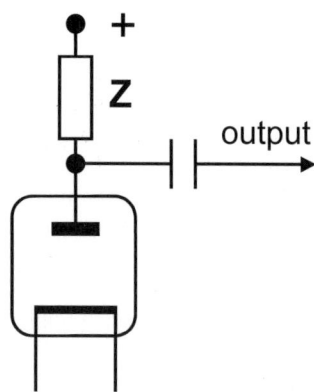

Fig. 10.4. Vacuum diode loaded by an impedance Z may serve as a source of calibrated noise voltage

Since noise voltages from independent sources are statistically independent (uncorrelated), the mean square of the sum of such voltages equals to the sum of the mean squares of the voltages. This evident rule is important for all measurements of electrical fluctuations.

With an averaging time of 1 s, temperatures of up to 1770 K were measured to within a statistical uncertainty of 0.3%. Pepper and Brown (1979) summarized the advantages of their noise thermometer as follows:

- A simple measurement of the DC current of the vacuum diode provides the difference between the absolute temperatures of the reference and sensor resistors.
- The result does not depend on the knowledge of the capacitance of the circuit.
- The need for accurate equalisation or measurement of the reference and sensor resistors is reduced.
- The small magnitude of resistance used ($R < 100\ \Omega$) renders the effect of lead capacitances between the resistors and the switches negligible.
- The difference in loss of the noise signals from the reference and sensor circuits due to the restricted frequency response of the amplifier can be made negligible.

Crovini et al. (1992) described a high-temperature noise thermometer designed at the IMGC. The authors pointed out main requirements to be fulfilled in noise thermometry. Thermal noise is of 'high quality' when it does not contain effects of interferences, of external noise, and of adverse and unstable products of the amplifier internal noise (e.g., small parasitic oscillations). The test consists in very sensitive spectral analyses and in the inspections of the stability of the noise. The noise must be stable and exhibit a white spectrum in the frequency range used. A careful inspection should not reveal any correlated component in the form of sine oscillations or peaks. If these conditions are not achieved, grounding, shielding, and input arrangement have to be improved. The authors formulated some basic rules for achieving a 'high-quality' noise:

- As far as possible, star ground connections, with each amplifier stage, shield, and chassis directly grounded to a common point, should be used.
- If a ground bus is used, this must be as short as possible and of very low resistance, and the connection of the various branches must follow given sequence avoiding unwanted feedback to the input.
- Shielding continuity must be provided everywhere, and a particular shielding symmetry must be realised at the input.
- Connections to computers must provide full isolation with ground circuit separation.
- Both electric and magnetic shielding is required for the temperature sensor.

Shepard et al. (1992) reported on a tuned-circuit noise thermometer developed for a space nuclear electric power system. The system requires temperature measurement at 1400 K in space for ten years, of which seven are expected to be at full reactor power. Brixy et al. (1992) described noise thermometry for industrial and metrological applications developed at Forschungszentrum Jülich. In particular, it was applied as a reference and calibration method in a nuclear power plant. Under such conditions, the thermopower of thermocouples may drift rapidly, as well as the resistance of a noise thermometer. When monitoring this resistance, correct temperature measurements are possible through the thermal noise. The authors pointed out that noise thermometry is in practice the only method of contact thermometers, which is not affected by the unavoidable changes of the sensor at high temperatures caused by the recrystallisation or interaction with the environment. Any changes in the resistance of the sensor generating thermal noise are measured and taken into account. The electronic system is divided into the preamplifier and the main device. The preamplifier is installed as close to the sensor as ambient conditions permit. The signals from the preamplifier can then be transmitted over long distances, up to several hundred meters. A computer controls the whole automated measuring procedure. The most severe problem for noise thermometry is the avoidance or rejection of electro-magnetic interferences. For the noise thermometer developed by the authors, the error of measurements was highly reduced. The uncertainty for a single measurement of freezing point of zinc was 10^{-4}.

10.2 Noise Correlation Thermometer

When measuring temperature of wire samples through their thermal noise, the noise of low-resistance samples may appear to be comparable with the inherent noise of amplifiers. The correlation method of measurements is useful in such cases.

10.2.1 Correlation Amplifier

In a correlation device, an electric signal U_c (e.g., the thermal noise generated by a resistor) is fed to the inputs of two similar amplifiers (Fig. 10.5).

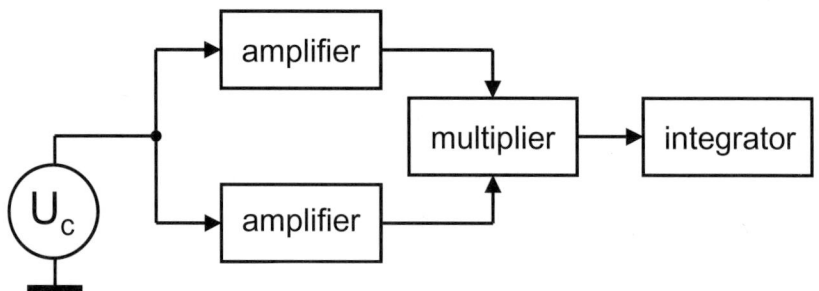

Fig. 10.5. Block diagram of correlation amplifier. After integration, the output voltage does not depend on the inherent noise of the channels

Their output signals contain the amplified common input voltage U_c (correlated) and inherent noise voltages U_1 and U_2. The signals proceed to a multiplier and then to an integrating circuit. The result of the multiplication is

$$V = K_1 K_2 (U_c^2 + U_c U_1 + U_c U_2 + U_1 U_2), \tag{10.7}$$

where K_1 and K_2 are total gains of the two channels. Since the inherent noise voltages of the amplifiers, $U1$ and $U2$, are uncorrelated with each other and with the common input signal U_c, the corresponding products vanish after averaging. The mean output voltage of the integrator is thus proportional to the square of the correlated input signal U_c:

$$<V> = K_1 K_2 <U_c^2>, \tag{10.8}$$

but it is sensitive to the gains of the channels K_1 and K_2.

Brophy et al. (1965) developed a correlation amplifier for very weak signals. The amplifier is capable of detecting correlated signals fed to both inputs in the presence of uncorrelated noise two to three orders of magnitude greater. The correlation technique was employed in noise thermometers (e.g., Shore and Williamson 1966; Storm 1970).

10.2.2 Compensation Method

The proposed correlation amplifier (Kraftmakher and Cherevko 1972a) differs in that it employs a compensation method (Fig. 10.6).

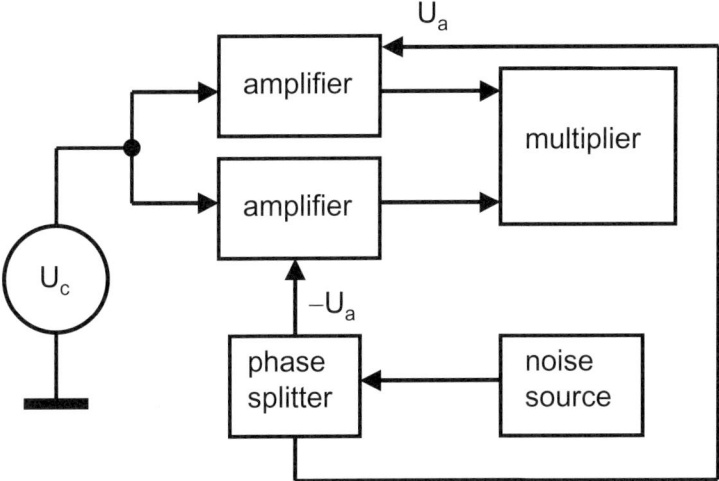

Fig. 10.6. Correlation amplifier with compensation. Anticorrelated voltages, U_a and $-U_a$, being added to the correlation voltage, U_c, nullify the output signal of the multiplier regardless of the gain of the channels (Kraftmakher and Cherevko 1972a). A noise source and a phase splitter provide the anticorrelated voltages

The noise voltage to be measured, U_c (correlated), is fed to two amplifiers, and the amplified voltages proceed to a multiplier. For the compensation, two anticorrelated voltages, U_a and $-U_a$, are also fed to the amplifiers. These voltages are taken from an independent noise generator and a phase splitter. Now the signal after averaging equals

$$<V> = K_1 K_2 (<U_c^2> - <U_a^2>). \qquad (10.9)$$

When the RMS value of the anticorrelated voltage equals that of the voltage to be measured, the output voltage becomes zero, regardless of the gains of the channels.

The correlation amplifier with compensation was employed as a noise thermometer for measurements on a 50-µm platinum wire heated electrically in vacuum (Kraftmakher and Cherevko 1972b). To simplify the calculations, two identical samples were used, with resistances of about 10 Ω at 273 K. They were connected in series relative to a DC source and in parallel relative to the inputs of the correlation amplifier. At various DC currents, the thermal noise was measured in a 60–150 kHz frequency band along with the resistance of the samples. In the range 1200–1900 K, the results show good agreement between the temperature derived from the noise measurements and that calculated from the resistance of platinum taking into account the vacancy contribution (Fig. 10.7). The scatter of the experimental points was about 1%, while the systematic deviation was several times smaller.

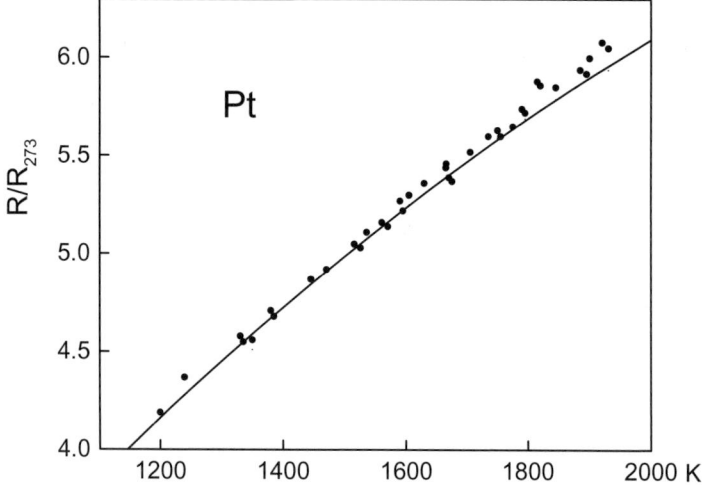

Fig. 10.7. Resistance ratio for platinum wire measured by noise correlation thermometer (Kraftmakher and Cherevko 1972b): • experimental points, ⎯⎯ quadratic approximation. The increase in the R/R_{273} ratio is in agreement with non-linear contribution due to vacancy formation in crystal lattice

10.2.3 Temperature Derivative of Resistance

The noise correlation thermometer with compensation is also efficient in a regime of periodically changing the temperature of the sample (Kraftmakher and Cherevko 1974). The changes in the thermal noise and in the electrical resistance are measured to calculate the temperature derivative of the resistance and the heat transfer coefficient of the sample. A DC current heating the sample is modulated with a period sufficiently long to achieve thermal equilibrium (Fig. 10.8).

The changes in the resistance of the samples corresponding to certain changes in the heating current were determined beforehand with a bridge circuit. This was sufficient to evaluate the changes in the resistance corresponding to the temperature changes. It is easy to see that the temperature derivative of the resistance obeys the relation

$$R' = R_1/[\Delta(RT)/\Delta R - T_2] = R_2/[\Delta(RT)/\Delta R - T_1], \qquad (10.10)$$

where R_1 and R_2 are the resistances of the sample at the initial, T_1, and the final, T_2, temperatures, $\Delta(RT)$ is a quantity available from the measured changes in the thermal noise, and ΔR is the corresponding change in the resistance. The heat transfer coefficient of the sample is obtainable in similar way:

$$P' = R_1\Delta P/[\Delta(RT) - T_2\Delta R] = R_2\Delta P/[\Delta(RT) - T_1\Delta R], \qquad (10.11)$$

where ΔP is the increment in the heating power necessary to change the temperature of the sample from T_1 to T_2.

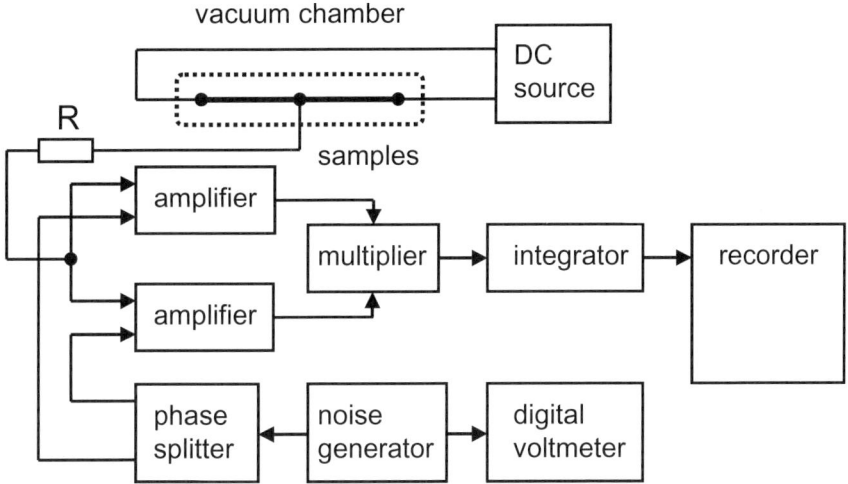

Fig. 10.8. Noise correlation thermometer operating under periodic changes of the temperature of the sample (Kraftmakher and Cherevko 1974). The resistor R at room temperature generates additive thermal noise for calibration

The measurements were performed on 8-μm tungsten wires placed in an evacuated and sealed-off glass bulb. The resistance of the two samples connected in parallel was about 11 Ω at 273 K. The modulation period was 120 s, and the amplitude of the temperature changes about 50 K. The system was calibrated by means of resistors at room temperature periodically connected to the inputs of the correlation amplifier in series with the samples. The quantity $\Delta(RT)$ was calculated as an average from 10 to 20 periods of the temperature variations.

A more difficult task is to apply noise measurements to direct determinations of small temperature oscillations in a wire sample.

10.2.4 Nyquist's Formula for High Current Densities

In vacuum, the current density necessary to heat up a wire sample of diameter d to a certain temperature is proportional to $d^{-\frac{1}{2}}$. Measurements on thin samples allow verification of the Nyquist formula under nonequilibrium conditions, where the current density is of the order of 10^8 A.m^{-2}. The results of the measurements on 8-μm tungsten wires confirmed the validity of the Nyquist formula under such conditions.

11 Electronics for Modulation Measurements

Modulation techniques require numerous but not sophisticated electronic instruments. Fortunately, the necessary equipment is now readily available, and the requirements specific for the modulation measurements are quite moderate.

11.1 Electronic Instruments

A list of electronic instruments and of their features important for modulation measurements is given below. This equipment can be used to assemble modulation set-ups and to further develop the modulation techniques.

DC sources are necessary for heating the samples to a desired mean temperature. Their main features are upper limits of the voltage and current, smallness of pulsations in the output voltage, stability, and convenience of regulation. Modern stabilised sources are very suitable, and any device with corresponding limits of the voltage and current is usable.

Low-frequency *oscillators* provide modulation of the heating power. The main requirements for them are the stability of the frequency and amplitude of the output voltage, small distortions, sufficient output power, and compatibility with a load, i.e., a sample or an electric heater. The maximum power is achievable when the load resistance equals the internal resistance of the source.

Power amplifiers supply high currents to heat thick samples or resistive heaters. They must provide sufficient output power, stability, small distortions, and to be compatible with a load.

Meters of electric current and voltage are necessary to measure oscillations of the heating power. The inaccuracy of modern digital meters usually is in the range 0.01% to 0.1%.

In some set-ups, *calibrated variable resistors and capacitors* are needed. As a rule, their inaccuracy does not exceed 0.1 %.

Frequency meters measure the modulation frequency. Usually, the calibration of oscillators is not sufficiently accurate, and one has to measure the frequency by an additional digital frequency meter. At low frequencies, it is preferable to determine the period of the oscillations, because in this case the desired accuracy is obtainable in a shorter time. An external frequency meter is unnecessary when a digital frequency meter is incorporated into the oscillator.

Modern digital *phase meters* provide wide frequency range and inaccuracy less than 0.1°, which is sufficient for modulation measurements.

Oscilloscopes are used for observing electric signals and as null indicators in compensation circuits. Modulation frequencies are relatively low, whereas the signals after amplification are sufficiently large. The requirements for the oscilloscopes are therefore quite moderate.

Light sources are needed for the modulated-light heating. They must provide the necessary radiation power and ensure its stability. Usually, incandescent lamps and lasers serve for this purpose. In some cases, optical fibres are used to guide the modulated light to the samples.

Photoelectric sensors detecting temperature oscillations include photomultipliers, photodiodes, and photoresistors. Their main features are the spectral response, stability, time constant, and inherent noise. Photomultipliers are very sensitive to visible and near-infrared radiation but are inconvenient because of high operating voltage. Photodiodes and photoresistors provide more favourable spectral response and are capable of detecting radiation from samples of lower temperatures (Fig. 11.1). The main advantage of photomultipliers, the high internal gain, is of no importance owing to availability of sensitive amplifiers.

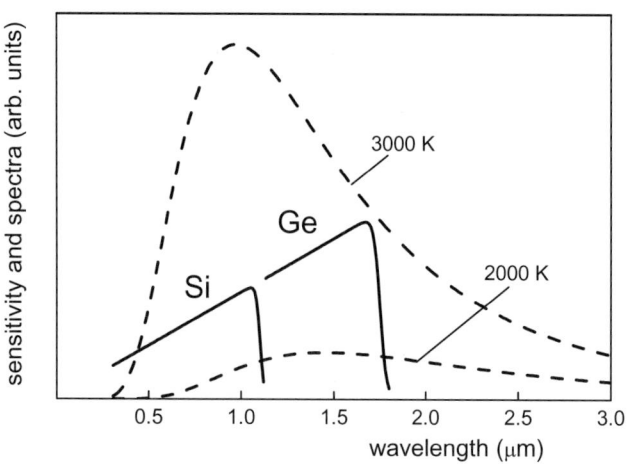

Fig. 11.1. Spectral response of silicon and germanium photodiodes. Spectra of perfect blackbody radiation for two temperatures are also shown

Selective amplifiers serve for measuring temperature oscillations or as null detectors in compensation circuits. Their main features are the frequency range, sensitivity, selectivity, stability, inherent noise, and convenience of tuning. For example, an amplifier PAR model 124A (Princeton Applied Research) operates as a selective nanovoltmeter in a frequency range 1 Hz–100 kHz. A digital switch sets the operating frequency. The quality factor Q, adjustable up to 100, defines the selectivity. A 1:100 input transformer allows a 100-fold enhancement of the sensitivity when the internal resistance of the source of the signal does not exceed 10 Ω. The output voltages, proportional to the input voltage, are a DC voltage (up to 10 V at a 1-kΩ load) and an amplified AC voltage (up to 0.1 V at a 600-Ω load).

The device is usable also as a wide-band amplifier or a lock-in amplifier. The stability of the gain is better than 1%, so that the amplifier may serve for direct measurements.

Lock-in amplifiers are capable of measuring periodic signals of selected frequency even much weaker than signals of other frequencies or noise. Their main features are the frequency range, sensitivity, stability, and convenience of tuning. For example, a Stanford Research Systems model SR 830 dual-channel lock-in amplifier employs digital signal processing and provides full-scale sensitivity 2 nV to 1 V (Fig. 11.2).

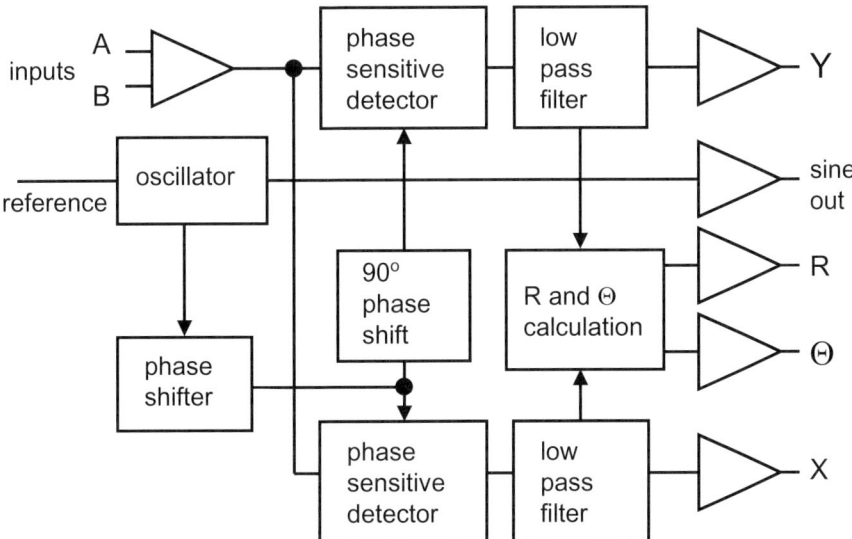

Fig. 11.2. Simplified block diagram of SR 830 dual-channel lock-in amplifier. Lock-in amplifier is an excellent tool for modulation measurements

The frequency range is 1 mHz to 100 kHz. Input impedance of each channel is 10 MΩ, with 25 pF in parallel. Gain accuracy is 1%, and the absolute phase error is less than 0.01°. Orthogonality of the channels is 90° ± 0.001°. The common-mode-rejection ratio at 100 Hz is 90 dB. The time constant is adjustable up to 30 s for frequencies above 200 Hz and up to 30000 s for lower frequencies. An internal oscillator provides output voltages of frequencies in the operating range. The inaccuracy in the frequency is 25 ppm + 3×10^{-5} Hz. The output impedance of this source is 50 Ω, and the maximum voltage on a 50-Ω load is 2.5 V (RMS). The result of the measurements is displayed, in digital form, as X and Y or R and Θ. Here X is the component of the signal coinciding in phase with the reference, Y is the quadrature component, R is the modulus of the signal, and Θ is the phase angle between the signal and the reference. The noise immunity of a lock-in amplifier is limited by the so-called dynamic reserve. This means the capability of measuring weak signals of the selected frequency in the presence of much stronger noise or periodic signals of other frequencies. The dynamic reserve of the SR 830 amplifier is more than 100 dB.

Data-acquisition systems are the best tools to control the measurements and accumulate experimental data.

11.2 Examples of Modulation Set-Ups

As an example of electronic instrumentation for modulation calorimetry, one may consider a calorimeter designed by Smaardyk and Mochel (1978). The apparatus is intended for high-resolution ($1:10^4$) studies on small quantities (0.1 mg) of organic liquids. The sample is confined between the Pyrex and Mylar sheets in direct contact with both the thermometer and heater (Fig. 11.3). Most of the heat flow is normal to this sandwich and into a thin layer of exchange gas below the Pyrex. The heater is a thin film of chromel, 7.5 nm thick, vacuum evaporated directly onto the Mylar sheet. Aluminium leads, 100 nm thick, are also evaporated on the sheet. The temperature oscillations are measured with a thin-film copper-bismuth thermocouple, 250 nm thick, evaporated directly onto the Pyrex sheet. A 150-nm layer of SiO prevents electrical contact with the heater. Copper wires are attached to the heater and thermocouple tabs with silver micropaint.

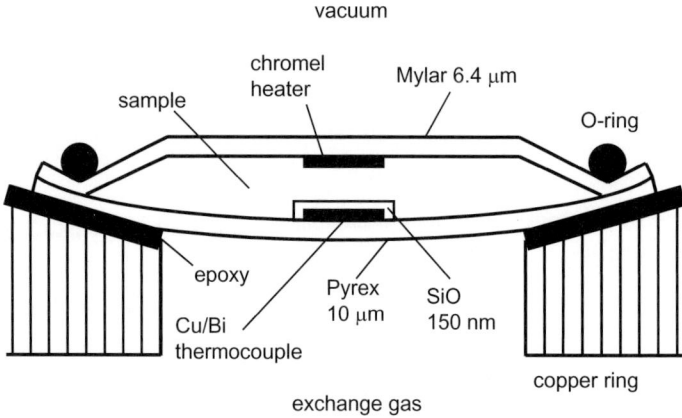

Fig. 11.3. Calorimeter cell of a high-precision calorimeter for organic liquids (Smaardyk and Mochel 1978)

A Hewlett-Packard model 3300A function generator provides the AC heating (Fig. 11.4). A Keithley model 148 nanovoltmeter measures the temperature rise of the sample above the heat sink. A PAR model AM-1 transformer amplifies the thermocouple signal, which is measured by a PAR model 128 lock-in amplifier. A chart recorder monitors the output voltage of the amplifier. In order to measure small changes in the temperature oscillations, a precision square-wave voltage is subtracted from the signal through a PAR model 113 preamplifier. The temperatures of the heat sink and of the vacuum chamber are controlled with thermistors. The heat sink temperature can also be swept at constant heating or cooling rates with periods as long as 14 h using a Wavetek model 133 signal generator.

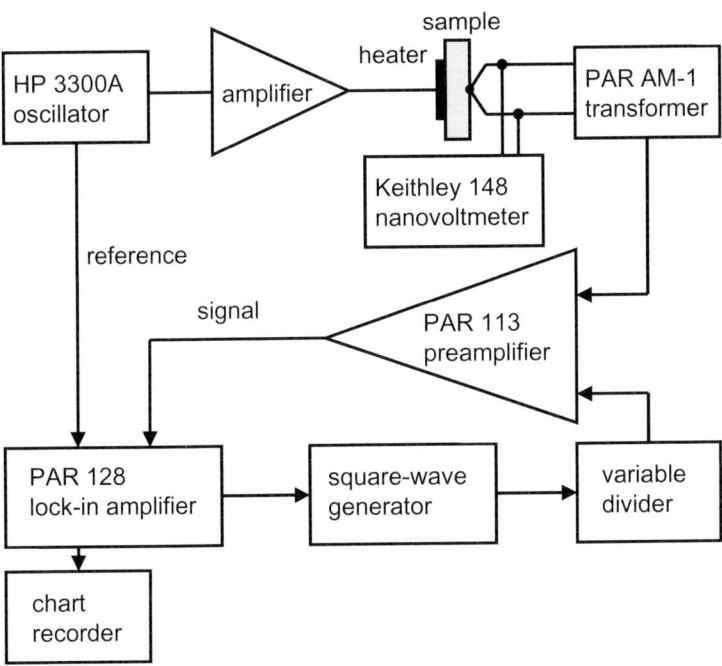

Fig. 11.4. Simplified block diagram of the electronics of a high-precision calorimeter described by Smaardyk and Mochel (1978)

Finotello et al. (1997) described a calorimeter for studying liquid crystals. The calorimeter cell incorporates a resistive heater and a thermometer, and is weakly anchored to a regulated bath (Fig. 11.5). A low-frequency oscillator feeds the heater, while a resistance thermometer measures the temperature oscillations in the calorimeter cell. By gradually changing the temperature of the bath, the heat capacity of the sample is obtainable versus temperature. The temperature stability of the bath is of about 0.1 mK at room temperature and of a few microkelvins at liquid helium temperatures. A commercially available AC resistance bridge or a temperature controller may serve for this purpose, as well as a homemade AC bridge.

The authors stressed that the success in the temperature control rests on using a high-resolution fast-responding thermometer. The power dissipated by the thermometer is 100–1000 times smaller than that by the heater. The DC component of the voltage drop across the thermometer is directly measured and averaged by a digital meter, yielding the mean temperature of the cell. A selective amplifier amplifies the AC component. Then the signal proceeds to a dual-channel lock-in amplifier operating at a long time constant for better signal averaging. The amplifier operates also in the so-called $2f$ mode, by creating a double-frequency reference signal, and provides both in-phase and quadrature outputs. The amplitude and the phase of the temperature oscillations are measured by a digital meter and stored by a computer. The computer stores also the mean temperature of the sample and controls the temperature of the bath.

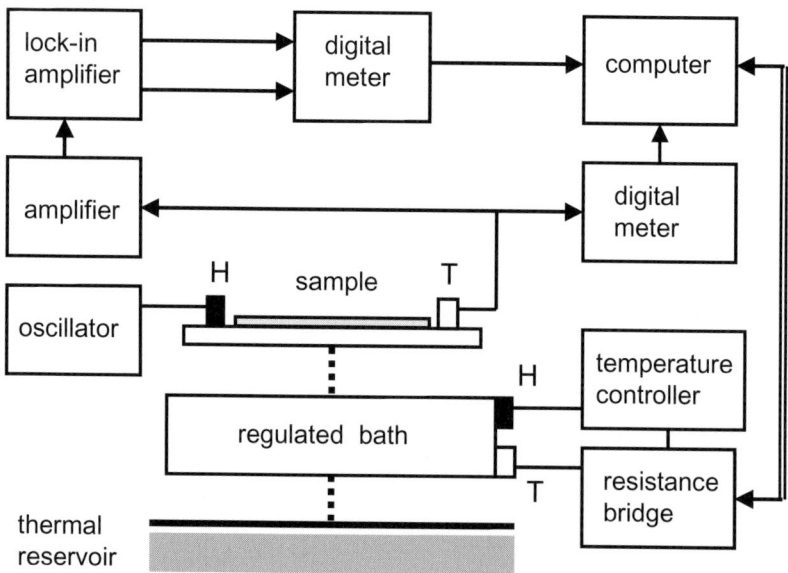

Fig. 11.5. Simplified diagram of modulation calorimeter described by Finotello et al. (1997). H – heaters, T – thermometers

Mehta et al. (1999) described in detail their set-up for calorimetric measurements of confined films of ^{4}He (Fig. 11.6). An AC voltage from a synthesizer, HP 3320B, is applied to a heater. This voltage was found to be stable to within 0.08% over 24 hr. The power used is of the order of 1 µW, which corresponds to temperature oscillations of a few microkelvins. The heater is evaporated at the bottom of the calorimeter cell. A resistance thermometer $T1$ measures the temperature oscillations, and a battery-operated source provides a DC current for it. The signal from the thermometer is of the order of 100 nV. It proceeds to a preamplifier and then to a tuned amplifier, PAR 124A, and a dual-channel lock-in amplifier, SR 530. A computer stores the output voltage. The regulation of the temperature of the cell is done with a thermometer $T2$. The computer acquires the heat capacity signal over a time interval of 100–300 s. This time is chosen to reduce the random noise after averaging to less than 0.1%.

Fraile-Rodríguez et al. (2001) developed a calorimeter for studies of phase transitions in solids ranging from 4.2 to 400 K. The typical mass of the samples is 1 to 25 mg. The calorimetric set-up is constituted by two concentric Dewars, which can be filled with liquid nitrogen and liquid helium (Fig. 11.7). The main chamber of the calorimeter contains the sample and a copper block, which acts as the thermal reservoir. This chamber is placed inside a secondary chamber providing good thermal isolation. Both chambers can be independently evacuated, or helium gas may serve to increase the thermal coupling. Chopped light from a halogen lamp is focused on a glass rod, which guides the light to the sample. The surface of the sample is black-painted to stabilise the absorption of light. The lighting system is fixed on an optical bench placed inside a metallic box. The reference signal for a lock-in amplifier is delivered by a phototransistor sensing the chopped light. A constantan heater heats the copper block. A platinum thermometer with an absolute accuracy of 1 mK is placed at the bottom of the copper block with the leads thermally anchored to. A multimeter, HP 3457a, measures its resistance. A temperature controller, which uses an independent platinum thermometer, regulates the temperature. The modulation frequency should be selected to make the thermal wavelength greater than the largest dimension of the sample. Two 25-μm iron-constantan thermocouples measure the DC and AC temperature components. Their junctions are attached to the rear surface of the sample and to the copper block. The electrical contact between the thermocouples should be avoided. The leads of the thermocouples support the sample. The AC thermocouple signal proceeds to a low-noise transformer (PAR model 190, gain = 100) and then to a lock-in amplifier, SR 850. This signal can also be amplified and observed with an oscilloscope.

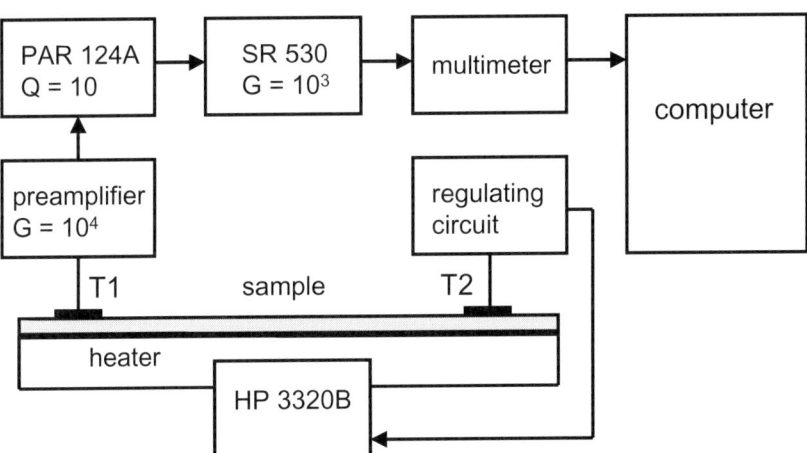

Fig. 11.6. Simplified diagram of set-up for measuring heat capacity at liquid helium temperatures (Mehta et al. 1999). The high gain and selectivity of the measuring channel makes it possible to measure temperature oscillations of the order of 10^{-6} K

The authors developed a program for the automation of the calorimeter using two procedures. The first procedure provides temperature jumps, which may be as small as 25 mK. Every time the approach to the thermal equilibrium is observed until the selected equilibrium criterion is satisfied. Typical temperature drifts are below 4×10^{-5} K.s^{-1}. The second procedure allows for continuous heating or cooling with the scanning rate from 0.001 to 0.1 K.s^{-1}. The authors examined the features of the calorimeter by measuring the specific heat of a triglycine sulphate (TGS) ferroelectric crystal undergoing a second-order phase transition at 49°C.

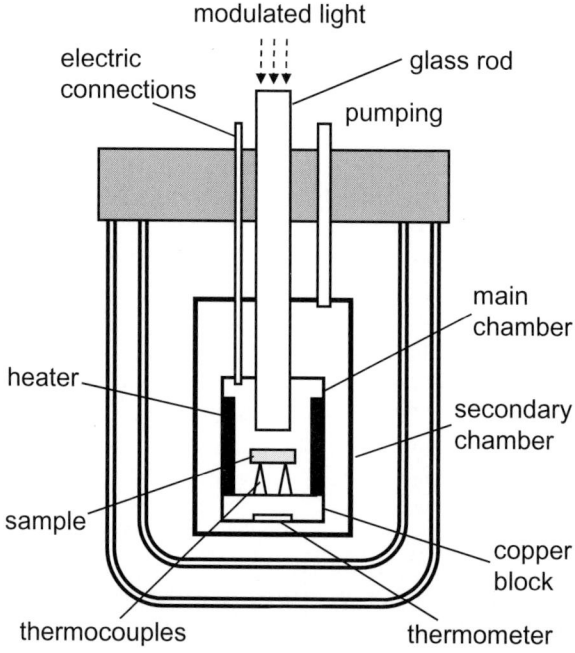

Fig. 11.7. Simplified diagram of calorimeter employing modulated-light heating (Fraile-Rodríguez et al. 2001)

12 Accuracy of Modulation Measurements

As in other cases, errors of modulation measurements fall into two categories: errors arising from differences between the theoretical model and experimental conditions, and instrumental errors arising from the inaccuracy of measuring instruments.

12.1 Requirements for Modulation Experiments

The theoretical model of modulation calorimetry includes the following assumptions:

(1) The mean temperature is the same over the entire calorimeter cell, as well as the amplitude and the phase of the temperature oscillations. This means that when a separate heater and a thermometer are used, the time of equilibration between them and the sample is much shorter than the period of the modulation.

(2) The modulation frequency is sufficiently high to meet the criterion of adiabaticity: changes in the heat losses from the calorimeter cell caused by the oscillations of its temperature are much smaller than the oscillations in the heating power. The conditions (1) and (2) thus pose opposing requirements for the modulation frequency.

(3) The heat capacity of the addenda (the substrate, the heater, and the thermometer) is much smaller than that of the sample. This requirement is not so strict because of the high resolution of modulation measurements. In many cases, the heat capacity of the sample, which was much smaller than that of the addenda, was successfully determined.

The theoretical model is well satisfied when a conducting sample, in the form of a wire, a foil, or a thin rod, is heated by an electric current, while the temperature oscillations in the sample are detected through its resistance or radiation from it. In this case, the requirement (3) is excluded and the assumption (2) is easy to meet by increasing the modulation frequency. To exclude cold-end effects, the measurements are performed on a central portion of the sample, where axial temperature gradients are small. Radial temperature differences are insignificant because of the high thermal conductivity and/or small thickness of the samples. Therefore, the requirement (1) is also satisfied when using direct electric heating.

With other methods of heating, the assumption (1) is met by decreasing the thickness of the sample and the modulated power. When the modulated power is fed into the sample, the mean temperature of the sample increases. This increase is much larger than the temperature oscillations and may result in temperature gradients inside the sample. To reduce these gradients, only a small modulated power is supplied to the sample, while a furnace controls the mean temperature. This method of heating is especially important for studies of phase transitions, where good temperature resolution is one of the main requirements.

The thickness of the samples of low thermal conductivity usually does not exceed 0.2 mm. A small thickness satisfies the assumption (1) but contradicts the requirement (3). To circumvent this difficulty, the heater and the thermometer are prepared as thin films, 10–100 nm thick. To decrease the modulation frequency, one can employ a nonadiabatic regime. The second possibility is to reduce AC heat losses from the sample by means of an active thermal shield. This method is usable when the heat exchange occurs through radiation. With other mechanisms of heat exchange, the modulation frequency should be chosen carefully. The simplest way is to experimentally find a frequency range, where the quantity $\omega\Theta_0$ remains constant. This means that both requirements (1) and (2) are satisfied. It is difficult to meet them simultaneously only in one case, when samples of low thermal conductivity are studied under high pressures provided by a dense medium. As a rule, significant corrections are necessary to determine the specific heat in such measurements.

A slab sample heated at one its side must be sufficiently thin in comparison to the thermal wavelength. Hatta and Minakov (1999) calculated the maximum thickness necessary to keep the error of measurements of the heat capacity within 1%. The thickness depends on the thermal diffusivity D of the sample and on the modulation frequency ω. To satisfy the above requirement, the thickness of the sample must not exceed $0.6(D/\omega)^{1/2}$.

An additional question concerns to contributions from leads and suspensions, which are unavoidable in calorimetric systems. Greene et al. (1972) considered this problem. The heat capacity of a wire, C_w, makes a contribution to the total heat capacity given by $\Delta C = C_w F(X)$, where $X = (\omega C_w/2K_w)^{1/2}$, and K_w is the thermal conductance of the wire. The function $F(X)$ has the limiting forms $F(X) \approx 1/3$ for $X \leq 0.5$ and $F(X) \approx 1/2X$ for $X \geq 2$. As the modulation frequency increases, a smaller fraction of the wire contributes to the total heat capacity.

Suzuki et al. (1982) found the heat capacity of the addenda of their calorimeter cell to decrease with increasing frequency. This was explained by contributions from leads and suspensions, which were expected to vary as $\omega^{-1/2}$ (the case of $X \geq 2$). To check this conjecture, the authors plotted the measured heat capacity of the addenda versus $\omega^{-1/2}$. A linear relation was obtained for modulation frequencies above 200 Hz. The slope of the line appeared to be in good agreement with calculations based on thermal diffusivities and cross-sectional areas of the wires involved.

Thus, in almost all cases the experimental conditions of modulation calorimetry are quite adequate to the theoretical model. In modulation measurements of thermal expansivity, electrical resistance, and thermopower, it is much easier to achieve such conformity because an adiabatic regime of the measurements is not required. The main sources of errors in modulation calorimetry are the determinations of the oscillations in the heating power and in the temperature of the sample. Accurate determinations of the heating power are feasible with direct electric heating or separate resistive heaters. With other methods of heating, the situation is more complicated. Thermocouples and resistance thermometers are the best tools for accurate measurements of the temperature oscillations. With other methods (temperature dependence of the resistance of the sample, photoelectric sensors), the errors depend on the reliability of the data used or on the accuracy of calibration. Employment of blackbody models in the samples greatly increases the reliability of the use of photoelectric sensors.

Table 12.1 shows estimates of inherent errors of modulation measurements. The term 'imprecision' refers to differences of single data points from smoothed values, whereas 'inaccuracy' refers to total errors including random and systematic errors. Only rough estimations of the inaccuracy are available.

Table 12.1. Estimates of errors in modulation measurements (standard deviations)

Source of error	Imprecision [%]	Inaccuracy [%]
Mean temperature of the sample		0.01–1[a]
Mean temperature from thermal noise		1[a]
Mass of the sample		0.1–5
Modulation frequency		< 0.01
Oscillations of heating power:		
Direct electrical heating		0.2–1
Modulated-light heating	0.1	2–5
Electron bombardment		1–2
Separate heaters		0.1
Induction heating		4–6[b]
Peltier heating	0.5	1–3[a]
Temperature oscillations:		
Supplementary-current method	1	1[c]
Third-harmonic method	0.1	1[c]
Equivalent-impedance method	0.1	1[c]
Photoelectric detectors	0.1	1[d]
Pyroelectric sensors	0.1	1[d]
Thermocouples	0.01–0.1	0.1–1[a]
Resistance thermometers	0.01–0.1	0.1–1[a]
Oscillations of sample length, resistance, and thermal EMF	0.01–0.1	0.1–1

[a] depends on temperature,
[b] according to Filippov and Makarenko (1968),
[c] depends on the accuracy of dR/dT values,
[d] depends on the accuracy of calibration

12. 2 Comparison with Recommended Values

Contrary to a widespread opinion, modulation measurements may provide data, whose absolute accuracy is comparable with that from other techniques. This relates, in particular, to calorimetric measurements with direct electric heating or separate heaters. Below, a comparison is given of results of some modulation measurements performed many years ago and of recommended values recently updated. The preparation of recommended values is one of the tasks of CODATA (Committee on Data for Science and Technology). A paper "Thermophysical properties of some key solids: An update" by White and Minges (1997) serves here as a source of recommended values. This paper considers the specific heat of Cu, Fe, W, and Al_2O_3; the thermal expansion of Cu, Si, W, and Al_2O_3; the electrical resistivity of Cu, Fe, Pt, and W; the thermal conductivity of Al, Cu, Fe, and W; the absolute thermopower of Pb, Cu, Pt, and W. We will compare the results of modulation measurements of the specific heat of copper, iron, and tungsten, of the thermal expansivity of tungsten, and of the electrical resistivity of platinum with the recently updated recommended values.

The second source of recommended data is a review paper by Sabbah et al. (1999) entitled "Reference materials for calorimetry and differential thermal analysis". We can compare the recommended values of the specific heat of molybdenum and platinum with the results of modulation measurements.

Finally, the thermal expansivity of platinum from modulation measurements will be compared with data recommended by Kirby (1991).

12.2.1 Specific Heat of Tungsten, Copper, and Iron

Modulation measurements of the specific heat of tungsten were carried out using the equivalent-impedance technique. In this case, one has to use data on the electrical resistance of the sample and its temperature derivative. Such data for high-melting-point metals were taken from the literature. It should be stressed once again that the accuracy of these measurements of the specific heat completely depends on the reliability of the resistance data. The numerical values of the specific heat of tungsten measured in the range 1500–3600 K are available from polynomial fits (Kraftmakher and Strelkov 1962) or from tables of the data (Kraftmakher 1966c, 1973a). The results are in reasonable agreement with the recommended values in the 1500–3100 K range: the difference is less than 3% (Fig. 12.1). However, it rapidly increases at higher temperatures amounting to about 10% at 3400 K. For other high-melting-point metals, the situation is not so good. Significant differences are seen between results of the modulation measurements on tantalum, molybdenum, and niobium and data from pulse and dynamic techniques.

In measurements of the specific heat of copper in the range 550–1250 K (Kraftmakher 1967c), a thermocouple served to determine the mean temperature of the sample and the temperature oscillations in it. The numerical data are available from the equation fitting the experimental data taking into account the vacancy contribution to the specific heat. The results are close to the recommended values (Fig. 12.2). The maximum difference is less than 1% in the range 550–1100 K and grows to about 2% at 1250 K.

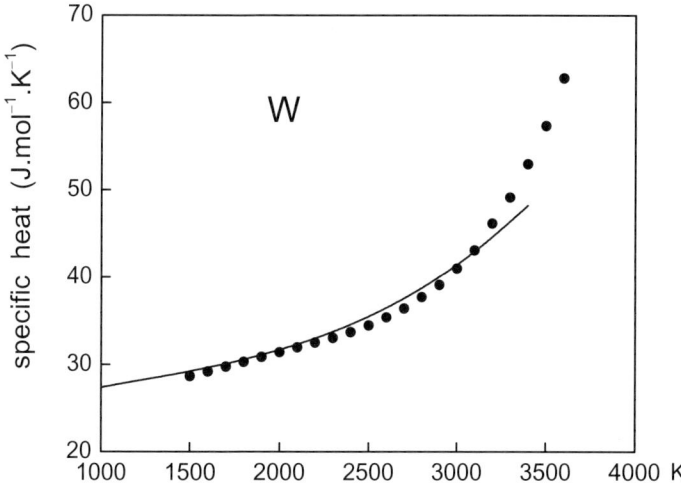

Fig. 12.1. Specific heat of tungsten: ● modulation measurements (Kraftmakher and Strelkov 1962), —— recommended values (White and Minges 1997). Reasonable agreement of the data is seen up to 3100 K

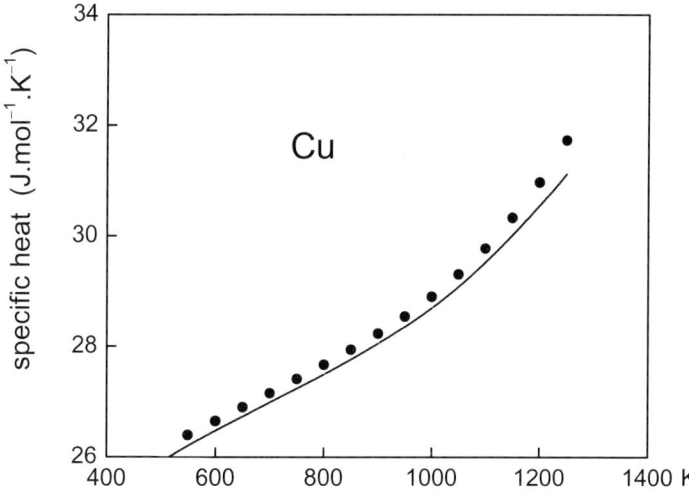

Fig. 12.2. Specific heat of copper: ● modulation measurements (Kraftmakher 1967c), —— recommended values (White and Minges 1997). Good agreement exists in almost all the range of the measurements

The specific heat of iron in the range 600–1250 K was measured employing electron heating and a thermocouple for determining the temperature of the sample and the temperature oscillations in it (Varchenko et al. 1978). Reasonable agreement with the recommended values exists in the 600–1100 K range, but above 1100 K a significant discrepancy appears, especially in the γ-phase (Fig. 12.3). This discrepancy may be caused by the lack of thermodynamic equilibrium in the sample used in the modulation measurements.

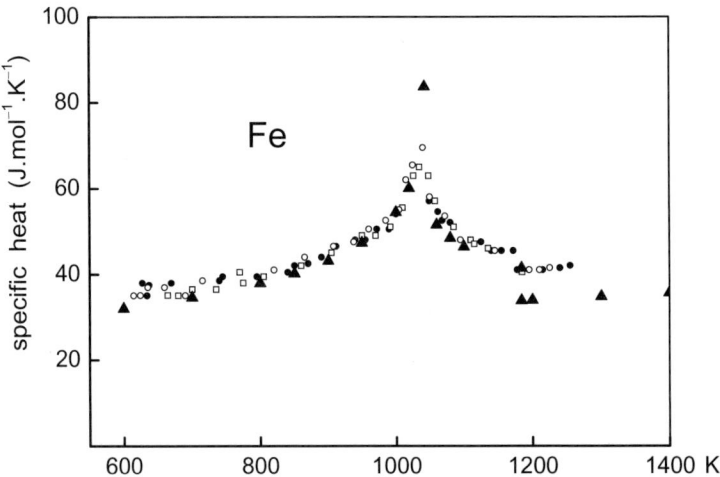

Fig. 12.3. Specific heat of iron: ○, ●, □ modulation measurements (Varchenko et al. 1978), ▲ recommended values (White and Minges 1997). The significant discrepancy for the γ-phase may be caused by the lack of equilibrium in the sample used in the modulation measurements

12.2.2. Specific Heat of Molybdenum and Platinum

First direct measurements of the high-temperature specific heat of molybdenum were undertaken by Rasor and McClelland (1960ab) by means of pulse technique. Then the equivalent-impedance technique was employed in the range 1300–2500 K (Kraftmakher 1964). Later, many measurements were carried out with dynamic calorimetry. The dynamic-calorimetry data constituted the base of recommended values (Sabbah et al. 1999). The comparison of the results of the modulation measurements with the recommended values (Fig. 12.4) shows good agreement up to 2300 K (the difference is less than 1%), but the data diverge at higher temperatures. This contradiction becomes very significant when one extrapolates the results of modulation calorimetry to the melting point.

For platinum, the recommended values are given only up to 1500 K (Fig. 12.5). In this range, they are close to the results of modulation measurements (Kraftmakher and Lanina 1965). In the range 1000–1400 K, the difference is less than 2%; it increases to 3% at 1500 K.

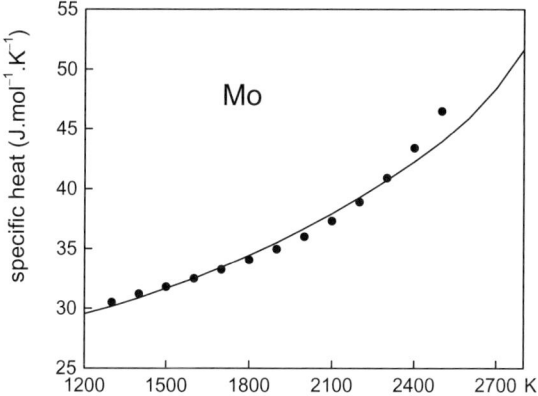

Fig. 12.4. Specific heat of molybdenum: • modulation measurements (Kraftmakher 1964), —— recommended values (Sabbah et al. 1999)

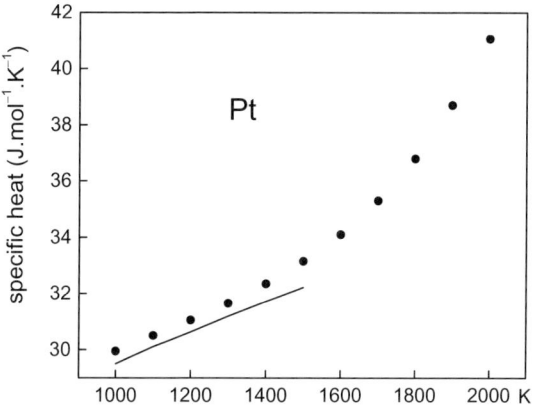

Fig. 12.5. Specific heat of platinum: • modulation calorimetry (Kraftmakher and Lanina 1965), —— recommended values (Sabbah et al. 1999)

12.2.3 Thermal Expansivity of Tungsten and Platinum

Modulation measurements immediately provide values of linear thermal expansivity. The measurements on tungsten were performed in the range 2000–2900 K (Kraftmakher 1972). The results depend on the accepted temperature dependence of the electrical resistance and the specific heat of tungsten. The numerical experimental values of the thermal expansivity were given in a review paper (Kraftmakher 1973b). The data are in good agreement with recommended values (Fig. 12.6) based mainly on measurements by Miiller and Cezairliyan (1990) at the National Bureau of Standards. In all the range of the modulation measurements, the difference does not exceed 4%.

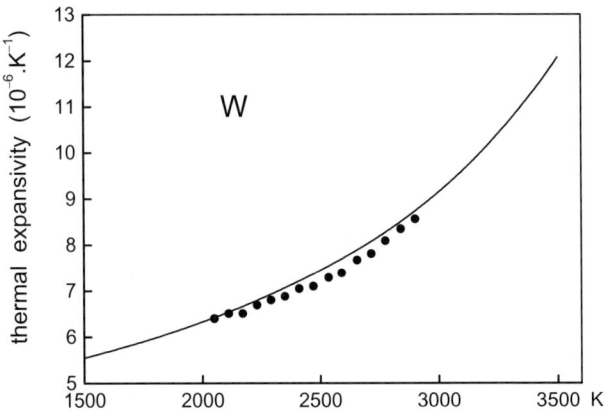

Fig. 12.6. Thermal expansivity of tungsten: • modulation measurements (Kraftmakher 1972), —— recommended values (White and Minges 1997). The strong non-linear increase in the thermal expansivity observed by modulation measurements was confirmed by dynamic measurements at NBS (Miiller and Cezairliyan 1990)

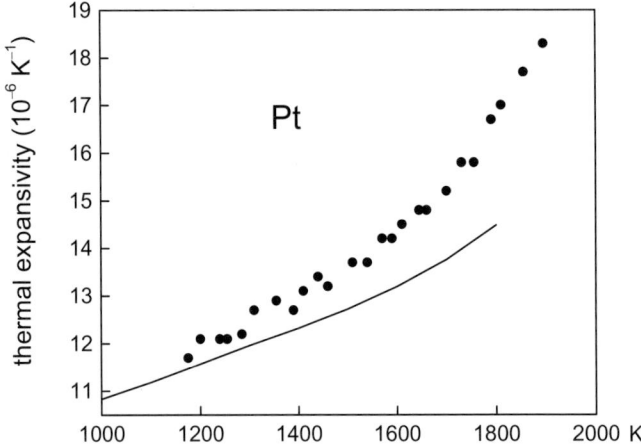

Fig. 12.7. Thermal expansivity of platinum: • modulation measurements (Kraftmakher 1967b), —— recommended values (Kirby 1991)

The results of modulation measurements of the thermal expansivity of platinum can be compared with data recommended by Kirby (1991). Five previous investigations, which used precise experimental techniques, were the base for establishing analytical expressions for the thermal expansivity of platinum from absolute zero to 1800 K. The modulation measurements were carried out in the range 1200 to 1900 K (Kraftmakher 1967b). There exists a significant discrepancy between the data (Fig. 12.7), which grows with increasing temperature, from 2% at 1200 K to about 15% at 1800 K.

12.2.4 Electrical Resistivity of Platinum

Platinum is widely used as a material for resistance thermometers (platinum thermometer). It was known for a long time that the temperature dependence of the electrical resistance of platinum, in a wide temperature range except low temperatures, obeys a quadratic relation $R = A + BT + CT^2$, where $C < 0$. This dependence was reliably established. However, an additional contribution arises at high temperatures due to vacancy formation in the crystal lattice. This fact was seen from the measurements of the specific heat of platinum (Kraftmakher and Lanina 1965). Later, the temperature derivative of the electrical resistance (not the resistivity) was measured directly with the modulation technique (Kraftmakher and Sushakova 1974). From these measurements, the deviation from the above equation became evident. To compare the results of the modulation measurements with the recommended resistivity values, it is necessary to correct the data for the thermal expansion. Clearly,

$$(1/\rho_{273})d\rho/dT = (1/R_{273})dR/dT \times l/l_{273} + (R/R_{273}) \times \alpha, \qquad (12.1)$$

where ρ is the electrical resistivity, R is the resistance of the sample, l/l_{273} corresponds to the thermal expansion of the sample, and α is the linear thermal expansivity of platinum.

The correction increases with temperature and amounts to about 5% at 1850 K. The corrected values appeared to be in good agreement with the recommended data (Fig. 12.8).

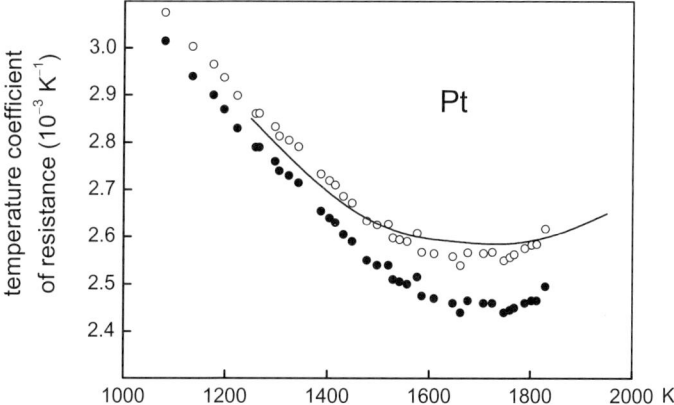

Fig. 12.8. Temperature coefficient of resistivity of platinum: ● original experimental points, uncorrected for thermal expansion (Kraftmakher and Sushakova 1974), ○ experimental points corrected for thermal expansion, —— recommended values (White and Minges 1997). The linear temperature dependence below 1500 K corresponds to the commonly accepted quadrature temperature dependence of resistance

13 Studies at High Temperatures

This chapter considers some topics related to the activity of the author, namely: (i) equilibrium point defects in metals, (ii) direct determination of the temperature derivative of specific heat, (iii) an unexpected premelting anomaly in specific heat, and (iv) observations of temperature fluctuations and determination of the ratio of isobaric and isochoric specific heats of solids.

13.1 Equilibrium Point Defects in Metals

Many years ago, Frenkel (1926) predicted the formation of point defects in solids. Some atoms in a crystal lattice acquire energies much larger than the mean energy. At high temperatures, the energy of such atoms may become sufficient to leave their regular sites in the lattice and occupy interstitial positions. A vacancy and an interstitial thus appear simultaneously, the so-called Frenkel pair (Fig. 13.1). Later, Wagner and Schottky (1930) pointed out how vacancies may arise without formation of interstitials: atoms leaving their lattice sites occupy positions at the surface or at internal imperfections of the crystal. Such vacancies are often called Schottky defects.

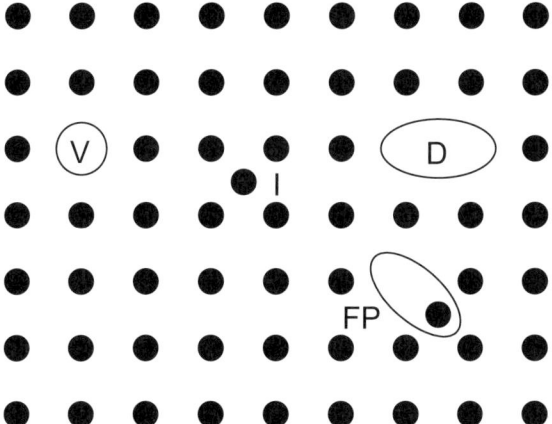

Fig. 13.1. Point defects in crystal lattice. V – vacancy, I – interstitial atom, D – divacancy, FP – Frenkel pair. The equilibrium defect concentrations are still under debate

The equilibrium vacancy concentration obeys the relation

$$c_v = \exp(-G_F/k_B T) = \exp(S_F/k_B)\exp(-H_F/k_B T) = A\exp(-H_F/k_B T), \qquad (13.1)$$

where G_F is the Gibbs free energy of vacancy formation, H_F and S_F are the corresponding enthalpy and entropy, k_B is the Boltzmann constant, and T is the absolute temperature. The entropy S_F does not include the configurational entropy, but relates to changes in vibration frequencies caused by softening of the atom binding near vacancies. Frenkel considered melting of solids as a result of softness of the lattice caused by point defects. He concluded that equilibrium concentrations of point defects may reach values of the order of 1%. The first experimental observations of vacancies in metals were made in the 1950s when an extra resistivity of quenched samples was discovered. Annealing at appropriate temperatures recovers the resistivity. Defect-induced resistivity was also found in measurements at high temperatures. During five decades, many experimental and theoretical investigations of point defects in metals have been carried out. Point defects strongly influence many physical properties of metals at high temperatures: mechanical properties, enthalpy, specific heat, thermal expansion, electrical resistivity, and positron annihilation. Many data confirm that the predominant equilibrium point defects in metals are vacancies.

The vacancy mechanism is the most probable one for self-diffusion. Point defects may appear also under nonequilibrium conditions, due to deformation or irradiation. Various theoretical calculations of the formation enthalpies of point defects are in reasonable agreement. However, equilibrium concentrations of the defects depend also on the formation entropies, theoretical predictions for which are ambiguous. Methods of studying point defects in metals may be divided into three groups:

- Studies of equilibrium defects through physical properties of metals at high temperatures. Results of such measurements do not depend on the history of the samples and are in satisfactory agreement. However, properties of a hypothetical defect-free crystal are unknown and cannot be calculated precisely. It is therefore impossible to reliably separate the point-defect contributions.
- Studies of samples, in which extra concentrations of point defects were created by quenching, deformation, or irradiation. The main disadvantage of the equilibrium measurements is completely excluded here because the samples under study are compared with well-annealed samples, defect concentrations in which are negligible. Unfortunately, it is difficult to reveal the equilibrium defect concentrations. During quenching, many vacancies have time to annihilate or form clusters. Therefore, vacancy concentrations after quenching are much smaller than equilibrium ones. This discrepancy grows with approaching melting points.
- Observations of relaxation phenomena caused by the point-defect equilibration: properties of samples at high temperatures are studied under such rapid temperature changes that the defect concentration cannot follow them. In this case, one measures properties corresponding to a defect-free crystal. This statement relates only to properties depending on changes in the defect concentra-

tions during the measurements: specific heat, thermal expansivity, and temperature derivative of electrical resistance. Such measurements are capable of unambiguous separation of the point-defect contributions to physical properties (Sect. 15.3).

It is commonly accepted now that equilibrium point defects are to be studied under equilibrium conditions. Criteria for the choice of a suitable physical property for such experiments are quite clear: (i) the magnitude of the defect contribution and reliability of separating it, (ii) the accuracy of the measurements, and (iii) knowledge of parameters entering relations between the contributions and concentrations of the defects. The reasons that the most suitable property is the specific heat are as follows:

- From theoretical calculations, high-temperature specific heat of a defect-free crystal depends weakly on temperature.
- As a rule, the defect contributions are much larger than the inaccuracy of the measurements of specific heat.
- The contribution of the defects to specific heat directly relates to their equilibrium concentration.
- Measurements of the specific heat under rapid temperature oscillations, necessary to separate the defect contributions, are simpler than corresponding measurements of other properties.

Modulation techniques appeared to be very useful for studying equilibrium point defects in metals (Kraftmakher and Strelkov 1966b, 1970; Kraftmakher 1971b, 1974, 1977, 1994a, 1996, 1997, 1998, 2000, 2001ab).

13.1.1 Specific Heat

In the 1960s, specific-heat data obtained by the equivalent-impedance technique were used to evaluate the enthalpies and entropies of vacancy formation in refractory metals. There exists a more or less extended temperature range, in which the specific heat of metals increases linearly with temperature. By extrapolating this dependence, the non-linear contribution was separated and attributed to vacancy formation (Kraftmakher and Strelkov 1962; Kraftmakher 1963ab, 1964, 1966c). Now the non-linear increase in the high-temperature specific heat of metals is evident (Fig. 13.2). This phenomenon was observed by all calorimetric techniques known at present. The excess molar enthalpy and specific heat caused by the vacancy formation are

$$\Delta H = NH_F c_v = NH_F A \exp(-H_F/k_B T), \qquad (13.2a)$$

$$\Delta C = (NH_F^2 A/k_B T^2) \exp(-H_F/k_B T), \qquad (13.2b)$$

where N is the Avogadro number.

A plot of $\ln T^2 \Delta C$ versus $1/T$ should be a straight line with a slope $-H_F/k_B$. After determination of the formation enthalpy, the coefficient A and the equilibrium vacancy concentration become available.

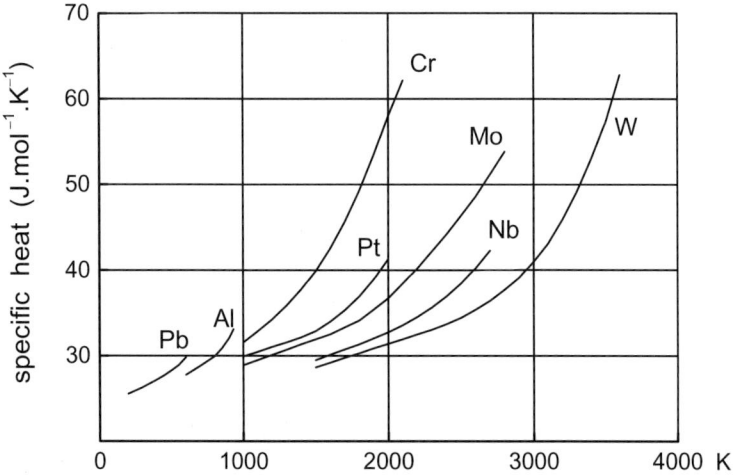

Fig. 13.2. Specific heat of some metals. Pb, Al – adiabatic calorimetry (Kramer and Nölting 1972); W, Pt – modulation calorimetry (Kraftmakher and Strelkov 1962; Kraftmakher and Lanina 1965); Cr – drop method (Kirillin et al. 1967); Mo, Nb – dynamic calorimetry (Cezairliyan et al. 1970; Righini et al. 1985). The difference between low- and high-melting-point metals is clearly seen: the latter manifest much larger non-linear increase of specific heat. However, the origin of the non-linear contribution to high-temperature specific heat of metals remains a disputable question

Table 13.1. High-temperature specific heat of metals from modulation measurements. The approximation takes into account the vacancy formation:
$C = a + bT + cT^{-2}\exp(-d/T)$ J.mol^{-1}.K^{-1}

Metal	a	$b \times 10^5$	$c \times 10^{-10}$	$d \times 10^{-3}$	Reference
W	20.5	545	740	36.5	Kraftmakher and Strelkov 1962
Ta	24.35	285	220	33.6	Kraftmakher 1963a
Mo	22.05	650	170	26	Kraftmakher 1964
Nb	20.7	525	30	23.7	Kraftmakher 1963b
Rh	22.5	91	80	22.05	Glazkov 1988
Zr	20.5	420	35	20.3	Kanel' and Kraftmakher 1966
Pt	24.5	545	25	18.6	Kraftmakher and Lanina 1965
Ti	24.05	440	46	18	Shestopal 1965
Ni	27.2	530	50	16.2	Glazkov 1987
Cu	23.65	500	5	13.2	Kraftmakher 1967c
Au	23.85	525	2.5	11.6	Kraftmakher and Strelkov 1966a
La	24.7	650	37	11.6	Akimov and Kraftmakher 1970

A more rigorous determination of the vacancy parameters consists in fitting the experimental data by a polynomial taking into account the vacancy formation:

$$C = a + bT + cT^{-2}\exp(-H_F/k_BT). \tag{13.3}$$

The coefficients of this equation are obtainable by the least-squares method. All possible values of H_F are tried, and for each value the standard deviation of experimental points is determined. A plot of the standard deviation versus the assumed value of H_F shows the most probable formation enthalpy and its uncertainty. For tungsten, tantalum, molybdenum, and niobium, the parameters of equilibrium vacancies were first deduced from the specific-heat data.

Table 13.1 presents polynomial fits for the high-temperature specific heats of metals obtained from modulation measurements.

13.1.2 Thermal Expansion

The vacancy formation causes an increase in the volume of the sample. After creation of a vacancy, relaxation of the lattice occurs. The vacancy volume is thus smaller than the atomic volume Ω and equals $\gamma\Omega$ ($\gamma < 1$). In an isotropic case, the changes in the volume, $\Delta V/V$, and in linear thermal expansivity of the sample, $\Delta\alpha$, are as follows:

$$\Delta V/V = \gamma c_v = \gamma A \exp(-H_F/k_BT), \tag{13.4a}$$

$$\Delta\alpha = (\gamma H_F A/3k_BT^2)\exp(-H_F/k_BT). \tag{13.4b}$$

The relative increase in linear expansivity, $\Delta\alpha/\alpha$, is much larger than that in the volume, $\Delta V/V$. Direct measurements of the expansivity are therefore preferable. The results of such measurements (Kraftmakher 1967b, 1972) are in reasonable agreement with Eq. (13.4b). A strong non-linear increase in the thermal expansivity of high-melting-point metals was observed by various techniques and now causes no doubts (Fig. 13.3). However, the derived vacancy concentrations appeared to be somewhat smaller than those from the specific-heat data. Besides an uncertainty in γ values (they were assumed to be 0.5), there exists another reason for this disagreement. The main sources and sinks for vacancies are internal imperfections in the crystal lattice (voids, grain boundaries, dislocations and, probably, vacancy clusters). Therefore, vacancy formation may partly occur without an increase of the outer volume of the sample.

Differential dilatometry is considered the most reliable method for determining equilibrium vacancy concentrations. This technique consists of simultaneously measuring the macroscopic dilatation, $\Delta l/l$, and the relative change in the lattice parameter, $\Delta a/a$, versus temperature. These quantities coincide at low temperatures. Due to vacancy formation, a difference arises between the two quantities. This difference grows rapidly with temperature. Under certain assumptions, the following relation should be valid in an isotropic case:

$$c_v - c_i = 3(\Delta l/l - \Delta a/a), \tag{13.5}$$

where c_i is the equilibrium concentration of interstitials, which is expected to be much smaller than the equilibrium vacancy concentration c_v.

This expression was used to determine vacancy concentrations in many metals. In all the cases, low concentrations were obtained, smaller than 0.1% at the melting points. However, such measurements have not been performed on high-melting-point metals. There are only few data on the lattice parameter of these metals at high temperatures, and further measurements are necessary to clarify the situation.

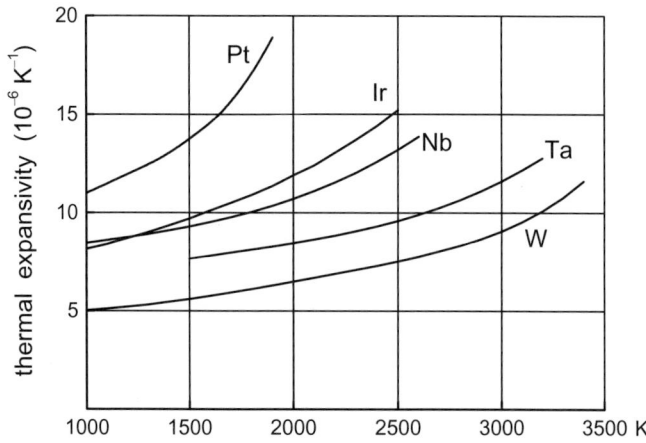

Fig. 13.3. Linear thermal expansivity of high-melting-point metals. Pt – modulation method (Kraftmakher 1967b); Ir – traditional dilatometry (Halvorson and Wimber 1972); Nb, Ta – dynamic technique (Righini et al. 1986b; Miiller and Cezairliyan 1982); W – recommended values (White and Minges 1997)

13.1.3 Electrical Resistivity

Point defects cause an additional scattering of conduction electrons. The extra resistivity due to vacancies, $\Delta\rho$, is proportional to their concentration:

$$\Delta\rho = \rho_v c_v = \rho_v A \exp(-H_F/k_B T), \tag{13.6}$$

where ρ_v is a proportionality coefficient.

The influence of point defects on resistivity is similar to that of impurities. As a rule, the values of ρ_v for various metals range from 1 to 10 μΩ.cm per 1% vac. The influence of interstitials is several times larger. However, significant deviations from Matthiessen's rule were observed, i.e., the coefficient ρ_v depends on temperature. This must be taken into account when comparing extra resistivities from equilibrium and quenching experiments.

The vacancy-induced extra resistivity is relatively small. Its separation therefore strongly depends on extrapolation of the data from medium temperatures. The situation becomes more favourable when one directly measures the temperature derivative of the resistivity. The increase in this derivative due to the vacancies is

$$\Delta(d\rho/dT) = (\rho_v H_F A/k_B T^2)\exp(-H_F/k_B T). \quad (13.7)$$

Modulation measurements of the temperature derivative of resistance were carried out on aluminium and platinum (Kraftmakher and Sushakova 1972, 1974). At medium temperatures, the derivative changes linearly with temperature, in accordance with the quadratic dependence of the resistance. A significant non-linear increase is seen at high temperatures, from which it is easy to deduce the vacancy contribution (Fig. 13.4). For both metals, plausible formation enthalpies were obtained. The extra resistivities are close to the results of some quenching experiments. For the two metals no contradiction thus exists between equilibrium and quenching data, despite the difficulties mentioned above.

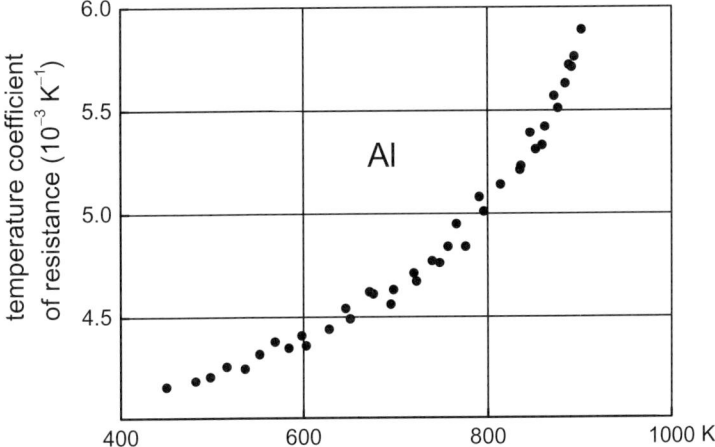

Fig. 13.4. Temperature coefficient of resistance of aluminium. The non-linear increase was attributed to vacancy formation (Kraftmakher and Sushakova 1972)

13.1.4 Equilibrium Concentrations of Point Defects

Surprisingly, until today we have no reliable knowledge of equilibrium point-defect concentrations in metals. The question is very important, especially for high-temperature applications. Two opposing viewpoints on equilibrium point defects in metals are as follows:

• Defect contributions to physical properties of metals at high temperatures are small and cannot be reliably separated from the effects of anharmonicity. The methods suitable for studying equilibrium vacancies are positron-annihilation spectroscopy, which provides the enthalpies of vacancy formation, and differential dilatometry, which provides the equilibrium vacancy concentrations. The equilibrium vacancy concentrations at melting points range from 10^{-4} to 10^{-3}.

Reasonable values of the formation enthalpies deduced from the non-linear increase in high-temperature specific heat of metals are accidental, and the derived defect concentrations are improbably large, so that this approach is generally erroneous.

• In many cases, the defect contributions to specific heat of metals are much larger than the non-linear effects of anharmonicity and can be separated without crucial errors. This approach is quite adequate for determination of the defect parameters, especially, of equilibrium defect concentrations. These concentrations at melting points are of the order of 10^{-3} in low-melting-point metals and of 10^{-2} in high-melting-point metals. The strong non-linear effects in high-temperature specific heat and thermal expansivity of metals are caused by the point-defect formation. Examination of these effects rules out anharmonicity as a possible origin of this phenomenon. Important arguments supporting this viewpoint have appeared in the last decades. It may turn out that just calorimetric determinations provide the most reliable data on equilibrium vacancy concentrations in metals.

The equilibrium vacancy concentrations from differential dilatometry and from calorimetric measurements are in strong contradiction (Fig. 13.5). Parameters of vacancy formation in metals derived from modulation measurements of specific heat are given in Table 13.2.

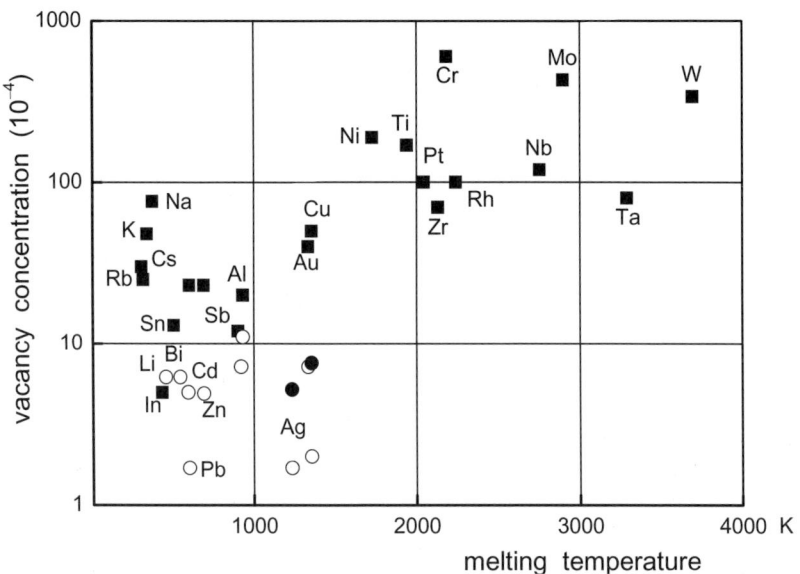

Fig. 13.5. Equilibrium vacancy concentrations in metals at melting points obtained from calorimetric measurements (■) and from differential dilatometry (○). The concentrations are of the order of 10^{-3} in low-melting-point metals and of 10^{-2} in high-melting-point metals. New differential-dilatometry data (●) on copper and silver (Kluin and Hehenkamp 1991; Mosig et al. 1992) are several times larger than those accepted earlier

Table 13.2. Parameters of equilibrium point defects in metals derived from modulation measurements of their specific heat

Metal	H_F [eV]	S_F/k_B	c_{mp} [%]	Reference
W	3.15	6.5	3.4	Kraftmakher and Strelkov 1962
Ta	2.9	5.45	0.8	Kraftmakher 1963a
Mo	2.24	5.7	4.3	Kraftmakher 1964
Nb	2.04	4.15	1.2	Kraftmakher 1963b
Rh	1.9	5.25	1.0	Glazkov 1988
Zr	1.75	4.6	0.7	Kanel' and Kraftmakher 1966
Pt	1.6	4.5	1.0	Kraftmakher and Lanina 1965
Ti	1.55	5.15	1.7	Shestopal 1965
Ni	1.4	5.4	1.9	Glazkov 1987
Cu	1.05	3.7	0.5	Kraftmakher 1967c
Au	1.0	3.15	0.4	Kraftmakher and Strelkov 1966a
La	1.0	5.8	1.2*	Akimov and Kraftmakher 1970

*at the phase-transition point

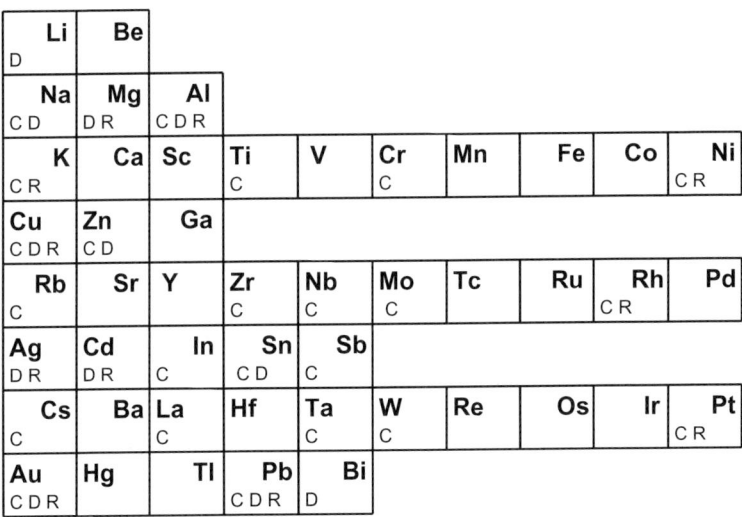

Fig. 13.6. Equilibrium point-defect concentrations are determined for only a half of metals. Techniques providing such data are indicated as follows: *C* – calorimetry, *D* – differential dilatometry, *R* – resistometry

Until now, the question remains under debate. Determinations of equilibrium point-defect concentrations from non-linear contributions to high-temperature specific heat are supported by many considerations, namely:

● There exists more or less wide temperature range, in which specific heat linearly depends on temperature. A linear extrapolation takes into account all linear contributions to specific heat, including linear anharmonic and electronic terms.

- Other non-linear contributions to specific heat (anharmonic or electronic) must be smaller than the linear terms.
- The temperature dependence of the extra specific heat well corresponds to calculations assuming point-defect formation, and reasonable values of the formation enthalpies are obtained.
- The non-linear effects in specific heat correlate with those in thermal expansivity.
- In aluminium and platinum, the specific-heat data do not contradict the extra electrical resistivity.
- Relaxation phenomenon in the specific heat of tungsten and platinum is in agreement with the non-linear part of their specific heat (Sect. 15.3.2).
- Results of rapid-heating determinations of the enthalpy of tungsten at the melting point well correspond to the defect contribution calculated from the non-linear increase in the specific heat (Sect. 15.3.5).

Equilibrium point-defect concentrations are determined for only a half of metals (Fig. 13.6).

13.2 Temperature Coefficient of Specific Heat

The direct determination of the temperature coefficient of specific heat (TCSH) rests on a strictly sine modulation of the heating power. The temperature dependence of specific heat brings about a deviation of the temperature oscillations from the sine form. A second-harmonic component in the temperature oscillations appears, which depends on TCSH (Kraftmakher and Tonaevskii 1972). The heat-balance equation, with the temperature dependence of the specific heat being taken into account, is

$$mc(1 + \alpha\Theta)\Theta' + Q + Q'\Theta + Q''\Theta^{'2}/2 = p_0 + p\cos\omega t. \tag{13.8}$$

Here $\alpha = (1/c)dc/dT$ is the TCSH, and Q, Q', and Q'' are the power of heat losses from the sample and its temperature derivatives. Since $\alpha \ll 1$, this non-linear equation is solvable by successive approximations. With high-order terms neglected, the solution is

$$\Theta = \Theta_1\cos(\omega t - \phi_1) + \Theta_2\cos(2\omega t - \phi_2), \tag{13.9a}$$

$$\Theta_1 = (p/mc\omega)\sin\phi_1, \tag{13.9b}$$

$$\Theta_2 = \alpha\Theta_1^2/4\cos\phi_2, \tag{13.9c}$$

$$\tan\phi_1 = mc\omega/Q', \tag{13.9d}$$

$$\tan\phi_2 = Q''/2mc\omega\alpha. \tag{13.9e}$$

At sufficiently high modulation frequencies, $\phi_1 \cong 90°$, $\phi_2 \cong 0$, and the above expressions become simpler:

$$\Theta_1 = p/mc\omega, \tag{13.10a}$$

$$\Theta_2 = \alpha\Theta_1^2/4. \tag{13.10b}$$

The TCSH is thus available from the fundamental component and the second harmonic of the temperature oscillations. Above the Debye temperature, α is usually of the order of 10^{-4} K^{-1}. It may increase several times due to the point-defect contribution. Close to a second-order phase transition, α may appear to be in the range 10^{-2}–10^{-1} K^{-1}, and one may succeed in measuring the second harmonic even with temperature oscillations smaller than 1 K. In all such cases, a high selectivity is necessary to measure the second harmonic in the presence of a much stronger fundamental signal.

A strictly sine modulation of the heating power was supposed here but it is difficult to attain such a mode in practice. It is convenient to heat wire samples by an AC current. When the internal resistance of the AC source is sufficiently high for all harmonics, no changes of the current are caused by the oscillations of the resistance of the sample. However, a second harmonic in the heating-power oscillations arises because the current flows through the oscillating resistance. The right side of Eq. (13.8) should be written as $I^2R(1 + \beta\Theta)\cos^2(\omega t/2)$, where I is the amplitude of the heating current, R is the resistance of the sample, $\omega/2$ denotes the frequency of the current, and $\beta = R'/R$ is the temperature coefficient of the resistance. Since $\beta \ll 1$, one obtains, instead of (13.10b),

$$\Theta_2 = \Theta_1^2(\alpha - \beta)/4. \tag{13.11}$$

Of great importance is the method of measuring the temperature oscillations. If they are detected through a parameter non-linearly depending on temperature, the results may be markedly distorted. For example, one should not measure the temperature oscillations through the radiation from the sample. It is better to use the temperature dependence of its resistance. To calculate the necessary correction, the increment in the resistance should be expanded in a series with the first two terms retained:

$$\Delta R = R\beta\Theta + R\beta'\Theta^2/2, \tag{13.12}$$

where $\beta' = d\beta/dT$. The temperature oscillations at the fundamental frequency result therefore in the resistance oscillations also at the doubled frequency. Finally,

$$\Theta_2 = \Theta_1^2(\alpha - \beta - \beta'/\beta)/4. \tag{13.13}$$

Hence, when heating a sample by an AC current and measuring the temperature oscillations through its electrical resistance, Eq. (13.13) is no longer as simple as Eq. (13.10b). However, such a method is quite acceptable in many cases. As an example, the quantities entering Eq. (13.13) were calculated for tungsten. When approaching the melting point, the main part is played by α due to the strong non-linear increase in the specific heat, while β prevails at low and medium temperatures. The correction associated with β′/β remains relatively small.

A set-up with two gas-filled incandescent lamps with tungsten filaments was used to test the method. The lamps formed a part of a bridge fed by 50-Hz mains current (Fig. 13.7). This frequency is sufficiently high to ensure the validity of Eqs. (13.10a) and (13.10b). Since the expected Θ_2/Θ_1 ratio is 10^{-3} to 10^{-2}, the content of harmonics in the heating current must be small. The harmonics of the temperature oscillations are measured with the frequency-conversion method. A small supplementary current from a low-frequency oscillator passes through the samples. The frequency of the supplementary current is close to that of the harmonic to be measured. This results in the appearance of a voltage across the sample at the difference frequency, of about 0.1 Hz. This voltage is proportional to the magnitude of the oscillations in the resistance of the sample and to the supplementary current. It is fed through an RC filter to an amplifier and then recorded. By tuning the frequencies of the supplementary current close to ω and 2ω, both components of the temperature oscillations are measured in turn.

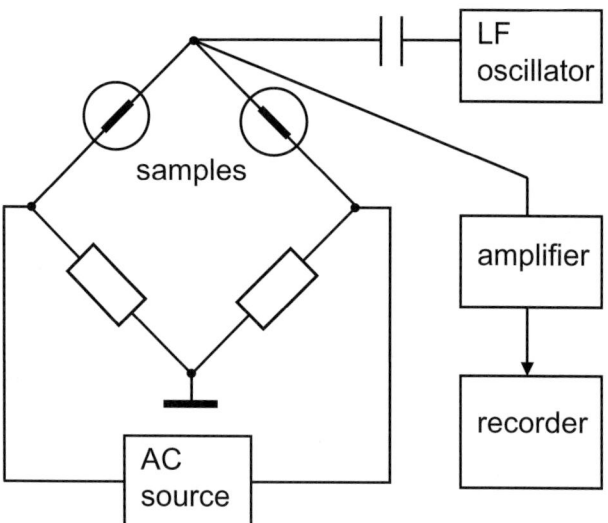

Fig. 13.7. Arrangement for direct measurement of the temperature coefficient of specific heat (Kraftmakher and Tonaevskii 1972). The harmonics of the temperature oscillations are measured with the frequency-conversion method

The method is capable of measuring the second harmonic in the presence of a much stronger signal of fundamental frequency (Fig. 13.8). The strong increase of TCSH corresponds well to the non-linear increase of the specific heat. It is not evident whether this method is applicable for studying second-order phase transitions. In this case, the temperature oscillations must be kept small to ensure good temperature resolution. A thermocouple is the most proper tool to measure the temperature oscillations in such studies.

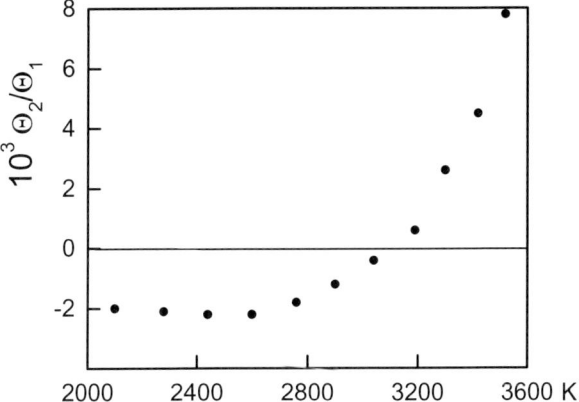

Fig. 13.8. Temperature dependence of Θ_2/Θ_1 ratio for tungsten. The strong increase above 2800 K reflects the vacancy formation. Negative values below 3200 K show predominant contribution of β (Kraftmakher and Tonaevskii 1972)

13.3 Unexpected Premelting Anomaly

An unexpected anomaly was observed in the specific heat of platinum near its melting point (Kraftmakher 1978). Wire samples, 0.02–0.1 mm in diameter, were heated in vacuum by passing through them a DC current with a small AC component. The light from a small central portion of the sample was projected onto a photosensor, a photomultiplier or a photodiode. The modulation frequency was in the range 20–2000 Hz, and the temperature oscillations ranged from 0.1 to 1 K. Just before the destruction of the samples, the oscillations in the radiation from the sample rapidly increased. Usually, the increase was 20–30 times but in several samples it reached 50 times. To reduce the amplitude of the oscillations, it was necessary to decrease the AC component of the heating current. Under a gradual increase of the heating current, after reaching the point C (Fig. 13.9) and even at somewhat higher temperatures, it was possible to return to the point A and repeat such cycles several times. The destruction of the samples occurred in the region CD.

The phenomenon was observable only in a central portion of the samples, shorter than 1 mm. It was impossible to directly determine the temperature region of the anomaly, so that only rough estimations could be made. The estimates correspond to approximately 10 K. After the first observations, the phenomenon was confirmed when platinum samples were heated by electron bombardment (Fridman 1983). Further details were reported later (Kraftmakher 1991). Unsuccessful attempts were undertaken to find an anomaly in the electrical resistivity of the samples. The resistance of the entire sample was measured, so that a contribution from a small portion of the sample could not be detected. However, by measuring the oscillations of the radiation from the entire sample the anomaly was still observable: the oscillations exhibited an increase by 1.5–2 times. The increase of the oscillations thus cannot be a simple result of a local increase of the resistivity.

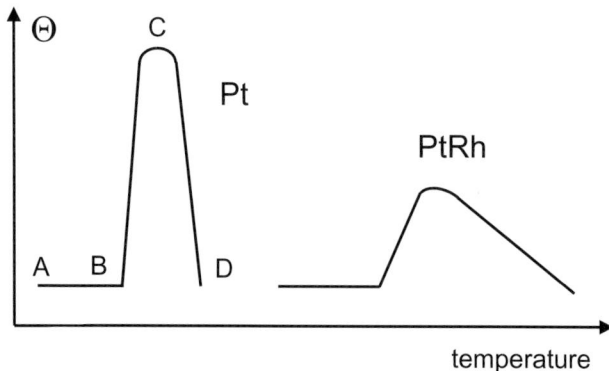

Fig. 13.9. Unexpected premelting anomaly in specific heat of platinum and PtRh alloy, schematically (Kraftmakher 1991). The phenomenon is still waiting for confirmation and explanation

A possible origin of the phenomenon might be an anomaly in the emittance of the samples. To check this conjecture, temperature oscillations of two frequencies, low and infra low, were created in the samples simultaneously. It turned out that in the region where the low-frequency oscillations of the radiation rapidly increase, the infra low oscillations (0.1 and 0.2 Hz) exhibit no anomaly. In addition, a change in the phase of the temperature oscillations was observed at 20 Hz. This confirms that the phenomenon is caused by an anomaly in the specific heat rather than in the emittance of the sample. The results described above were obtained on samples 20 mm long. To lengthen the region of the anomaly, longer samples were also investigated. However, in this case the destruction of the samples occurred earlier, in the region BC. Probably, the equilibrium melting corresponds to the point B, while the anomaly relates to a superheated (metastable) state. Such a conclusion was made because the specific heat in the anomalous region is too small to be attributed to any equilibrium solid or liquid state. The anomaly was also detected by the equivalent-impedance technique, which provides the ratio of the heat capacity of the sample to the temperature derivative of its resistance. An anomalous behaviour of this quantity was observed just before the destruction of the samples. However, such measurements relate to the entire sample, while the anomaly relates to a small portion of it.

The strong temperature dependence of the specific heat in the anomalous region causes significant distortions of the temperature oscillations. The distortions were observable directly by an oscilloscope and by detecting the second-harmonic signal. Probably, Bezemer and Jongerius (1976) observed just this phenomenon when determining the melting point of platinum. The authors attributed the distortions to changes in the emittance of the sample during melting and did not consider the specific heat.

In PtRh alloys (10% and 30% Rh), the anomaly is more moderate and its temperature region seems to be wider. The phenomenon was observed also in nickel and palladium, whereas no anomaly was seen in tungsten, tantalum, and niobium.

13.4 Isochoric Specific Heat of Solids

Observations of temperature fluctuations under equilibrium conditions offer unique opportunity to determine the isochoric specific heat of solids. Such observations were carried out on thin tungsten wires (Kraftmakher and Krylov 1980ab). Modulation calorimetry allowed us to circumvent the main difficulty of the measurements, the necessity to determine absolute values of the temperature fluctuations. With the method proposed, the ratio of the isobaric and isochoric specific heats is available.

13.4.1 Temperature Fluctuations

The mean square of the temperature fluctuations in a sample, $<\Delta T^2>$, and their spectral density, $<\Delta T_f^2>$, are given by

$$<\Delta T^2> = k_B T^2/mC_v, \qquad (13.14)$$

$$<\Delta T_f^2> = 4k_B T^2/Q'(1+x^2), \qquad (13.15)$$

where k_B is Boltzmann's constant, Q' is the heat transfer coefficient, $x = mC_v\omega/Q'$, and m and C_v are the mass and isochoric specific heat of the sample, respectively (Milatz and Van der Velden 1943; Landau and Lifshitz 1980).

It is very important that the isochoric specific heat C_v, rather than the isobaric specific heat C_p, enters the above expressions. The reason is that the temperature fluctuations and fluctuations of the volume of the sample are uncorrelated. In contrast to liquids, it is impossible to directly measure the isochoric specific heat of solids. Measurements of the temperature fluctuations provide such opportunity. The main difficulty is the smallness of the temperature fluctuations, even in very small samples. To observe markedly different values of the two specific heats, C_p and C_v, the measurements are to be performed at high temperatures.

To verify the theory, Chui et al. (1992) observed temperature fluctuations at low temperatures. The authors pointed out that the central point of the problem is the applicability of the relation between fluctuations of two fundamental quantities in thermodynamics, the energy U and the temperature T. One point of view consists in accepting the relation $\Delta U = C\Delta T$, where C is the heat capacity of a subsystem thermally connected to a reservoir. Energy exchanges between them lead to the above formulas for the temperature fluctuations. Another approach rests on the statement that the temperature of a canonical ensemble is constant and does not fluctuate. Temperature fluctuations in a microcanonical ensemble are possible but in this case $<\Delta T^2> = k_B T^2/MC$, where MC is the total heat capacity of the system and the reservoir. Considering the problem, Kittel (1988) and Mandelbrot (1989) presented two opposite conclusions.

The experiment (Chui et al. 1992) was aimed at a choice between the two theories. The authors measured the temperature-dependent magnetisation of a paramagnetic salt, copper ammonium bromide, in a fixed magnetic field. A SQUID coupled to a superconducting coil wound around the salt pill detected the changes in the magnetisation. The temperature fluctuations were measured at temperatures near 2 K, above the Curie point of the salt (1.8 K). To exclude another origin of the phenomenon, fluctuations in two samples were measured simultaneously, and no correlation between them was found. The high sensitivity of the SQUID, amounted to $(5-14) \times 10^6$ V.K^{-1}, allowed the authors to measure the spectral densities of the temperature fluctuations of the order of 10^{-20} K^2.Hz^{-1}. The spectral density appeared to be in agreement with expectations based on the known heat capacity and the relaxation time of the samples. Below 0.1 Hz, the spectral density was about 10^{-19} K^2.Hz^{-1}. Estimates were made of the fluctuations in the magnetisation of the samples, which could be erroneously considered as being the temperature fluctuations. This effect causes an effective noise of about 2×10^{-23} K^2.Hz^{-1}, much smaller than the fluctuations observed. The authors concluded that the relation $\Delta U = C \Delta T$ is applicable to the fluctuations in U and T to an accuracy of 20%. The heat capacity of the reservoir was 1700 times larger than that of the sample, and the authors claimed that the results confirm the fluctuation theory leading to Eqs. (13.14) and (13.15). Later, Day et al. (1997) discussed the fluctuation-imposed limit for temperature measurements.

13.4.2 Experimental

At high temperatures, photoelectric sensors are the best tools to detect extremely small temperature oscillations and temperature fluctuations. A simple set-up was built for the measurements (Kraftmakher and Krylov 1980a). The signal caused by the temperature fluctuations was expected to be comparable with the inherent noise of a photosensor. To suppress this noise, a correlation method was used (Fig. 13.10). The set-up includes two identical channels, each consisting of a photodiode, a preamplifier, a selective amplifier, and a power amplifier. An electrodynamometer serves for accurately multiplying the signals fed into its movable coil and one of the fixed coils. The displacement of the movable coil is registered by means of an additional low-frequency current passing through it. A voltage induced in the second fixed coil depends on the orientation of the movable coil. This voltage is measured by a lock-in amplifier with a long time constant and then recorded. An averaging of the signal over several hours is therefore possible.

When detecting small temperature oscillations in wire samples, a low-frequency oscillator provided the modulation of the heating power. Only one amplification channel operated, while the oscillator supplied the second signal for the multiplier. In this case, the electrodynamometer operated as a lock-in detector. With this technique, temperature oscillations in the range 10^{-6}–10^{-5} K are measurable (Fig. 13.11).

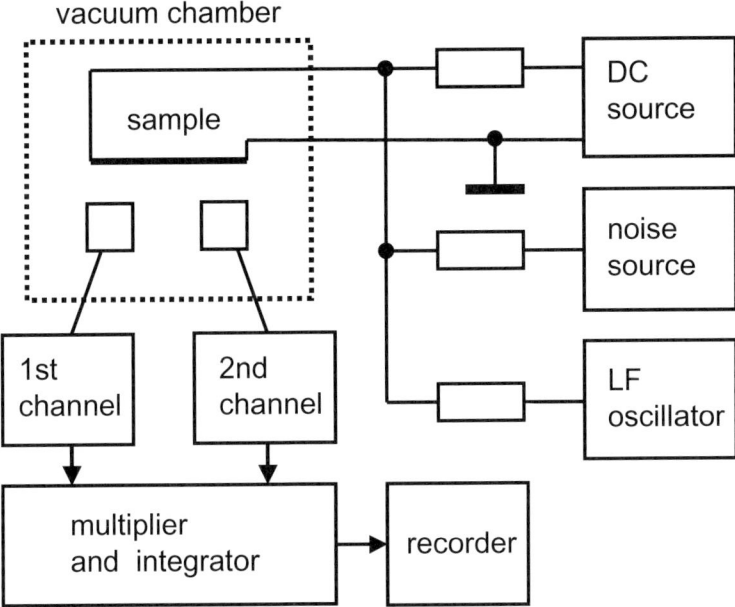

Fig. 13.10. Arrangement for measuring extremely small temperature oscillations and temperature fluctuations at high temperatures (Kraftmakher and Krylov 1980a). Nowadays, a computer-based technique might be used for obtaining more accurate data

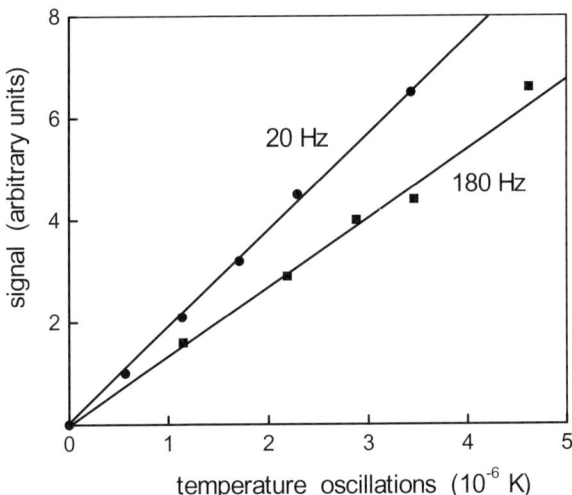

Fig. 13.11. Results of measurements of small temperature oscillations of two frequencies, 20 and 180 Hz, at high temperatures (Kraftmakher and Krylov 1980a). Temperature oscillations of the order of 10^{-6} K are measurable

13.4.3 Determination of the C_p/C_v Ratio

Probably, it were a hopeless venture to accurately measure all the quantities entering the Eqs. (13.14) and (13.15), which are necessary to deduce the isochoric specific heat. However, modulation calorimetry helps to solve the problem. When the same sample is subjected to a modulated heating power, the temperature oscillations in it are governed by the isobaric specific heat. This allows one to exclude all the quantities that cannot be accurately measured and to determine the specific-heat ratio C_p/C_v. To deal with spectral densities in both cases, a noise generator provides the modulation, and the temperature oscillations in the sample are of the same character as the temperature fluctuations. The only difference is that the oscillations are governed by the isobaric specific heat. The mean temperature of the sample is the same, and the measurements are carried out using the same measuring system. To determine the specific-heat ratio, it is sufficient to compare the frequency dependence of the spectral densities of temperature fluctuations and of the temperature oscillations caused by the noise modulation. The spectral density of the temperature oscillations in the sample is given by

$$<\Theta_f^2> = <p_f^2>/Q'^2(1+y^2), \qquad (13.16)$$

where $<p_f^2>$ is the spectral density of the square of the heating-power oscillations, and $y = mC_p\omega/Q'$.

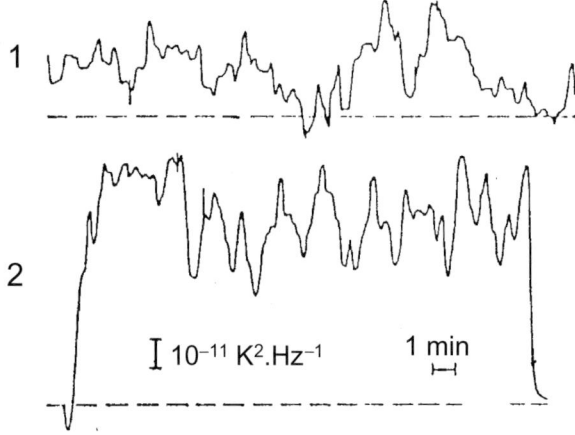

Fig. 13.12. Records of output voltage of the integrating system. 1 – temperature fluctuations in the sample without modulation of the heating power, 2 – temperature fluctuations plus oscillations caused by noise modulation of the heating power (Kraftmakher and Krylov 1980a)

The ratio of the spectral densities (13.15) and (13.16) may be presented in a form convenient for the determination of the specific heat ratio $\gamma = C_p/C_v$:

$$<\Delta T_f^2>/<\Theta_f^2> = A(1+z^2)/(1+z^2/\gamma^2), \qquad (13.17)$$

where $A = 4k_BQ'/\langle p_f^2 \rangle$ is constant at a given temperature, $z = f/f_0$, f_0 is the modulation frequency, at which the temperature oscillations become $\sqrt{2}$ times smaller than at zero frequency, and $\gamma = C_p/C_v$.

Temperature fluctuations were observed in thin tungsten wires, 3.5 μm thick and 1 mm long. An example of records of the output voltage of the integrator is reproduced here (Fig. 13.12). At 2200 and 2400 K, the spectra of the temperature fluctuations and of the temperature oscillations due to the noise modulation appeared to be in good agreement with the theoretical prediction (Fig. 13.13). Equation (13.17) makes it possible to determine the C_p/C_v ratio by the least-squares method, even without evaluation of the temperature oscillations, the mass, and the heat transfer coefficient of the sample. The only requirement is that the spectral density of the noise used for the modulation, $\langle p_f^2 \rangle$, is independent of the frequency.

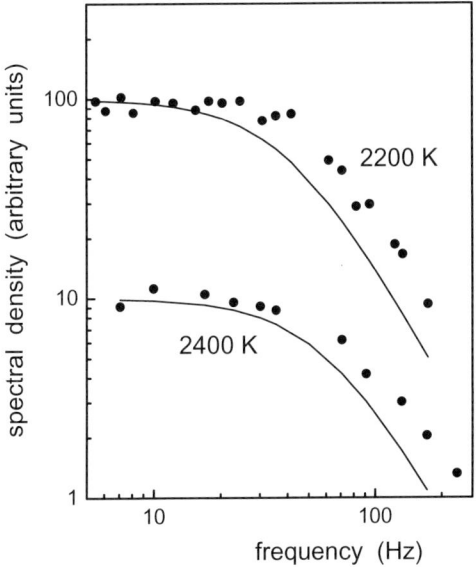

Fig. 13.13. Spectra of temperature fluctuations (●) and of temperature oscillations caused by noise modulation of heating power (——). After fit at low frequencies, the $\langle \Delta T_f^2 \rangle / \langle \Theta_f^2 \rangle$ ratio at higher frequencies shows the C_p/C_v ratio (Kraftmakher and Krylov 1980b)

Below 20 Hz, the spectral density of the temperature fluctuations equals approximately 3×10^{-11} K^2.Hz^{-1}, whereas at high frequencies it is inversely proportional to the frequency squared. The C_p/C_v ratio was found to be 1.4 ± 0.1 (Fig. 13.14). The isochoric specific heat at 2200 K is 24 ± 2.5 J.mol^{-1}.K^{-1}. The electronic contribution to the specific heat is available from low-temperature measurements (Waite et al. 1956), and it was believed to be linear with temperature. At 2200 K, the electronic contribution amounts to about 2.5 J.mol^{-1}.K^{-1}. The isochoric specific heat of the crystal lattice is thus somewhat smaller than the classical limit $3Nk_B = 24.9$ J.mol^{-1}.K^{-1}. This result does not contradict the theory because the lattice specific heat may decrease with increasing temperature due to a negative anharmonic contribution. Further determinations of the C_p/C_v ratio unquestionably deserve efforts. Nowadays, employment of a data-acquisition system and a computer could significantly simplify such measurements and provide more accurate data.

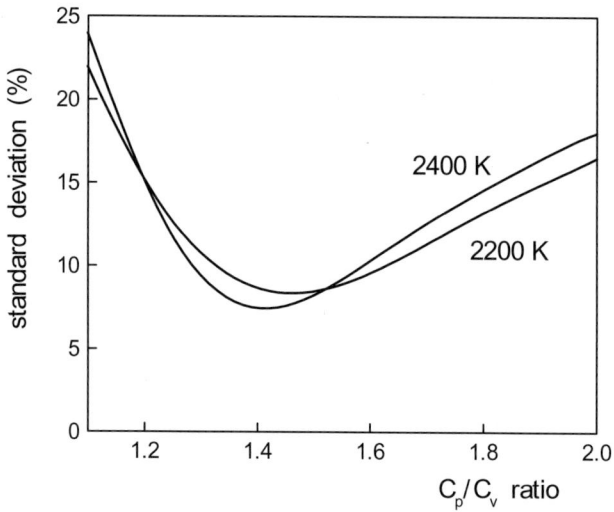

Fig. 13.14. Determination of C_p/C_v ratio for tungsten by the least-squares method. This ratio at 2200 and 2400 K was found to be 1.4 ± 0.1

14 Phase Transitions

In studies of phase transitions in solids, specific-heat measurements answer two important questions, namely: (i) is the transition either of the first or of the second order, and (ii) what is the temperature dependence of the specific heat near the transition point.

14.1 First- and Second-Order Phase Transitions

Usually, modulation calorimetry does not allow measurements of the latent heat of a first-order phase transition. This feature may be considered to be an advantage because specific heats of both phases are not plagued by the latent heat, as when using other calorimetric techniques. This was already seen in the first such studies (Holland 1963; Kraftmakher and Romashina 1965; Zaitseva and Kraftmakher 1965). In modulation measurements, only a discontinuity is seen in the point of a first-order phase transition (Fig. 14.1).

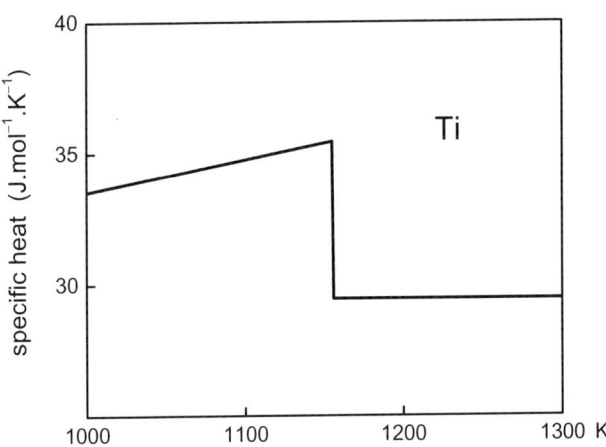

Fig. 14.1. Specific heat of titanium near the first-order phase transition from modulation measurements (Zaitseva and Kraftmakher 1965)

However, Garnier and Salamon (1971) determined a small latent heat accompanying the phase transition in chromium. In these measurements, the temperature oscillations involved both phases. The absorption and emission of the latent heat during a cycle of the temperature oscillations modify the form of the temperature changes, and a plateau was found due to the latent heat of the transition (Fig. 14.2).

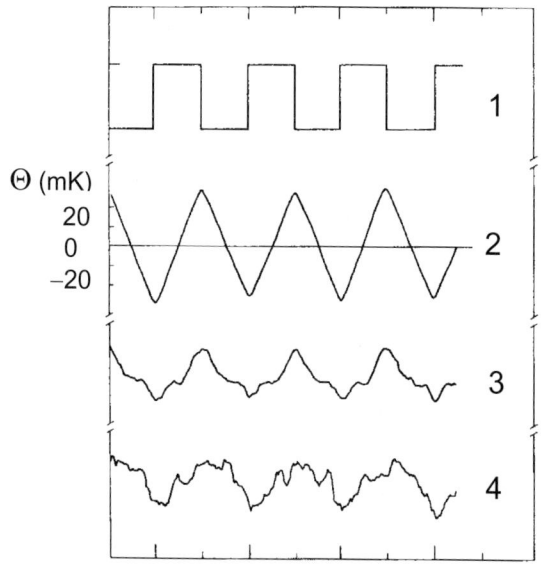

Fig. 14.2. Temperature oscillations of a chromium sample. 1 – input heat pulses at 11 Hz, 2 – triangular oscillations at $T = T_N + 0.4$ K, 3 – appearance of a plateau when the oscillations span the first-order transition (average of sixteen passes), 4 – single pass spanning T_N (Garnier and Salamon 1971)

Saruyama (1992) used modulation calorimetry to study the melting transition of an n-paraffin, $C_{20}H_{42}$. Mizuno et al. (1992) studied the ferroelectric phase transition in the vinylidenefluoride-trifluoroethylene (VDF-TrFE) random copolymers. In the 52% VDF copolymer, they observed a frequency-dependent peak close to the transition point. The transition shows a heat-capacity peak even after excluding the first-order component. The authors explained this result as evidence that the transition has the nature of a high-order transition, along with a latent heat. Ema et al. (1993) pointed out that in such measurements the speed of the phase transition should be faster than the AC heating rate and superheating (or supercooling) should not be significant. Such conditions are unlikely to be fulfilled in most first-order transitions in insulating materials.

In studies of first-order phase transitions, modulated differential scanning calorimetry (MDSC) provides a more straightforward and adequate method. This technique combines the advantages of modulation calorimetry and of differential scanning calorimetry. Hatta and Nakayama (1998) used modulation calorimetry and MDSC for studies of first-order phase transitions. They measured the specific heat of $BaTiO_3$ and $NaNO_2$ at their first-order ferroelectric-to-paraelectric phase transitions. In the MDSC measurements, the underlying heating rate was 0.1 K.min^{-1}, and the amplitude of the temperature oscillations was 0.1 K for modulation periods of 100 and 60 s. The specific heat of octyloxycyanobiphenyl (8OCB) was determined by both techniques in a range including the first-order phase transition from the crystalline to the smectic-A phase. By comparison of the results obtained by MDSC with those by modulation calorimetry, the conclusions by the authors were as follows:

- Heat capacity measurements at a first-order phase transition should employ temperature oscillations as small as possible. From such measurements, one obtains the detailed behaviour of the heat capacity except for the latent heat.
- The latent heat is available, with high accuracy, from MDSC measurements.
- When studying the kinetics of a phase boundary, the MDSC measurements should be carried out under conditions excluding a cooling process during the measurements.

Diosa et al. (2000b) studied the phase transitions in $LiNH_4SO_4$ single crystals by means of MDSC and usual modulation measurements. The transitions were observed at 28, 225, 285, and 460 K. The transitions at 285 and 460 K are of first order. Using modulation calorimetry, Fritsch and Lüscher (1983) and Fritsch et al. (1982, 1984) studied the behaviour of the specific heat of some low-melting-point metals close to their melting points.

Solid Electrolytes

A number of ionic solids have highly conducting phases, with conductivities comparable to those of liquid electrolytes. Vargas et al. (1976) reported on simultaneous measurement of the ionic conductivity σ and the specific heat near the 208-K order-disorder phase transition of $RbAg_4I_5$. They applied modulated-light heating to simultaneously measure the specific heat and temperature derivative of conductivity of single-crystal samples. Single crystal slices were thinned to 0.1 mm and connected to silver wires for four-terminal resistance measurements. An AC current (12 μA at 10 kHz) passed through the samples, while they were heated by light chopped at 1.5 Hz. The induced temperature oscillations were measured using a 25-μm thermocouple. From dR/dT and $R(T)$ data, the logarithmic derivative $d(\ln\sigma)/dT$ was calculated. It turned out that this quantity is a linear function of the specific heat. The data cover three decades of reduced temperature above and below the transition point. The linear relation holds up to smallest values of $|T/T_c - 1|$, of the order of 10^{-4}. The critical indices of the specific heat, below and above the transition point, are 0.15 ± 0.02. Vargas et al. (1977) measured the electrical conductivities and specific heats of other MAg_4I_5 family solid electrolytes (M = K and NH_4) in the range including the order-disorder transition. Lederman et al. (1976) and Jurado et al. (1997) also studied the transition in $RbAg_4I_5$.

Table 14.1 presents some studies of phase transitions in solids. A list of studies of phase transitions in ferro- and antiferroelectrics is given in Table 14.2.

Table 14.1. Studies of phase transitions in solids

Item	Reference
Ti, α–β transition	Holland 1963; Zaitseva and Kraftmakher 1965
CuZn	Ashman and Handler 1969
CuZn, specific heat and dR/dT	Simons and Salamon 1971
$SrTiO_3$	Garnier 1971; Hatta et al. 1977; Gallardo et al. 2002
NH_4Cl	Schwartz 1971
$2H$-$TaSe_2$, $2H$-TaS_2	Craven and Meyer 1977
SnTe	Hatta and Kobayashi 1977
$Sn_xGe_{1-x}Te$	Hatta and Rehwald 1977
$TiSe_2$	Craven et al. 1978a
$NH_4Br_xCl_{1-x}$	Lushington and Garland 1980; Yoshizawa et al. 1983
$Pt_6(NH_3)_{10}Cl_{10}(HSO_4)_4$	Inoue et al. 1982
$C_{24}Rb$ intercalate	Suematsu et al. 1980
Co, Zr	Boyarskii and Novikov 1981
$CsBi(MoO_4)_2$	Stokka and Samulionis 1981
$Pb_{1-x}Ge_xTe$	Sugimoto et al. 1981
$(CH_3NH_3)_2FeCl_4$	Goto et al. 1982
C_6Li intercalate	Robinson and Salamon 1982
$CsPbCl_3$	Stokka et al. 1982
NaN_3	Hirotsu et al. 1983
CuV_2S_4	Sekine et al. 1984
$(C_nH_{2n+1}NH_3)_2FeCl_4$ (n = 1, 2)	Yoshizawa et al. 1984
Quartz, α–β transition	Matsuura et al. 1985; Hatta et al. 1985; Yao and Hatta 1995
Graphite-ICl intercalates	Tashiro et al. 1985, 1990
$AgCrS_2$	Kawaji et al. 1989
C_{60}	Chung et al. 1992
(Rare earth)$_{1.85}Ce_{0.15}CuO_{4-\delta}$	Hwang et al. 1992
$ZrTe_3$, $K_{0.3}MoO_3$	Chung et al. 1993b
K_2ZnCl_4, K_2SeO_4, Rb_2ZnCl_4	Haga et al. 1995ab
$CuGeO_3$	Kuo et al. 1995, 1996
Cs_2ZnI_4	Melero et al. 1995
C_{70}	Sekine et al. 1995
$(NH_4)_2HPO_4$	Vargas et al. 1995
$(ND_4)_2DPO_4$	Vargas and Diosa 1997
NaV_6O_{11}	Akiba et al. 1998
$KHSO_4$, $LiKSO_4$, $LiNH_4SO_4$	Diosa et al. 2000abc
$(TMTSF)_2PF_6$	Powell et al. 2001
$(DMe$-$DCNQI)_2M$ (M = Ag, Li)	Nakazawa et al. 2003

TMTSF = tetramethyltetraselenafulvalene,
DMe-DCNQI = 2,5-dimethyl-N,N'-dicyanoquinonediimine

Table 14.2. Studies of phase transitions in ferro- and antiferroelectrics

Item	Reference
$LiTaO_3$	Glass 1968
$NaNO_2$	Hatta and Ikushima 1971, 1973; Hatta et al. 1995
$BaTiO_3$	Hatta and Ikushima 1972, 1976; Tura et al. 1998
$Gd_2(MoO_4)_3$	Cheung and Ullman 1974
TGSe	Ema et al. 1977
$K_2Pt(CN)_4Br_{0.3} \cdot 3H_2O$	Hatta et al. 1978
TGS	Ema et al. 1979; Fraile-Rodríguez et al. 2001
SbSI	Stokka et al. 1981
CsH_2PO_4, CsD_2PO_4	Kanda et al. 1982
TGS, TGSe, TGFB	Ema 1983
KH_2PO_4	Sandvold and Fossheim 1986; Moon et al. 1996
K_2ZnCl_4	Gesi 1992
VDF-TrFE copolymers	Mizuno et al. 1992
$LaBGeO_5$	Onodera et al. 1993
TlH_2AsO_4	Irokawa et al. 1994
MHPOCBC liquid crystal	Ema et al. 1996c
$(CH_3)_2NH_2H_2PO_4$	Hatori et al. 1996
MHDOBBC liquid crystal	Puértolas et al. 1996
$Tb_2(MoO_4)_3$	Strukov et al. 1996
$BaTiO_3$, $NaNO_2$	Hatta and Nakayama 1998

TGSe = triglycine selenate, TGS = triglycine sulfate, TGFB = triglycine fluoberyllate, VDF-TrFE = vinylidene fluoride-trifluoroethylene

14.2 Ferro- and Antiferromagnets

For studies of second-order phase transitions, modulation calorimetry provides unique temperature resolution and precision. Both features are crucial for accurate determinations of the temperature dependence of specific heat. In this case, absolute values of specific heat are not so important, and all variants of modulation calorimetry are usable. The theory of second-order phase transitions predicts a power-law temperature dependence of specific heat near the transition points:

$$C^-(t) = A^-(|t|^{-\alpha^-} - 1)/\alpha^- + B^- \quad \text{for } T < T_c, \quad (14.1a)$$

$$C^+(t) = A^+(t^{-\alpha^+} - 1)/\alpha^+ + B^+ \quad \text{for } T > T_c, \quad (14.1b)$$

where $t = T/T_c - 1$, T_c is the transition temperature, and α^- and α^+ are the so-called critical indices of specific heat for $T < T_c$ and $T > T_c$, respectively.

Generally, A^- and A^+ do not coincide, as well as B^- and B^+. Experimental data below and above the transition point are treated separately, and different values of T_c may provide best fits the two sets of data. Usually, at the final stage one accepts fits with the same T_c for both branches of the specific-heat curve. An additional requirement often posed is the same critical index below and above the transition point, i.e., $\alpha^+ = \alpha^-$. Connelly et al. (1971) used this procedure to derive the transition temperature and the critical indices for nickel. The dependence of α^+ and α^- on T_c allowed the authors to find the transition temperature for which $\alpha^+ = \alpha^-$. This temperature was about 0.13 K above the point of the maximum in the specific heat. Taking into account the uncertainties in the parameters fitting the experimental data, a region of probable values of T_c and α appears (Fig. 14.3). The authors concluded that $\alpha^+ = \alpha^- = -0.10 \pm 0.03$.

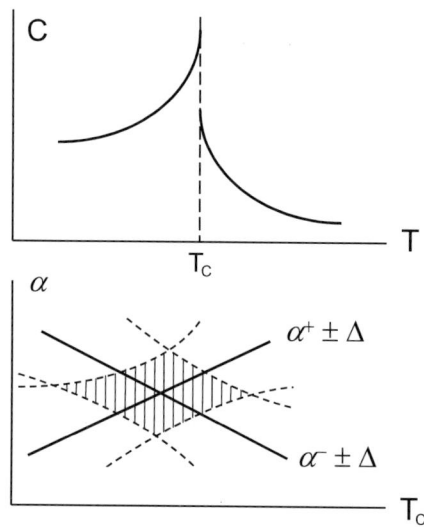

Fig. 14.3. Determination of transition temperature and of critical index of specific heat, schematically. The dashed area shows probable limits for T_c and α compatible with the assumption that both quantities are the same below and above the transition point (Connelly et al. 1971)

The first modulation measurements of the specific heat of ferromagnets near the Curie points were made in the 1960s on iron (Kraftmakher and Romashina 1965), nickel (Kraftmakher 1966a), and cobalt (Kraftmakher and Romashina 1966). The singularity in the Curie point was found to be of logarithmic type, which means that $\alpha^+ = \alpha^- = 0$. Many workers accepted such dependence for specific heat. It seemed preferable because only four fitting parameters, B^-, B^+, $A^\pm$, and T_c, are necessary. The coefficients of the polynomial fits for the three ferromagnets are given in Table 14.3. This approximation means that the specific heat behaves symmetrically below and above the transition point ($A^+ = A^-$), but the high-temperature branch is shifted down ($B^+ < B^-$). Usually, the transition temperature is chosen to provide the accepted temperature dependence of the specific heat in a widest temperature range (Fig. 14.4). However, only upper and lower limits for the critical indices are obtainable from experimental data. In the measurements mentioned above, it was concluded that $|\alpha| < 0.2$.

Table 14.3. Specific heat of iron, cobalt, and nickel near their Curie points. The experimental data are fitted by polynomials $C^{\pm}/3Nk_B = -A\log|T - T_c| + B^{\pm}$, where $A = A^+ = A^-$

Metal	A	B^-	B^+	ΔB
Iron	1.65	9.15	8.15	1
Cobalt	0.7	6.95	6.3	0.65
Nickel	0.46	4.72	4.4	0.32

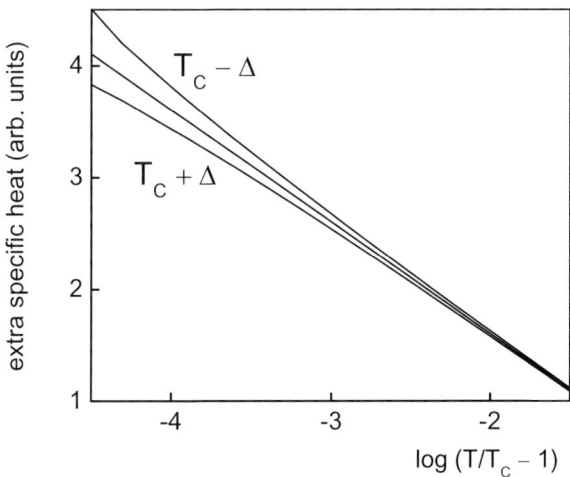

Fig. 14.4. Presentation of data accepting logarithmic dependence of specific heat, schematically. The graph depends on the choice of the transition temperature. With these coordinates, the best value of T_c should provide a linear dependence in a widest temperature range. The data relate to $T < T_c$

Ikeda et al. (1976) observed a symmetric logarithmic singularity in the specific heat of two-dimensional Ising-like antiferromagnets K_2CoF_4 and Rb_2CoF_4 near their Néel points. Ikeda et al. (1981) reported on a similar singularity for the two-dimensional random antiferromagnets $Rb_2Co_xNi_{1-x}F_4$ (x = 0, 0.5, 0.65, and 0.8). Hatta and Ikeda (1980) found a symmetric logarithmic divergence of the specific heat of several antiferromagnets of the K_2NiF_4 family. However, in many other cases the specific heat of ferro- and antiferromagnets manifests a more complex behaviour. The specific-heat anomaly is not symmetric, and the critical indices depend on temperature. For instance, Lederman and Salamon (1974) observed a change of the critical indices of the specific heat of dysprosium near the Néel point.

For $|t| \approx 5\times10^{-5}$, $\alpha^+ = \alpha^- = -0.02 \pm 0.01$ for the outer region, while for the inner region $\alpha^+ = \alpha^- = 0.18 \pm 0.08$. Hiraka and Endoh (1999) measured the specific heat of a CoS_2 single crystal in magnetic fields up to 5 T over 30–250 K temperature range including the Curie point ($T_c = 121$ K).

In 1960s, a question arose about the temperature dependence of the electrical resistivity of a ferromagnet near its Curie point. The modulation technique immediately provides data on the temperature derivative of the resistance and thus is more suitable to answer this question (Kraftmakher 1967a). For nickel, it turned out that this derivative behaves like the specific heat, consisting of a jump and a logarithmic singularity. In a wide temperature range, $T_c \pm 50$ K, the data obey a logarithmic dependence $(1/R_c)dR/dT = B - A\log|T-T_c|$. Fisher and Langer (1968) developed a theory that predicts such behaviour above the Curie point.

Simons and Salamon (1971) measured the specific heat and the temperature derivative of resistance of β-brass near the order-disorder transition (Fig. 14.5). The authors found that the anomalous part of the temperature derivative of resistance is proportional to that of the specific heat.

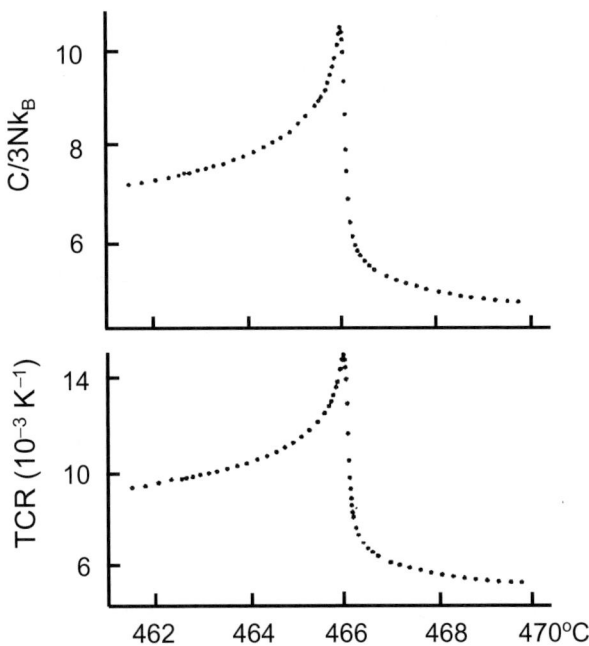

Fig. 14.5. Specific heat and temperature coefficient of resistance (TCR) of β-brass near the order-disorder transition (Simons and Salamon 1971)

A precise determination of the critical indices is difficult because of imperfection of the samples (impurities, grain boundaries, dislocations). This causes rounding of experimental points in the close vicinity of the transition point. The maximum of the specific heat does not coincide with the transition temperature.

With the modulation method of measuring the thermopower, a discontinuity was found in the Curie point of iron, nickel, and cobalt (Fig. 14.6). Modulation studies of phase transitions in ferro- and antiferromagnets are displayed in Tables 14.4 and 14.5.

Table 14.4. Studies of phase transitions in ferromagnets

Item	Reference
Fe, specific heat	Kraftmakher and Romashina 1965; Lederman et al. 1974; Varchenko et al. 1978
dR/dT	Kraftmakher and Romashina 1967; Kraftmakher and Pinegina 1974; Shacklette 1974
thermopower	Kraftmakher and Pinegina 1970
Ni, specific heat	Kraftmakher 1966a; Handler et al. 1967; Connelly et al. 1971; Maszkiewicz 1978; Maszkiewicz et al. 1979
dR/dT	Kraftmakher 1967a
specific heat and thermopower	Papp 1984
Co, specific heat	Kraftmakher and Romashina 1966
thermopower	Kraftmakher and Pinegina 1971
Gd, specific heat	Lewis 1970; Wantenaar et al. 1977; Jeong et al. 1991; Jung et al. 1992; Bednarz et al. 1992, 1993; Glorieux and Thoen 1994; Glorieux et al. 1995
transition at 226 K	Salamon and Simons 1973
specific heat and dR/dT	Simons and Salamon 1974
EuO	Salamon 1973
CoS_2	Ogawa and Yamadaya 1974; Hiraka and Endoh 1999
$Fe_{75}P_{15}C_{10}$, amorphous	Schowalter et al. 1977
$Fe_{34}Pd_{46}P_{20}$, amorphous	Craven et al. 1978b
$Fe_{80-x}M_xP_{13}C_7$ (M = Ni, Mn, Cr)	Ikeda and Ishikawa 1980
$Eu_xSr_{1-x}S$	Haeiwa et al. 1988
$Hf_{1-x}Ta_xFe_2$, amorphous	Murayama et al. 1995
$YCo_{12}B_6$, $GdCo_{12}B_6$ (ferrimagnet)	Nahm et al. 1995
$NdNi_2$, $TbNi_2$, $DyNi_2$	Melero and Burriel 1996
$Ni_{90}Cr_{10}$	Jurado et al. 1997
$La_{0.7}Ca_{0.3}MnO_3$	Park et al. 1997
$Fe_xMn_{0.6-x}Al_{0.4}$ (0.2 <x < 0.6)	González et al. 2002
$La_{0.7}M_{0.3}MnO_3$ (M = Sr, Ca), $La_{0.66}(Pb,Ca)_{0.34}MnO_3$	Salamon and Chun 2003
Thermopower	
$Fe_xCo_{80-x}B_{20}$, $Fe_xNi_{80-x}B_{19}Si$	Kettler et al. 1982, 1984
$ErCo_2$	Resel et al. 1996

Table 14.5. Studies of phase transitions in antiferromagnets

Item	Reference
Cr, specific heat and dR/dT	Salamon et al. 1969
CoO	Salamon 1970
Cr	Garnier and Salamon 1971
K_2NiF_4	Salamon and Hatta 1971
$MnBr_2 \cdot 4H_2O$	Hempstead and Mochel 1973
K_2MnF_4, K_2NiF_4	Salamon and Ikeda 1973
Dy	Lederman and Salamon 1974
K_2CoF_4, Rb_2CoF_4	Ikeda et al. 1976
$CsMnF_3$	Ikeda 1977
MnF_2	Ikeda et al. 1978
$Rb_2Co_xMg_{1-x}F_4$	Suzuki and Ikeda 1978
K_2NiF_4 family	Hatta and Ikeda 1980
CoF_2, MnF_2, $KMnF_3$	Akutsu and Ikeda 1981
$Rb_2Co_xNi_{1-x}F_4$	Ikeda et al. 1981
YFe_2O_4	Tanaka et al. 1982
C_6Eu intercalate	Ohmatsu et al. 1983
$ZnCr_2Se_4$	Ershov et al. 1984
$Mn_{0.5}Zn_{0.5}F_2$	Ikeda 1986
$GdCu_2Si_2$, $GdNi_2Si_2$, $GdGa_2$, $GdCu_5$	Bouvier et al. 1991
Er, dR/dT	Terki et al. 1992
CoO, Cr_2O_3	Glorieux et al. 1994
Cr_2O_3, FeF_2, Cr	Marinelli et al. 1994b
Dy, magnetic phase diagram	Izawa et al. 1996
$Gd_2O_2SO_4$, magnetic phase diagram	Kratz et al. 1996
$R_{1.5}Ce_{0.5}Sr_2FeCu_2O_9$ (R = Y, Eu)	Felner et al. 1997
Cr_2O_3	Murtazaev et al. 2001

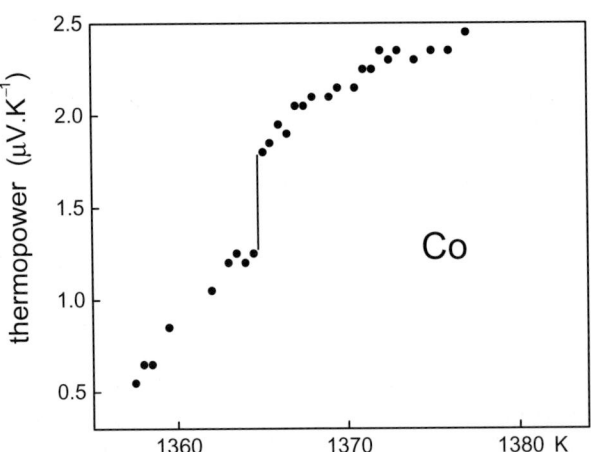

Fig. 14.6. Thermoelectric power of Co/PtRh thermocouple near the Curie point of cobalt (Kraftmakher and Pinegina 1971)

14.3 Superconductors

Modulation calorimetry provides excellent opportunity for measurements of specific heat of superconductors near their phase-transition points. Sullivan and Seidel (1968) carried out the first modulation studies on a superconductor. They measured the specific heat of indium versus an external magnetic field. The aim of these measurements was to demonstrate the important advantages of the new calorimetric technique. Since then, many studies of superconductors were performed, including bulk samples, thin films, fine particles, organic superconductors, pure metals and alloys, crystalline and amorphous.

Low-Temperature Superconductors

Greene et al. (1972) measured the specific heat of granular $Al-Al_2O_3$ films using modulated-light heating. Farrant and Gough (1975) studied niobium in a vicinity of the transition point at constant magnetic induction. The magnetic field was trapped inside the sample by allowing it to cool from a temperature well above T_c in a homogeneous magnetic field.

Tsuboi and Suzuki (1977) measured the specific heat of tin particles, whose average radius ranged from 13 to 110 nm. Thin oxide layers mutually insulated the particles. A substrate for the particles made of thin glass (5–10 μm) was mounted on a copper frame. A 20-nm manganin heater was deposited on the glass. A SiO film, 300 nm thick, was deposited on the heater, and then a 300-nm gold film to improve thermal diffusion was deposited over the region of the sample. A second SiO film served to insulate the gold film from a germanium thermometer. On the opposite side of the substrate, tin particles were deposited in vacuum on an area of 1×1 cm. After an amount of the particles of 0.5 mg or more was obtained, a 300-nm SiO film was deposited on the particles to protect them from oxidation in air. At 4.2 K, the contribution of the particles was at most 10% of the total heat capacity. The authors summarized their main results as follows:

- The temperature of the maximum in the specific heat of fine tin particles is lower than the transition temperature for bulk samples.
- As the particle size decreases, the peak of the specific heat shifts towards lower temperatures and becomes broader.

Similar measurements were performed under magnetic fields up to 3 T (Suzuki and Tsuboi 1977). Along with usual measurements of the specific heat versus temperature for several magnetic fields, the authors carried out field-sweep measurements. Fortune et al. (1990) performed calorimetric measurements under high magnetic fields on an organic conductor $(TMTSF)_2ClO_4$ (TMTSF = tetramethyltetraselenafulvalene), superconducting below 1.3 K. From the data, the B–T phase diagram of the sample was determined. Park et al. (2002) measured the specific heat of polycrystalline $Mg^{11}B_2$ from 5 to 45 K at magnetic fields up to 7 T. Park et al. (2003) measured the specific heat of a single crystal of the superconductor YNi_2B_2C as a function of the orientation of an external magnetic field.

Studies of low-temperature superconductors are listed in Table 14.6.

Table 14.6. Studies of low-temperature superconductors

Item	Reference
Specific heat	
In, in magnetic field	Sullivan and Seidel 1968
InPb alloy	Zoller and Dillinger 1969
BiSb, amorphous films	Zally and Mochel 1971, 1972; Krauss and Buckel 1975
Al, granular films	Greene et al. 1972
Ag-Pb-Ag sandwiches	Manuel and Veyssié 1973
V_3Ga, Nb_3Al	Viswanathan et al. 1974
Nb, in magnetic field	Farrant and Gough 1975
Al foils, Sn films	Manuel and Veyssié 1976
$LaRu_2$	Viswanathan et al. 1976
Sn, fine particles	Suzuki and Tsuboi 1977; Tsuboi and Suzuki 1977
$2H$-TaS_2	Garoche et al. 1978
In films	Gibson et al. 1979
V_3Si, in magnetic field	Huang et al. 1980
Zr_3Rh, amorphous	Garoche and Johnson 1981
$Bi_{84}Sb_{16}$ and $Ga_{95}Ag_5$ films	Kämpf et al. 1981
Nb_3Ge films	Rao and Goldman 1981
Al, thin films	Suzuki et al. 1982
CuZr, amorphous	Garoche and Bigot 1983
$BaPb_{1-x}Bi_xO_3$	Sato et al. 1983
Al	Machado and Clark 1988
$Ba_{0.6}K_{0.4}BiO_3$	Graebner et al. 1989
Au_xSn_{1-x}	Rieger and Baumann 1991
$PtGa_2$	Hsu 1994
Cu_xSn_{1-x}	Sohn and Baumann 1996
$Mg^{11}B_2$	Park et al. 2002
YNi_2B_2C, field-angle dependence	Park et al. 2003
Thermopower	
$LaRu_2$	Resel et al. 1996

TMTSF = tetramethyltetraselenafulvalene,
BEDT-TTF = bis(ethylenedithio)tetrathiafulvalene

High-Temperature Superconductors

Studies of specific-heat anomalies in high-temperature superconductors pose serious problems. First, the phonon specific heat near the transition points is large, and the contribution to be studied is of the order of 1%. Second, measurements on single crystals are desirable, and a small volume of the samples causes additional difficulties. Modulation calorimetry is therefore very attractive for such studies.

Inderhees et al. (1987) were the first to use modulation calorimetry in a study of $YBa_2Cu_3O_{7-\delta}$ (YBCO). The ceramic sample was a disc, 3 mm in diameter and 0.5 mm thick, with a mass of 18 mg. Modulated light provided the AC heating. A flattened thermocouple, made of 25-μm chromel and constantan wires, was attached to the backside of the sample with a small amount of varnish. The temperature oscillations, of about 5 mK, were measured and recorded by means of a lock-in amplifier and a computer. The same thermocouple recorded the enhancement of the mean temperature. A calibrated carbon-glass thermometer measured the temperature of the bath. Adiabatic heat-pulse measurements at 77 K were carried out to normalise the data. The electronic specific heat of YBCO was calculated from the Pauli paramagnetic susceptibility. The lattice contribution was fitted according to the Debye formula. The ratio $\Delta C/\gamma T_c$ was found to be 1.23 ± 0.08, which is close to 1.43 predicted by the Bardeen-Cooper-Schrieffer theory. Later, this group performed calorimetric measurements on small single crystals, including measurements in magnetic fields (Salamon et al. 1988, 1990, 1993; Inderhees et al. 1988, 1991; Ghiron et al. 1992). Results of such measurements are often presented as the ratio of specific heat to temperature. With increasing magnetic field, the sharpness of the specific-heat peak decreases. (Fig. 14.7).

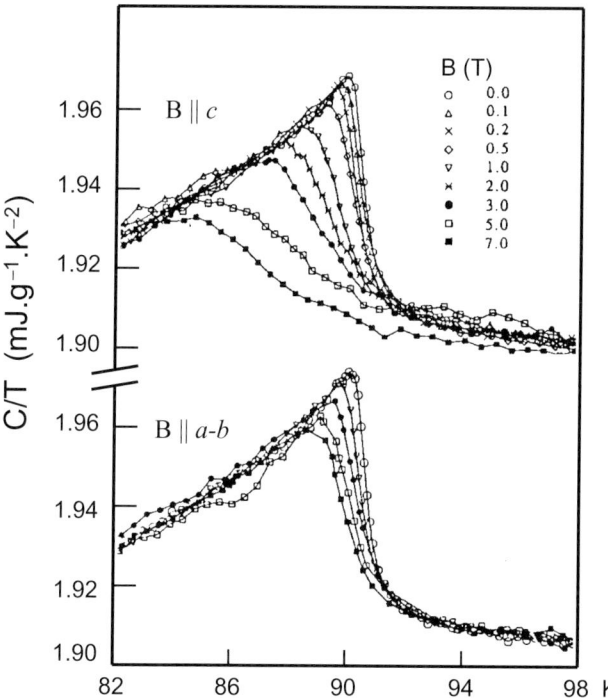

Fig. 14.7. Specific heat of an $YBa_2Cu_3O_{7-\delta}$ crystal in external magnetic fields applied parallel to the c axis and parallel to the a-b plane (Inderhees et al. 1991)

Chung et al. (1996) performed measurements of heat capacity of single whiskers of Bi-based cuprate superconductors, $Bi_2Sr_2CaCu_2O_x$ and $Bi_2Sr_2Ca_2Cu_3O_x$. The samples were typically 2 mm long, 0.2–0.3 mm wide, and 10–30 μm thick, with a calculated mass from 40 to 100 μg. One of the large sample surfaces was exposed to light chopped at a frequency typically below 10 Hz. This surface was coated with an evaporated 0.1-μm PbS film to eliminate possible effects of a temperature dependent absorptance. The temperature oscillations were measured with a 13-μm thermocouple attached to the opposite large surface of the sample with diluted GE 7031 varnish. The addenda contribution to the measured heat capacity was estimated to be less than 10%. The measurements were carried out in the range 25 to 280 K.

Carrington et al. (1996) employed a differential modulation calorimeter in studies of a 30-μg single crystal of $Tl_2Ba_2CuO_{6+\delta}$ ($T_c \cong 15$ K). The sample under study and a reference one were attached to chromel-constantan thermocouples and placed on opposite sides of a copper plate. The samples were heated independently, via optical fibres, by two light-emitting diodes. It was possible to monitor the temperature oscillations of each sample or to measure the difference between them. At a temperature near the transition, this difference was adjusted to be close to zero. Measurements in magnetic fields up to 7 T served to determine the normal background. In measurements on $HgBa_2Ca_2Cu_3O_{8+\delta}$ ($T_c \cong 111$ K), the sample was a small platelet crystal with dimensions $0.2\times0.1\times0.03$ mm and a mass of about 5 μg (Carrington et al. 1997). The differential method greatly improves the sensitivity and reproducibility of the measurements and effectively eliminates the effect of the magnetic field on the thermocouple calibration. To obtain a reference sample with a phonon contribution well matched to the sample under study, one of the samples was annealed at 450°C. After this treatment, the sample had no traces of superconductivity above 60 K.

Charalambous et al. (1999) studied the asymmetry of the critical indices in YBCO by means of a high-resolution microcalorimeter designed by Riou et al. (1997). The high precision of the calorimeter, of the order of 10^{-4}, allowed the authors to carry out the measurements on small detwinned single crystals. The asymmetry of the two branches of the specific heat, below and above the transition point, was clearly seen from the dC/dT data (Fig. 14.8). The authors concluded that the two critical indices are definitely different: $\alpha^- = -0.2 \pm 0.3$, while $\alpha^+ = 0.5 \pm 0.2$. Such a behaviour cannot be explained within the framework of a regular second-order phase transition. Deligiannis et al. (2000) observed hysteretic effects at the vortex-melting transition in high-purity $YBa_2Cu_3O_{7-\delta}$ single crystals. This means that this transition is of the first order.

Yu et al. (1988) measured the thermopower in the normal state of ceramic samples of the high-temperature superconductors $YBa_2Cu_3O_{7-\delta}$ and $La_{1.85}Sr_{0.15}CuO_4$. The transition temperatures of these are 92 and 35 K, respectively. The high-field experiments, up to 30 T, were carried out at the Francis Bitter National Magnet Laboratory at the Massachusetts Institute of Technology. Howson et al. (1989, 1990) measured the thermopower of YBCO crystals in the range from the transition point T_c to 250 K and in magnetic fields up to 2 T. A sharp peak close to T_c is very sensitive to the oxygen content of the samples. The authors argued that the peak might be due to fluctuation effects.

Table 14.7 lists studies of high-temperature superconductors.

Table 14.7. Studies of high-temperature superconductors

Item	Reference
Specific heat	
YBCO	Inderhees et al. 1987, 1988;
	Ishikawa et al. 1988;
	Kishi et al. 1988;
	Anshukova et al. 1989;
	Lægreid et al. 1989;
	Marinelli et al. 1990;
	Regan et al. 1991;
	Zhou et al. 1991;
	Marone and Payne 1997;
	Riou et al. 1997;
	Garfield et al. 1998;
	Charalambous et al. 1999
in magnetic field	Ginsberg et al. 1988;
	Calzona et al. 1990;
	Salamon et al. 1988, 1990, 1993;
	Ghiron et al. 1992;
	Overend et al. 1994, 1996;
	Kamilov et al. 1995
along H_{c2} line	Inderhees et al. 1991
transitions at 90 and 220 K	Lægreid et al. 1988;
	Vargas et al. 1989
$La_{1.85}Sr_{0.15}CuO_4$	Feng et al. 1988
$Bi_{0.7}Pb_{0.3}SrCaCu_{1.8}O_x$	Slaski et al. 1989
(Bi,Pb)-Sr-Ca-Cu-O	Okazaki et al. 1990
$Bi_2Sr_2CaCu_2O_y$	Nes et al. 1991
$Bi_{1.7}Pb_{0.3}Sr_2Ca_2Cu_3O_{10-\delta}$	Ausloos et al. 1994
$Tl_2Ba_2CuO_{6+\delta}$, in magnetic field	Carrington et al. 1996
$Bi_2Sr_2Ca_{n-1}Cu_nO_x$ whiskers (n = 2, 3)	Chung et al. 1996
$HgBa_2Ca_2Cu_3O_{8+\delta}$, in magnetic field	Carrington et al. 1997
$DyBa_2Cu_3O_{7-\delta}$, up to 8 T	Garfield et al. 1999
$La_2CuO_{4.093}$	Hirayama et al. 2000
Thermopower	
YBCO	Lowe et al. 1991; Aubin et al. 1993
in magnetic field	Yu et al. 1988;
	Howson et al. 1989, 1990;
	Oussena et al. 1992
$La_{1.85}Sr_{0.15}CuO_{4-\delta}$, up to 30 T	Yu et al. 1988
$Nd_{1.85}Ce_{0.15}CuO_{4+y}$	Mangelschots et al. 1992
$Nd_{2-x}Ce_xCuO_4$	Xu et al. 1992
$Tl_2Ba_2CuO_6$	Lin et al. 1993

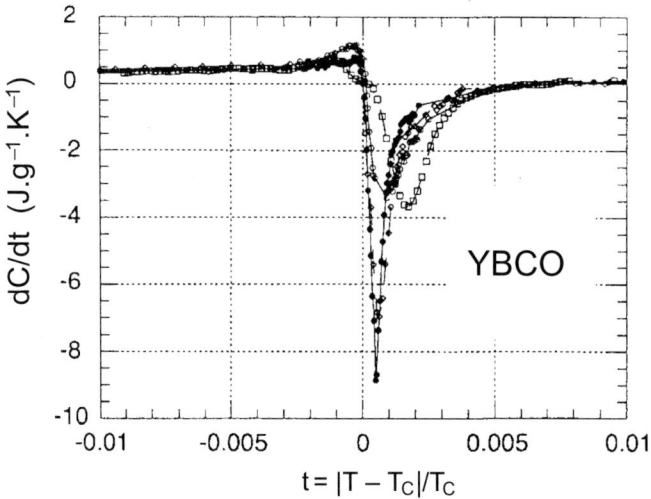

Fig. 14.8. Temperature derivative of specific heat of four YBCO samples (Charalambous et al. 1999). The asymmetry of the two branches of specific heat is clearly seen

14.4 Thin Films and Confined Systems

Last decades, many studies were devoted to phase transitions in thin films and physical systems subjected to a variety of confining conditions. Typically, such conditions are achieved by embedding the system in solid materials with cavities, either interconnected or isolated, of dimensions in the range 1 nm to 10 μm. To study confined systems, a small signal must be separated from a large background. Modulation calorimetry is regarded as the best tool for such measurements.

Birmingham et al. (1996) measured specific heat of hydrogen films quench-condensed on gold as a function of ortho concentration x ($0.28 < x < 0.75$). The measurements were performed for coverages between 24.3 Å^{-2} and 92.3 Å^{-2}, in the range 0.4–3 K. At room temperature, the ortho-para distribution in hydrogen gas is called normal hydrogen, n-H_2 ($x = 0.75$). After cooling to liquid helium temperatures, a slow conversion to para-hydrogen occurs. A calorimeter made out of a thin piezoelectric quartz plate operated simultaneously as a microbalance to measure the adsorbate coverage. The plate was suspended under tension by three nylon fibres. Four doped germanium thermistors and two nichrom resistive heaters on silicon substrates were epoxied to the backside of the calorimeter. Before the hydrogen is quench-condensed, a new gold surface was evaporated *in situ* on the front side of the plate from a gold-coated 0.25-mm tungsten wire. The coverage was monitored through the resonant frequency of the quartz substrate using an oscillator circuit. Coverage of 0.1 Å^{-2} results in a frequency shift of –0.35 Hz, while frequency shifts of about 0.02 Hz could be distinguished. The authors formulated their main results as follows:

- Two different types of orientational ordering for $x = 0.74$ were observed, for films condensed above 2.5 K and below 1.5 K.
- At 3 K, above the long-range order-disorder transition, the entropies for all the samples with $x = 0.74$ are comparable regardless of the condensation temperature and comparable with that for bulk H_2.
- As x decreases through the ortho-para conversion, the peaks or maxima in the specific heat shift to lower temperatures, just as in bulk H_2.
- For $x < 0.5$, when long-range ordering is excluded, the specific heat is not sensitive to the condensation temperature and hence to the film structure.

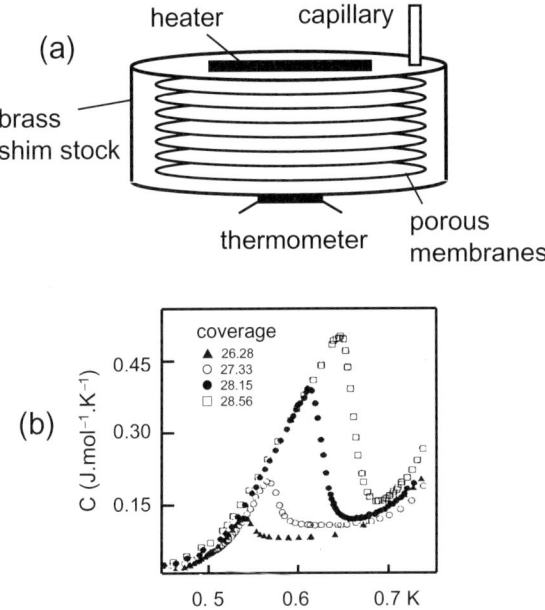

Fig. 14.9. (a) Simplified diagram of calorimeter cell for studying thin helium films condensed on porous membranes (Finotello et al. 1997). (b) Specific heat versus temperature for superfluid films in Millipore filter paper. Coverage is given in $\mu mol.m^{-2}$ (Steele et al. 1993)

Finotello et al. (1997) reviewed their studies of the superfluid-to-normal transition of helium films in Anopore membranes and in Xerogel porous glass and of phase transitions in liquid crystals, bulk and confined to Millipore filter paper or Anopore membranes. The calorimeter cell for helium films consists of inner and outer brass-shim cups, 21 mm in diameter and 50 μm thick, fitting snugly inside each other (Fig. 14.9). The substrates, onto which the helium films are adsorbed, aluminium oxide Anopore membranes, are discs 60 μm thick, with parallel cavities 0.2 μm in diameter. Several such discs are stacked, one above the other, within the brass cups. Helium is admitted to the cell through a coiled cupronickel or stainless steel capillary soldered to one face of the cell. A heater made of a thin high-resistivity wire is also glued to this face. A shaved down carbon resistor functions as the thermometer. The power dissipated in it was in the range 10^{-9}–10^{-12} W. To determine a proper modulation frequency, frequency scans were made at several temperatures, the lowest temperature was 0.1 K. In some measurements, data were obtained for two or more modulation frequencies. Before condensing a known amount of helium gas in the cell, the addendum heat capacity was measured. At 0.5 K, it was 4.5×10^{-6} J.K^{-1}.

Finotello et al. (1997) described also a calorimeter operating in the 25–125°C range for studying phase transitions in bulk liquid crystals and in those confined in porous substrates (Fig. 14.10). The calorimeter cell is placed inside a concentric cavity in a brass cylinder. Its upper part is sealed with a brass cap. The cylinder is provided with a manganin heater wrapped around it and with a platinum thermometer. A brass post, 3 mm in diameter and 4 cm long, supports the cylinder and provides thermal link to a larger, evacuated copper can. The latter rests on aluminium block partly submerged in a water bath. To achieve lower temperatures, a thermoelectric device is placed between the copper can and the aluminium block (not shown in the figure). The calorimeter cell consists of a 10 mm in diameter, 0.1 mm thick sapphire disc with a wire heater and a carbon flake thermistor attached on the lower face of the disc. Sapphire was chosen for its rigidity, flatness, high thermal conductivity, low heat capacity, and compatibility with many solvents used for cleaning the surface. The thermistor flake is less than 50 µm on side, which allows a short internal time constant. The heater and thermistor are glued to the disc with a small amount of varnish or epoxy. The internal time constant of the cell is about 1 s. Electrical leads are thermally anchored to the brass cylinder, and the external time constant depends on the length and diameter of the leads. The temperature oscillations were about 2 mK. About 10 mg of liquid crystal formed a droplet on the upper face of the sapphire disc.

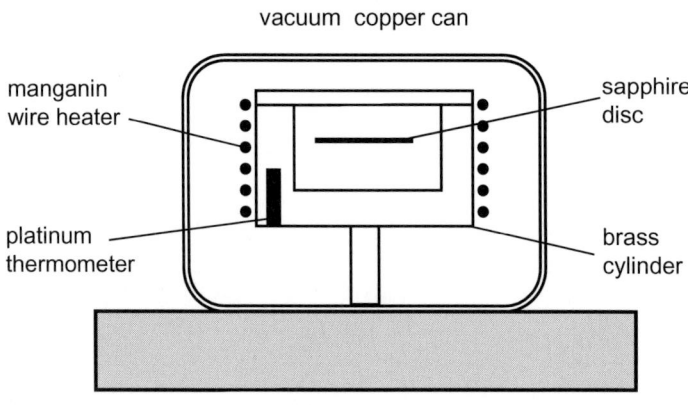

Fig. 14.10. Simplified diagram of calorimeter for studying phase transitions in bulk liquid crystals and in porous substrates (Finotello et al. 1997)

Table 14.8. Studies of phase transitions in thin films and confined systems

Item	Reference
N_2 monolayers on graphite	Chan et al. 1984
Ar submonolayers on graphite	Migone et al. 1984
^{4}He on graphite	Campbell and Bretz 1985
Ethylene and methane on graphite	Kim et al. 1986ab
^{4}He in porous glasses	Finotello et al. 1988
Ethane on graphite	Zhang and Migone 1988
^{4}He monolayer on graphite	Chae and Bretz 1989
Xe films	Steinmetz et al. 1989
^{4}He submonolayers on Ag and sapphire	Kenny and Richards 1990ab
Acetylene and butane on graphite	Alkhafaji and Migone 1992, 1993, 1996
Confined ^{4}He films	Steele and Finotello 1992
Confined liquid crystals	Iannacchione and Finotello 1992, 1993, 1994; Finotello and Iannacchione 1995; Wu et al. 1995; Qian et al. 1995, 1996, 1998; Iannacchione et al. 1995b, 1996, 1997, 1998a; Qian and Finotello 1997
H_2, Ne, O_2, Ar, melting in porous glasses	Molz et al. 1993
H_2, orientational ordering	Phelps et al. 1993; Birmingham et al. 1996
^{4}He superfluid films	Steele et al. 1993; Steele and Finotello 1994
^{3}He and ^{4}He films in disordered substrates	Yeager et al. 1994ab, 1995
4O.8, in silica aerogels	Kutnjak and Garland 1997
^{4}He films	Mehta and Gasparini 1997, 1998; Mehta et al. 1999; Kimball et al. 2000
^{4}He, superfluid transition in porous gold	Yoon and Chan 1997
8S5, in silica aerogels	Zhou et al. 1997a
Liquid crystals + silica aerosil particles	Haga and Garland 1997ab; Zhou et al. 1997b
^{4}He films on porous gold and on H_2	Csáthy et al. 1998

Mehta et al. (1999) measured the heat capacity of helium in a planar confinement. To test predictions of the correlation-length scaling, one requires a wide range of confinement sizes with high uniformity. The calorimeter cell was made by bonding together two 5 cm in diameter, 25-μm-thick silicon wafers at submicron separation. One of the wafers had a lithographically constructed SiO_2 triangular array of 0.125×0.125-mm squares, placed 0.5 mm apart. A 4-mm-wide circular border surrounded the array. The other wafer had a small central hole, 25–50 μm in diameter, to introduce helium into the cell. The SiO_2 squares maintained a uniform separation between the wafers, from 0.05 to 0.7 μm. The circular border served as a vacuum seal for the structure. After contact at room temperature, the wafers were annealed at 1100°C for about 6 hours. This procedure yielded defect-free cells possessing gap uniformity better than 1%. Following bonding, the wafers were assembled to complete the cell and then mounted on a cryostat. A uniform thin-film heater, in the form of a counter-wound spiral, was evaporated onto the wafers. Two doped-germanium thermometers measured the temperature oscillations and the mean temperature of the cell. Two thermally regulated stages served to change the temperature. One of these was maintained warmer than the cell, thus preventing condensation of helium away from the cell. The heat drain on top of the cell was via indium wires connected to a copper wire leading to the second stage.

Table 14.8 presents studies of phase transitions in thin films and confined systems. Calorimetric studies of organic conductors are listed in Table 14.9.

Table 14.9. Calorimetric studies of organic conductors

Item	Reference
TTF-TCNQ	Craven et al. 1974; Viswanathan and Johnston 1975
$(TMTSF)_2ClO_4$, in magnetic field	Brusetti et al. 1983; Fortune et al. 1990
κ-$(BEDT-TTF)_2Cu(NCS)_2$	Katsumoto et al. 1988; Graebner et al. 1990; Müller et al. 2002
$(TMTSF)_2ClO_4$, magnetic field up to 30 T	Fortune et al. 1990
α-$(BEDT-TTF)_2I_3$	Fortune et al. 1991
β-$(BEDT-TTF)_2I_3$	Fortune et al. 1992
$(TMTSF)_2ReO_4$, $(TMTSF)_2BF_4$	Chung et al. 1993a
κ-$(BEDT-TTF)_2Cu[N(CN)_2]Cl$, $(TMTSF)_2ReO_4$	Chung et al. 1993b
α-$(BEDT-TTF)_2MHg(NCS)_4$, M = K, Rb	Henning et al. 1995
TSF-TCNQ	Powell et al. 1997
$(DMET)_2BF_4$, $(DMET)_2ClO_4$	Akutsu et al. 1999
TTF-TCNQ, TSF-TCNQ	Saito et al. 1999b
κ-$(BEDT-TTF)_2Cu[N(CN)_2]X$, X = Br, Cl	Akutsu et al. 2000
$(DIMET)_2I_3$	Saito et al. 2000
$(DIMET)_2BF_4$	Saito et al. 2001a
κ-$(ET)_2Cu(NCS)_2$	Müller et al. 2003

TTF-TCNQ = tetrathiafulvalene-tetracyanoquinodimethane,
BEDT-TTF = bis(ethylenedithio)tetrathiafulvalene,
TMTSF = tetramethyltetraselenafulvalene,
DMET = dimethyl(ethylenedithio)diselenadithiafulvalene,
DIMET = sulphur analogy of DMET,
TSF-TCNQ = tetraselenafulvalene-tetracyanoquinodimethane

15 Relaxation Phenomena

Modulation calorimetry provides opportunity to vary in a wide range the frequency of the temperature oscillations in a sample. This makes possible a search for relaxation phenomena in specific heat, which appear when the modulation period becomes comparable to the relaxation time of a process contributing to the specific heat. Three such phenomena have been found by modulation calorimetry. First, supercooled organic liquids were studied near glass transitions. In this region, the specific heat becomes complex and frequency dependent (Birge and Nagel 1985, 1987; Birge 1986). Second, a relaxation phenomenon in the specific heat of tungsten and platinum was found at modulation frequencies of the order of 10^5 Hz (Kraftmakher 1985, 1990). The phenomenon was attributed to equilibration of point defects in crystal lattice. Finally, a relaxation was observed in the low-temperature specific heat of Mn_{12} acetate crystal (Fominaya et al. 1999b) and in some other cases.

Smith (1966) carried out the first calorimetric measurements with various modulation frequencies. He measured the ratio c/R' (R' is the temperature derivative of electrical resistance of the sample) of germanium whiskers by the third-harmonic technique and varied the frequencies of the heating current from 5.7 to 92 Hz. It turned out that at frequencies above 10 H the values obtained are up to 20% higher than at lower frequencies. The author concluded that the phenomenon is a result of changes in R' due to the generation of point defects in the crystal lattice. At higher frequencies, the defect concentration is unable to follow the temperature oscillations, which results in changes of R'. However, the difference between the low-frequency and high-frequency data practically did not depend on temperature. An analysis by Smith and Holland (1966) led to a serious contradiction with existing data on self-diffusion in germanium.

Further attempts to observe the relaxation in specific heat were undertaken with modulation frequencies up to 10^3 Hz. Measurements on platinum have shown no relaxation (Seville 1974). Modulation measurements on gold by Skelskey and Van den Sype (1974) revealed a frequency dependence of the quantity c/R'. The increase in this quantity at high frequencies is explainable if the relative vacancy contribution to the temperature derivative of electrical resistance is larger than that to the specific heat. The measurements were carried out at a single temperature, 1164 K, so that the result could not be supported by the temperature dependence of the phenomenon. Greene et al. (1972), Eno et al. (1977), and Suzuki et al. (1982) also performed measurements at various modulation frequencies. At high temperatures, high-frequency temperature oscillations were observed by means of photosensors (Kraftmakher 1981).

15.1 Formulas for Relaxation

Van den Sype (1970) calculated the relaxation in specific heat assuming it to contain two parts, the non-relaxing part C and the relaxing part ΔC. Designating the product of the angular frequency of the temperature oscillations and the relaxation time as $X = \omega\tau$, he obtained

$$|C(X)|^2 = (C_0^2 + C^2 X^2)/(1 + X^2), \qquad (15.1a)$$

$$\tan\Delta\phi = X \Delta C/(C_0 + CX^2), \qquad (15.1b)$$

where $C_0 = C + \Delta C$ is the equilibrium specific heat measured when $X^2 \ll 1$, and $\Delta\phi$ is the change in the phase of the temperature oscillations.

It is easy to deduce this formula considering complex specific heat $C(X) = C + \Delta C/(1 + iX)$. Fominaya et al. (1999b) also derived Eq. (15.1a), from which one obtains

$$X^2 = (C_0^2 - |C(X)|^2)/(|C(X)|^2 - C^2). \qquad (15.2)$$

The difference between the specific heats measured at a low and a high modulation frequency thus depends on C, ΔC, and X. The relaxation is also observable through changes in the phase of the temperature oscillations. In an adiabatic regime, the phase shift between the oscillations of the heating power and the temperature oscillations is close to 90°. The relaxation reduces this phase shift. The phase shift in the temperature oscillations depends nonmonotonically on X (Fig. 15.1). Measurements of the phase shift are very important because they provide an additional confirmation of the relaxation in specific heat.

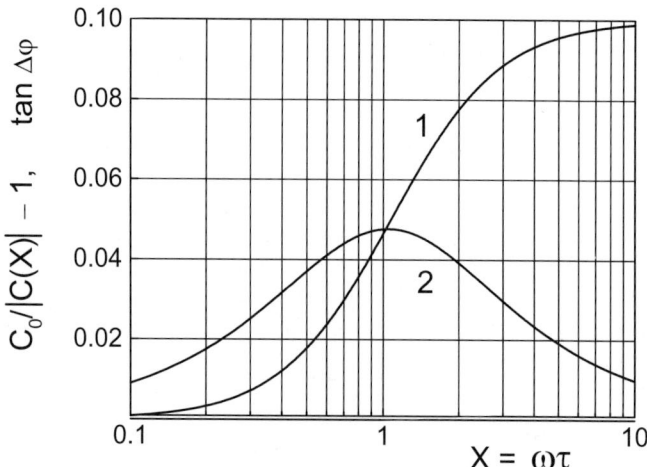

Fig. 15.1. Expected relaxation in specific heat versus $X = \omega\tau$, for $\Delta C/C = 0.1$. 1 – change in specific heat, 2 – shift in phase of temperature oscillations

The temperature dependence of the relaxation phenomenon was calculated for the vacancy equilibration in tungsten. The relaxation time decreases with increasing temperature. Accepting constant density of internal sources (sinks) for the vacancies, it is possible to evaluate the $C_0/|C(X)|$ ratio and the phase of the temperature oscillations (Fig. 15.2). For a constant modulation frequency, the ratio of the specific heats measured at a low and a high frequency of the temperature oscillations, $C_0/|C(X)|$, first increases with temperature, reaches a maximum, and then falls because the relaxation time decreases. The same relates to the change in the phase of the temperature oscillations.

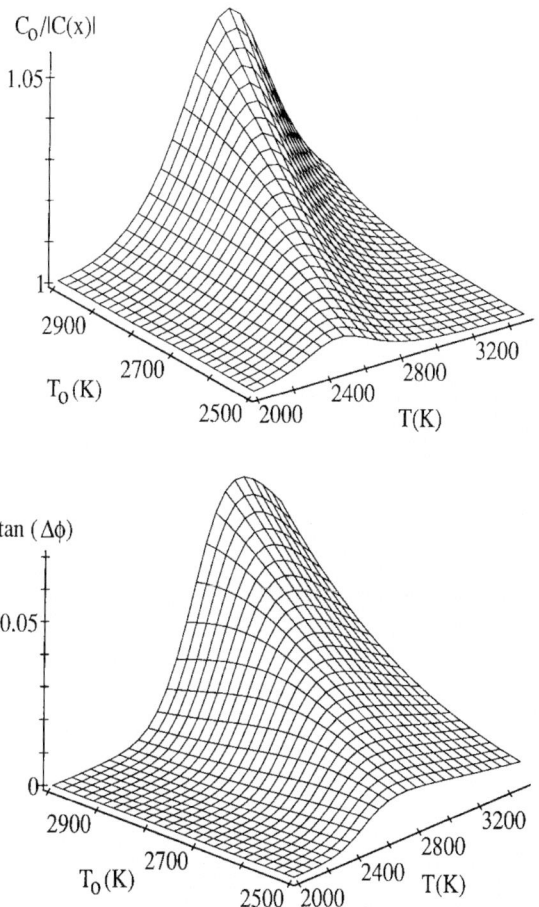

Fig. 15.2. Expected relaxation phenomenon in specific heat of tungsten. $C_0/|C(X)|$ is the ratio of specific heats measured at low and high modulation frequencies, and $\Delta\phi$ is the change in the phase of the high-frequency temperature oscillations. The accepted vacancy parameters are as follows: formation enthalpy $H_F = 3.15$ eV, formation entropy $S_F = 6.5k_B$, enthalpy of migration $H_M = 3$ eV. T_0 is a temperature for which $X = 1$ (Kraftmakher 2000)

15.2 Glass Transitions

An outstanding achievement of modulation calorimetry is the specific-heat spectroscopy (Birge and Nagel 1985, 1987; Birge 1986). It was aimed at measurements of the frequency dependence of specific heat of supercooled organic liquids in the region of the glass transitions. A thin nickel film evaporated onto a glass substrate is immersed in the liquid and generates temperature waves in it. The probe functions as a heater and a thermometer simultaneously (Sect. 7.4). A relaxation phenomenon was observed in many glass-forming liquids (Fig. 15.3).

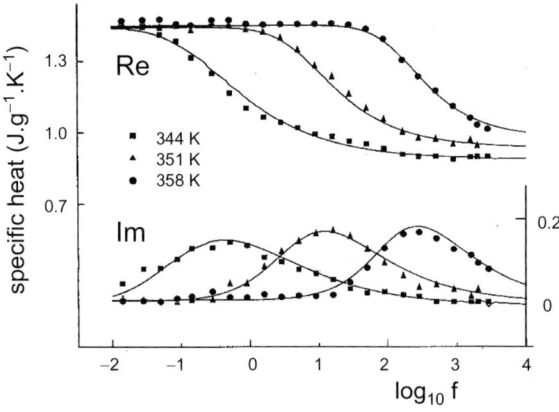

Fig. 15.3. Relaxation phenomenon in specific heat of supercooled $[(Ca(NO)_3)]_{0.4}(KNO_3)_{0.6}$ (CKN) in the region of glass transition (Jeong and Moon 1995; Moon and Jeong 2001)

Chirtoc et al. (2001) reported on photopyroelectric study of glass-forming liquids. The method was applied to measurements on glycerol at modulation frequencies up to 30 kHz. Later, the authors reported on the extension of the frequency range up to 100 kHz (Bentefour et al. 2003).

Minakov et al. (2001, 2003) developed a double-channel modulation calorimeter for simultaneous measurements of frequency-dependent heat capacity and thermal conductivity, in the frequency range 0.01 to 100 Hz. This technique is capable of distinguishing relaxation processes related to heat capacity and thermal conductivity. In an advanced version of the calorimeter, a disc-shaped sample is placed between two identical dielectric rods (Fig. 15.4). Metal heaters (Ni films, 30 nm thick) and temperature sensors (copper-constantan thermocouples, 100 nm thick) are sputtered on the polished ends of the rods, which are in direct thermal contact with the sample. A 1.5-μm dielectric layer of Al_2O_3 separates the heaters and the temperature sensors. The system can be described by the thermal effusivity of the rods, the cross-sectional area of the system, and the parameters of the sample: thickness, complex thermal effusivity and complex thermal diffusivity. The rods are 1 cm long and 4.5 mm in diameter. The second heater can be used to compensate for the temperature gradient in the sample or for 3ω measurements. The calorimeter was applied to study the glass transition in glycerol.

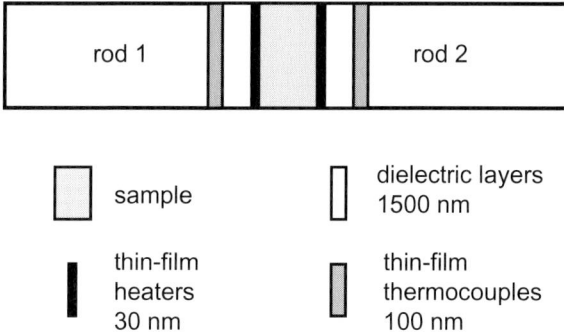

Fig. 15.4. Calorimeter cell of a double-channel calorimeter for simultaneous measurements of frequency-dependent heat capacity and thermal conductivity, in the frequency range 0.01 to 100 Hz (Minakov et al. 2003)

Table 15.1. Studies of glass transitions

Item	Reference
Glycerol, propylene glycol, 0.01–6000 Hz	Birge and Nagel 1985, 1987; Birge 1986
CKN, photoacoustic technique	Büchner and Korpiun 1987
(o-terphenyl)$_{1-x}$–(o-phenylphenol)$_x$	Dixon and Nagel 1988
Salol, 0.02–5000 Hz	Dixon 1990
Glycerol, propylene glycol, 10–10000 Hz	Inada et al. 1990
KDP, CKN, 0.01–5000 Hz	Jung et al. 1992; Jeong and Moon 1995; Moon et al. 1996; Jeong 1997
o-terphenyl, 2–6300 Hz, up to 0.1 GPa	Leyser et al. 1995
Polymers	Hensel et al. 1996
As$_x$Se$_{1-x}$, MDSC	Wagner and Kasap 1996
Di-n-butylphthalate, 0.004–8000 Hz	Menon 1996; Birge et al. 1997
Natural rubber, polystyrene, polymers, 0.2–2000 Hz	Korus et al. 1997
Polystyrene, 10^{-4}–2000 Hz	Weyer et al. 1997a
Poly(n-hexyl methacrylate), 0.2–2000 Hz	Beiner et al. 1998
(DMET)$_2$BF$_4$, (DMET)$_2$ClO$_4$	Akutsu et al. 1999
κ-(BEDT-TTF)$_2$Cu[N(CN)$_2$]Br	Saito et al. 1999a
Pd$_{40}$Ni$_{10}$Cu$_{30}$P$_{20}$, MDSC, 0.01–0.1 Hz	Hu et al. 1999
Pyroelectric sensor, frequencies up to 30 kHz	Chirtoc et al. 2001
Undercooled metallic liquids	Wilde 2002
Pyroelectric sensor, frequencies up to 100 kHz	Bentefour et al. 2003
Poly(vinyl acetate), 1–20 mHz	Castro and Puértolas 2003
Frequencies up to 30 kHz	Jung et al. 2003

CKN = [(Ca(NO$_3$)$_2$]$_{0.4}$(KNO$_3$)$_{0.6}$, KDP = KH$_2$PO$_4$, DMET = dimethyl(ethylenedithio), BEDT-TTF = bis(ethylenedithio)tetrathiafulvalene

Weyer et al. (1997a) studied polystyrene in the glass-transition region combining results from MDSC and those based on measurements of thermal effusivity by the third-harmonic method. The MDSC measurements were performed in the range 10^{-4} to 0.1 Hz, while the third-harmonic method provided data in the range 0.2 to 2000 Hz. The combination of the two techniques thus allows specific heat spectroscopy in a frequency range of seven orders of magnitude. Suga (2001) considered an adiabatic calorimeter as an ultra-low frequency spectrometer.

Table 15.1 lists studies of glass transitions with the specific-heat spectroscopy.

15.3 Point-Defect Equilibration in Metals

In studies of the point-defect equilibration in metals, the sample is subjected to such rapid temperature oscillations that the defect concentration cannot follow them. Under such conditions, the defect contribution to a given physical property is almost completely excluded. This statement relates only to properties that sense changes in the defect concentrations during the measurements: specific heat, thermal expansivity, and temperature derivative of electrical resistance. This is because the main contributions to the above properties are caused not by the presence of point defects, but by the temperature dependence of their concentration. When the defect concentration does not follow temperature oscillations and retains a mean value, these properties practically correspond to a hypothetical defect-free crystal.

The only obstacle for this approach arises from short equilibration times due to the high mobility of the defects at high temperatures and numerous internal sources and sinks for them. The relaxation is therefore observable only with sufficiently high modulation frequencies. The amplitude of the temperature oscillations is inversely proportional to their frequency, and such measurements require a sensitive technique. When measuring specific heat at a very high modulation frequency, the result should correspond to a defect-free crystal. At intermediate frequencies, it depends on the modulation frequency and the relaxation time, according to Eq. (15.1a).

15.3.1 Enhancement of Modulation Frequency

To search for relaxation phenomena in metals, a method of measuring high-temperature specific heat at frequencies of the order of 10^5 Hz has been developed (Kraftmakher 1981). A high-frequency current slightly modulated by a low-frequency voltage heats a wire sample (Fig. 15.5). Therefore, temperature oscillations of the two frequencies occur in the sample simultaneously. A photomultiplier detects these oscillations. The low-frequency component of its output signal proceeds to a lock-in amplifier. The high-frequency component selected by a resonant circuit is measured using frequency conversion and lock-in detection. An auxiliary frequency converter provides the reference voltage for the lock-in detector. A plotter records the signal proportional to the difference between the amplitudes of the high-frequency and low-frequency temperature oscillations. The measurements start at sufficiently low temperatures, where the point-defect concentration is negligible and no relaxation in specific heat is expected. At these temperatures, the recorded signal is adjusted to be zero. Then the difference between the specific heats corresponding to the two frequencies is directly measured at various temperatures of the sample.

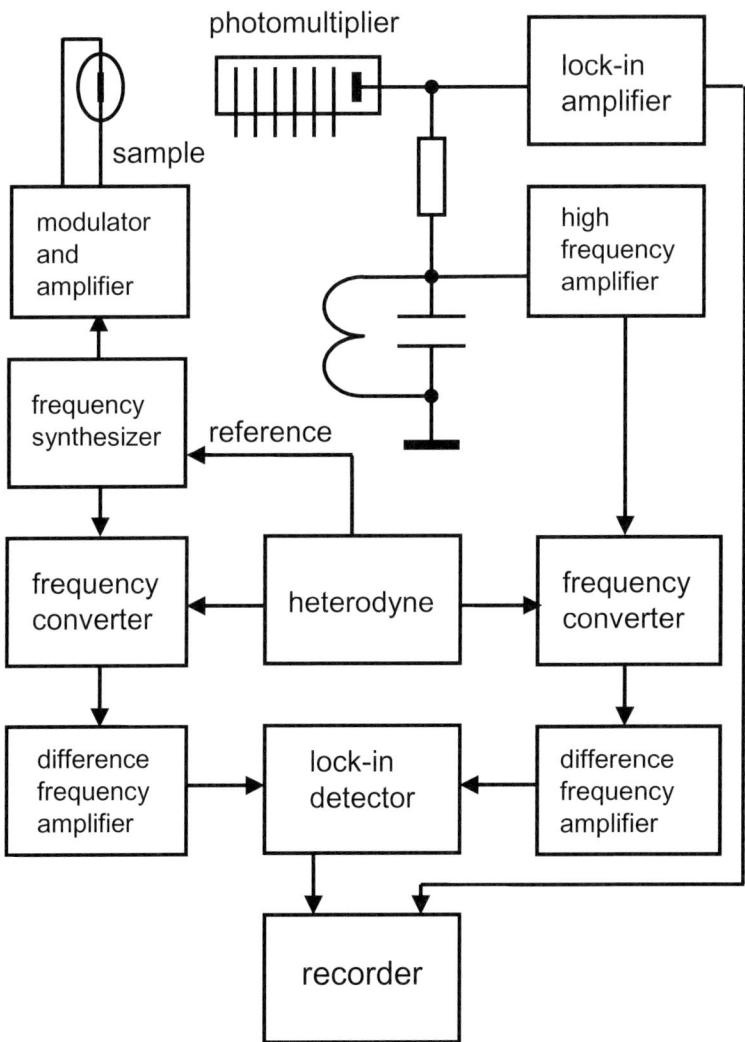

Fig. 15.5. Diagram of apparatus for observation of relaxation in high-temperature specific heat of metals and alloys (Kraftmakher 1981). Temperature oscillations of low and high frequencies occur in the sample simultaneously, and the ratio of corresponding specific heats is measured directly

15.3.2 Relaxation Phenomenon in Tungsten and Platinum

The measurements were carried out on commercial 8-µm tungsten wires and on vacuum incandescent lamps with 10–20 µm tungsten filaments (Kraftmakher 1985). The highest modulation frequency was 3×10^5 Hz. The temperature dependence of the relaxation observed was within the expectation (Fig. 15.6). However, this observation gained no recognition. Considering this work, Trost et al. (1986) concluded: "…on the basis of the information available at present we cannot exclude with certainty that the observed effect is partly or even entirely due to the experimental procedure and hence not intrinsic." Regretfully, the authors did not perform their own investigation of the phenomenon. No attempts were also reported on such measurements by other groups.

In the case of platinum, the high frequency was 5×10^4 Hz (Kraftmakher 1990). The samples were cut from a 10-µm platinum foil. Due to the lower melting temperature, the power heating the sample and the amplitude of the power oscillations becomes much smaller. This results in decrease of the temperature oscillations and, consequently, of the applicable modulation frequencies. The observed relaxation appeared to be in agreement with the non-linear increase in the specific heat. In both cases, the scatter of the experimental points increases at temperatures, where $X = \omega\tau$ is close to unity. This is quite explainable because only in this range the relaxation strongly depends on X. The latter is governed by the relaxation time, i.e., by the density of the sources (sinks) for the vacancies. At high temperatures, this density varies due to diffusion processes in the sample.

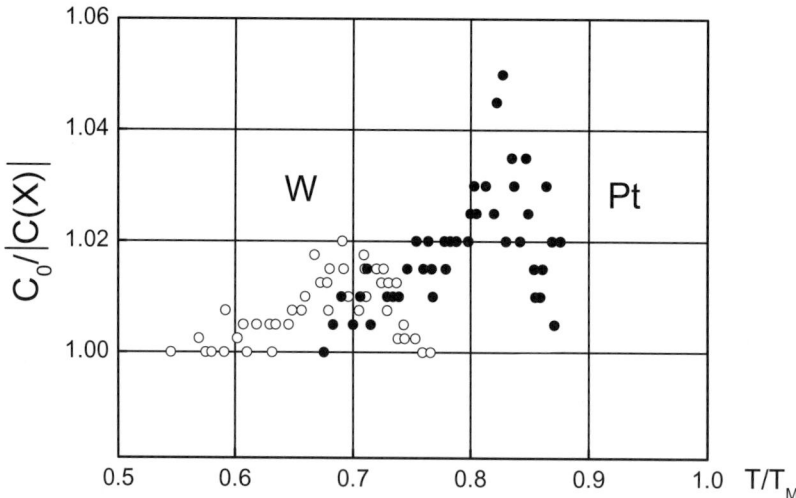

Fig. 15.6. Ratio of specific heats measured at low and high frequencies of temperature oscillations in tungsten and platinum (Kraftmakher 1985, 1990). The high frequency is 3×10^5 and 5×10^4 Hz, respectively. In both cases, the phenomenon observed is within expectations

15.3.3 Equilibration Times

A common opinion has been established that the quenched-in resistivity and changes in the positron-annihilation parameters are caused by vacancies. At the same time, the relation between the vacancy formation and the non-linear increase in specific heat is a disputable question. It seems therefore useful to compare results of relaxation experiments on various metals (Fig. 15.7). The relaxation time τ depends on the density of sources (sinks) for vacancies, i.e., on the quality of a sample. The difference between the relaxation times in gold obtained by measurements of the electrical resistivity (Seidman and Balluffi 1965) and by positron annihilation (Schaefer and Schmid 1989), which amounts to a factor of 50, is therefore quite understandable.

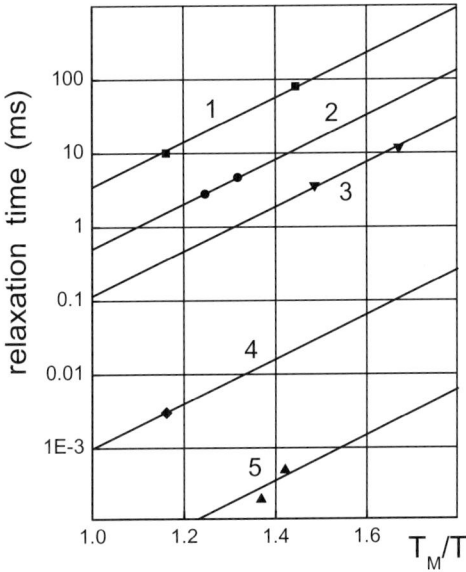

Fig. 15.7. Temperature dependence of equilibration times from various measurements: 1 – Au, resistivity (Seidman and Balluffi 1965); 2 – Al, current noise (Celasco et al. 1976); 3 – Au, positron annihilation (Schaefer and Schmid 1989); 4 – Pt, specific heat (Kraftmakher 1990); 5 – W, specific heat (Kraftmakher 1985). The straight lines correspond to $H_M = 7k_B T_M$ and constant densities of sources (sinks) for vacancies

For the comparison of relaxation times in various metals, the migration enthalpies, H_M, were considered to be proportional to the melting temperatures T_M ($H_M = 7k_B T_M$). The straight lines in the graph correspond to $\tau = B\exp(H_M/k_B T)$, where H_M is the migration enthalpy, and B is a factor different for various samples. Assuming a temperature-independent density of sources (sinks) for vacancies, the temperature dependence of the relaxation time is available even from a single measured value. In the tungsten sample, the relaxation times were found to be

5×10^{-7} s at 2600 K and 2×10^{-7} s at 2700 K. The short relaxation times are consistent with the well-known fact that dislocation densities in refractory metals are much higher than those in metals such as gold or platinum. The comparison of the data is thus favourable for the conclusion that all the relaxation phenomena considered here are of common nature.

15.3.4 Two Proposals for Studies of Vacancy Equilibration

A simple experimental approach was proposed to check whether the relaxation phenomenon observed in the specific heat of tungsten and platinum originates from the vacancy equilibration (Kraftmakher 2000). The relaxation should be observed during a period including a quench and subsequent anneal of the sample. The main sources and sinks for vacancies are dislocations and, probably, vacancy clusters. Their density dramatically increases after a quench, and a certain time is necessary to anneal the sample at the high temperature. If, while the relaxation is observed, to interrupt the heating current and then return the sample to the initial temperature, the phenomenon may disappear. It will appear again after a proper anneal of the sample and recovery of its structure (Fig. 15.8).

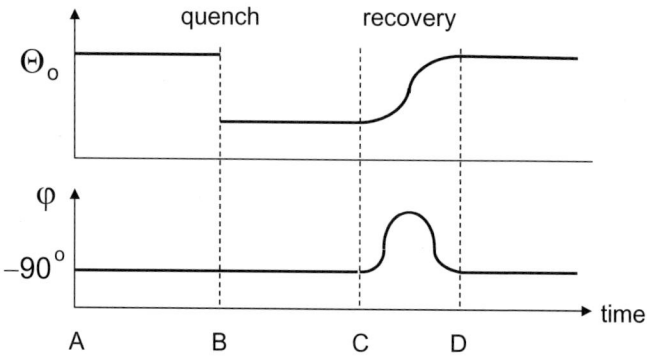

Fig. 15.8. Expected behaviour of amplitude and phase of temperature oscillations during quench and subsequent anneal of the sample, schematically. Observation of such behaviour would reliably confirm the vacancy contribution to specific heat

The amplitude and the phase of the high-frequency temperature oscillations should not alter when the sample is quenched from a low temperature, where $\Delta C \ll C$ or the relaxation time remains sufficiently long even after the quench. No changes are expected also at high temperatures, where the modulation frequency becomes insufficient to observe the relaxation. Only under very favourable conditions, when $X^2 \gg 1$ before the quench and $X^2 \ll 1$ after it, would the change in the amplitude of the high-frequency temperature oscillations reveal the true magnitude of the relaxation. Otherwise, this change corresponds to only a part of the phenomenon. Measurements with various modulation frequencies would reveal the temperature dependence of the relaxation. Along with providing data on equilibration, this approach would unambiguously prove the vacancy origin of the relaxation phenomenon in specific heat.

The directions of further investigations of the point-defect equilibration are as follows:

- Employment of pure and well-prepared samples with reduced densities of sources (sinks) for the vacancies, to observe the relaxation over a wider temperature range.
- Measurements on other metals, in which low dislocation densities are obtainable.
- Observations of changes in the relaxation time during a quench and subsequent anneal of the samples. A modified experimental set-up may appear to be useful for such measurements (Fig. 15.9).

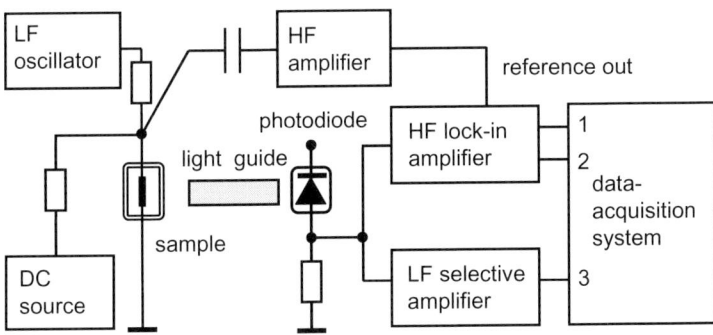

Fig. 15.9. Modified set-up for observing relaxation in specific heat of metals and alloys at high temperatures (Kraftmakher 1998). It may appear to be useful for observing changes in the relaxation phenomenon during quench and subsequent anneal of the sample

An additional approach exists to check the origin of the non-linear increase in high-temperature specific heat of metals. Under high pressures, equilibrium vacancy concentrations decrease because of increase in the Gibbs free energy of vacancy formation. Vacancy contributions to specific heat thus decrease under high pressures. Experimental data on the pressure dependence of the vacancy-induced resistivity show that the pressures to be employed are quite moderate, of the order of 0.1 GPa. Such a pressure is probably insufficient to markedly change other contributions to the specific heat. Dynamic calorimetry and modulation calorimetry seem to be most suitable for measurements under high pressures.

15.3.5 Questions to be Answered by Rapid-Heating Experiments

Two important questions may be answered by rapid-heating experiments. The first one relates to equilibrium vacancy concentrations in metals. Under very rapid heating, vacancies have no time to appear, and the enthalpy of a sample at a given premelting temperature should be smaller than that under a moderate heating rate.

For molybdenum and tungsten, the vacancy-related enthalpy at the melting point calculated from the non-linear increase in the specific heat amounts to about 10% of the total enthalpy. To check this conjecture, an examination of typical data now available was made (Kraftmakher 1996, 1997). The sources of the data included equilibrium measurements of the enthalpy and the specific heat, and rapid-heating determinations of the enthalpy at the melting points. From estimated times of the vacancy equilibration, only experiments with heating rates of the order of 10^8 K.s^{-1} or more may be expected not to contain the vacancy contribution.

It turned out that certainly different values of the enthalpy at the melting points were obtained under equilibrium and in rapid-heating measurements. To make a quantitative comparison, parts related to the non-linear increase in the specific heat were extracted from the results of equilibrium measurements. For this purpose, the experimental data were fitted by equations supposing vacancy formation. To fit the specific heat measured only at high temperatures, the enthalpy at 1500 K was taken to be 32.6 kJ.mol^{-1} for tungsten (Chekhovskoi 1981) and 33.5 kJ.mol^{-1} for molybdenum (Chekhovskoi and Petrov 1970). We thus obtained three sets of the enthalpies of tungsten and molybdenum at the melting points (Table 15.2): (i) the equilibrium enthalpies after subtracting the assumed vacancy contributions, H_1; (ii) the total equilibrium enthalpies including the vacancy contributions, H_2; and (iii) the enthalpies from rapid-heating measurements, H_3, which are expected to be close to H_1 rather than to H_2. With heating rates in the range 10^8–10^9 K.s^{-1}, the uncertainty in the enthalpy is within 3-5% (Pottlacher et al. 1993). The difference between H_1 and H_2 is therefore quite detectable.

Table 15.2. Enthalpy of solid tungsten and molybdenum at the melting points (kJ.mol^{-1}): H_1 – enthalpy not including the assumed defect contribution, H_2 – total enthalpy, H_3 – results of rapid-heating experiments

H_1	H_2	H_3	Reference
	Tungsten		
112	122		Kraftmakher and Strelkov 1962
112	116		Cezairliyan and McClure 1971
109	116		Chekhovskoi 1981
		112	Hixson and Winkler 1990
		111	Pottlacher et al. 1993
109	117		Righini et al. 1993
	Molybdenum		
81	91		Rasor and McClelland 1960b
84	92		Kraftmakher 1964
83	89		Chekhovskoi and Petrov 1970
		85	Seydel and Fischer 1978
85	89		Cezairliyan 1983
83	90		Righini and Rosso 1983
		87	Hixson and Winkler 1992
		87	Pottlacher et al. 1993

For tungsten, the results of the rapid-heating experiments, H_3, are close to the evaluated H_1 values and thus support the above conjecture. For molybdenum, the results of rapid-heating experiments lay between the two values, H_1 and H_2. Possible explanations of this may be as follows: (i) the heating is not fast enough to completely avoid the vacancy formation; (ii) a superheat of the samples under rapid heating leads to enhancement of the apparent melting point and of the corresponding enthalpy; (iii) vacancy formation accounts for only a part of the non-linear increase in the specific heat. A way to distinguish between these possibilities is to carry out the measurements with various heating rates and to determine the apparent melting temperature independently. To avoid a superheat, one may measure the enthalpy also at a selected premelting temperature. The rapid-heating technique is thus a very promising tool to reveal the vacancy-related enthalpy of metals.

The second question relates to the mechanism of melting. In some rapid-heating experiments, a significant superheat above equilibrium melting points was observed. This may be due to the lack of vacancies in the crystal lattice. To answer the question, it would be probably useful to compare results of two rapid-heating experiments: (i) starting at a temperature, where the vacancy concentrations are still negligible, and (ii) starting at a temperature, where they are sufficiently large.

15.3.6 Determination of Vacancy-Related Enthalpy – A Proposal

A straightforward approach was proposed (Kraftmakher 1997) to reveal vacancy contributions to high-temperature enthalpy of metals (Fig. 15.10). After heating the sample to a premelting temperature, the initial part of the cooling curve should depend on whether the vacancies had time to arise. If they had not, they will appear immediately after the heating.

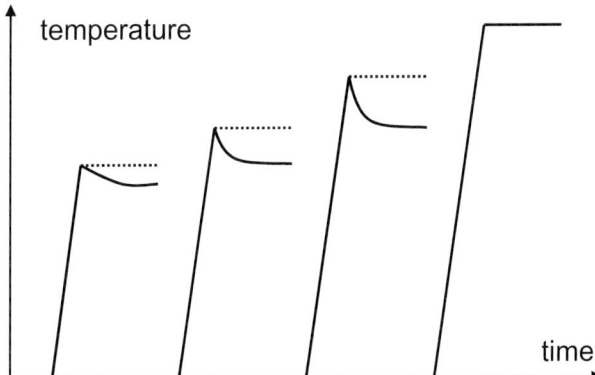

Fig. 15.10. Temperature traces expected after rapid heating of a sample to premelting temperatures (Kraftmakher 1997). At the highest temperature, vacancies have arisen during heating. Observation of this phenomenon would strongly confirm the vacancy nature of the non-linear increase in specific heat

At premelting temperatures, the equilibrium vacancy concentrations are set up in 10^{-4} to 10^{-2} s in low-melting-point metals and in 10^{-8} to 10^{-6} s in refractory metals. Under normal conditions, the temperature of the sample after the heating, in the time interval of interest, should remain nearly constant. Therefore, the decrease of the temperature of the sample immediately after the heating may be due only to the vacancy formation.

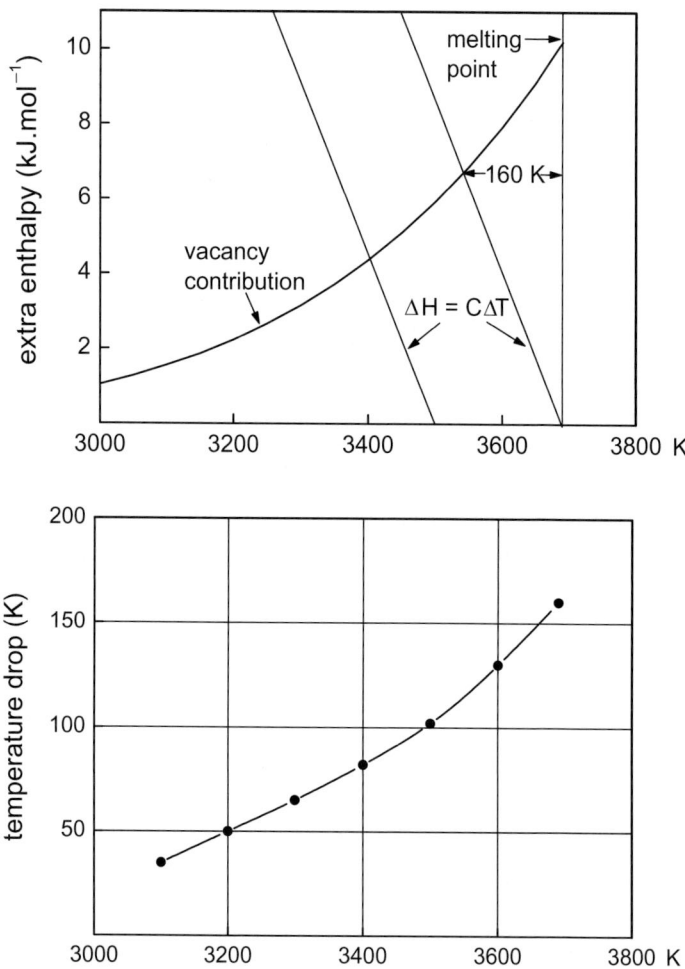

Fig. 15.11. Calculation of the temperature drop ΔT caused by vacancy formation after rapid heating of tungsten sample to various premelting temperatures (Kraftmakher 1998). If the basic assumption is correct, the temperature drop should be observable

When the vacancies arise after the heating, the initial cooling curve should depend on the vacancy contribution to the enthalpy and the relaxation time. Both quantities strongly depend on temperature. The enthalpy necessary to create the vacancies should be measurable from the temperature drop in the sample immediately after the heating. If the heating is not sufficiently fast, then the phenomenon could be studied under gradually increasing the upper temperature of the sample. The temperature drop has first to increase with the upper temperature, reach a maximum, and then fall because of decrease in the relaxation time.

To evaluate the expected temperature drop ΔT, it is necessary to consider the equilibrium vacancy concentration at the final temperature of the sample after the equilibration. The vacancy-related enthalpy equals $\Delta H = H_F \exp(-G_F/k_B T)$. For tungsten, the calorimetric data predict the vacancy-related enthalpy of about 10 kJ.mol^{-1} at the melting point. The heat-balance requirement is $C\Delta T = \Delta H$, where C is the specific heat of tungsten not including the vacancy contribution ($C \cong 40$ J.mol^{-1}.K^{-1}). The temperature drop due to the vacancy formation thus can be evaluated for any temperature achieved after a rapid heating (Fig. 15.11).

After heating up to the melting point, the maximum temperature drop amounts to about 160 K. This figure reduces to 105 K after heating up to 3500 K and to 50 K after heating up to 3200 K. For molybdenum, the situation is very similar. The phenomenon should be clearly seen if the equilibrium vacancy concentrations are of the order of 10^{-2}, but unobservable if they are less than 10^{-3}. From the measurements, the temperature dependence of the vacancy-related enthalpy could be determined.

The rapid-heating experiment to be made is similar to those reported earlier (Hixson and Winkler 1990, 1992; Pottlacher et al. 1991, 1993). Owing to the proposed approach, the measurements may be even simpler. Now there is no need to measure the heating current and the voltage drop across the sample. It is sufficient to rapidly heat up the sample, within 10^{-8} or 10^{-7} s, to a premelting temperature and to observe the initial part of the cooling curve. A very important point is to completely avoid a heating of the sample after reaching the selected temperature. Any uncontrollable heating will cause ambiguities in the vacancy-related enthalpy. The expected phenomenon should be clearly seen and amenable to quantitative treatment. Still more important, the nature of the phenomenon would be evident. This technique offers reliable determination of equilibrium vacancy concentrations in metals. It seems to be the simplest experiment that could be undertaken for this purpose.

The vacancy formation in high-melting-point metals might be also seen from the dilatation of the sample immediately after a rapid heating. It is easy to show that the temperature drop due to the vacancy formation in the sample should be accompanied by an increase in its volume. Such unusual behaviour would be the best confirmation of the vacancy origin of the phenomenon. The same approach is probably applicable to determinations of vacancy contributions to electrical resistivity. To be more informative, measurements of the extra resistivity should be made along with determinations of the vacancy-related enthalpy. However, the relative vacancy contribution to the electrical resistivity of refractory metals is much smaller than that to the enthalpy.

15.4 Other Examples of Relaxation in Specific Heat

There exist many other examples of relaxation phenomena in specific heat. Garoche et al. (1978) measured the specific heat of TaS_2 single crystals in the trigonal $2H$ phase between 0.3 and 3 K. In the superconducting state, below 0.6 K, the specific heat is frequency independent in the range 3–20 Hz. However, when the sample was restored to the normal state by applying an external magnetic field, the specific heat was found to decrease with increasing frequency. The authors explained this result by a frequency-dependent nuclear contribution to the specific heat. The phenomenon is governed by the rate, at which the equilibrium is established between the nuclear spins and the lattice.

Jeong et al. (1991) claimed that there is an indication that the specific-heat peak associated with the magnetic phase transition in gadolinium "may depend on the measuring frequencies." This conclusion was derived from a minor difference in the specific-heat data obtained with modulation frequencies 0.24 and 316 Hz. An increase of the upper frequency would clarify the question. Morilov and Ivliev (1995) reported on the relaxation behaviour of specific heat of gadolinium in the vicinity of the hcp-bcc phase transition. Saruyama (1997) observed a frequency dependence of the specific heat of polyethylene in the vicinity of the melting point. Minakov et al. (1999) discussed the complex heat capacity.

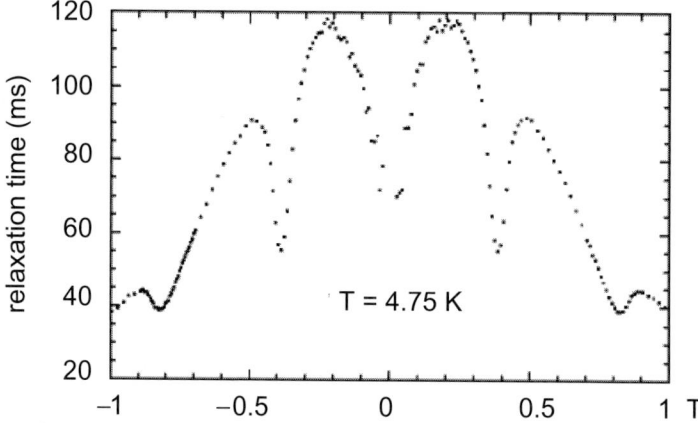

Fig. 15.12. Relaxation phenomenon in specific heat of Mn_{12}-acetate crystal (Fominaya et al. 1999b). Relaxation time at 4.75 K is shown as a function of external magnetic field

Fominaya et al. (1999b) found a relaxation phenomenon in the specific heat of Mn_{12} acetate. The crystals are arrangements of identical magnetic clusters in an organic matrix, without exchange coupling from one cluster to another. The magnetic contribution to the specific heat and the relaxation time depend on temperature and applied magnetic field. The magnetic field parallel to the easy axis of magnetisation was varied from −1 to 1 T. The modulation frequencies were in the range 4–20 Hz. From the measurements, the relaxation time was calculated for various temperatures and magnetic fields (Fig. 15.12). Fominaya et al. (1999a) observed similar behaviour of specific heat in magnetic field in Fe_8 crystals, but without any indication of frequency dependence in the range 35–700 Hz.

Akutsu et al. (1999) detected a frequency-dependent anomaly in specific heat of two organic conductors, $(DMET)_2BF_4$ and $(DMET)_2ClO_4$, around 110 K. This anomaly was attributed to a glass transition due to the freezing of the intramolecular motion of the ethylene group in the DMET molecule. Very moderate frequencies, in the range 1–16 Hz, were sufficient to reveal the frequency dependence of the specific heat. Saito et al. (1999a) found this effect in κ-$(ET)_2Cu[N(CN)_2]Br$ in the range 90–120 K. Similar relaxation phenomena were observed in κ-$(BEDT$-$TTF)_2Cu[N(CN)_2]Br$ and in κ-$(BEDT$-$TTF)_2Cu[N(CN)_2]Cl$ (Akutsu et al. 2000) (Fig. 15.13) and in deuterated salts (Sato et al. 2001).

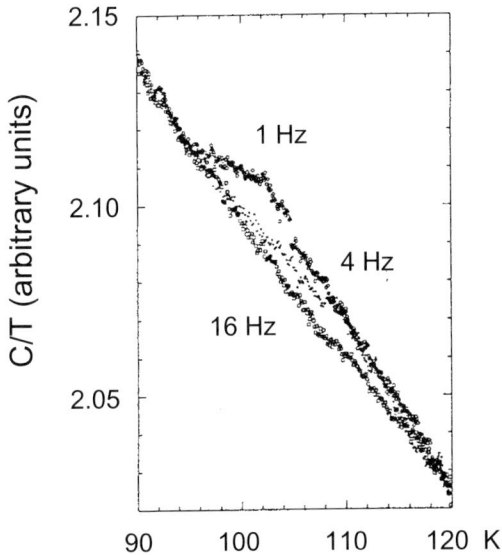

Fig. 15.13. Frequency dependent C/T ratio for κ-$(BEDT$-$TTF)_2Cu[N(CN)_2]Br$ (Akutsu et al. 2000)

16 Five Student Experiments

Five student experiments are briefly described below. The experiments relate to pulse calorimetry, modulation techniques, and noise thermometry. All the experiments do not require sophisticated equipment.

16.1 Pulse Calorimetry

A simple set-up is usable for acquaintance with the basic features of pulse calorimetry (Sect. 5.3). A low-power light bulb with a tungsten filament is included in a bridge circuit (Fig. 16.1). Unipolar electric pulses from a pulse generator feed the bridge, while the output voltage of the bridge is amplified by a differential amplifier and displayed by an oscilloscope or a data-acquisition system. Between the pulses, the output voltage is zero, regardless of the balance of the bridge.

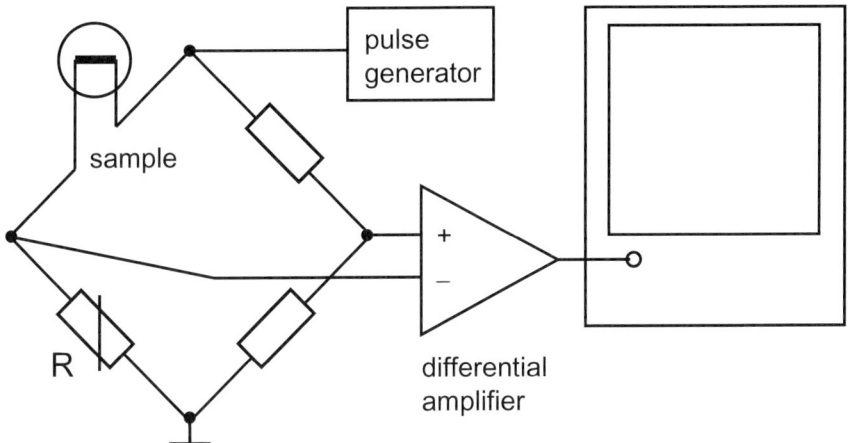

Fig. 16.1. Arrangement for pulse calorimetry experiment for students. A low-power light bulb is an excellent sample for the measurements

During the heating, the temperature of the filament increases, which is seen from an increase of its resistance (Fig. 16.2). The heating period is chosen to make this increase to be linear with time, which corresponds to an adiabatic regime. The heat-balance equation has a simple form

$$mc\Delta T = (P - Q)\Delta t, \tag{16.1}$$

where P is the power supplied to the filament, Q is the power of heat losses from it, and ΔT is the increase of the temperature of the filament during the interval Δt. In an adiabatic regime, $P - Q$ is independent of time because the changes in the heat losses from the sample are much smaller than the supplied power. The heating time Δt equals a half of the period. Equation (16.1) can be presented as

$$mc = (P - Q)R'\Delta t/\Delta R = (P - Q)R_{273}\beta\Delta t/\Delta R, \tag{16.2}$$

where ΔR is the increment of the resistance of the filament during the interval Δt, R_{273} is the resistance of the filament at 273 K, and $\beta = R'/R_{273}$ is the temperature coefficient of resistance.

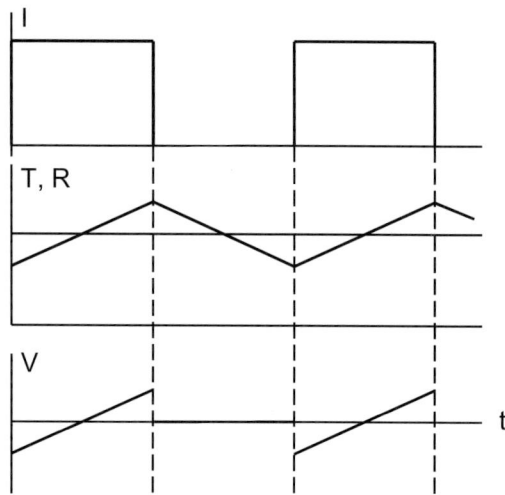

Fig. 16.2. Diagram of the pulse-heating technique. I is the heating current, T and R are the temperature and the resistance of the sample, and V is the output voltage of the bridge reflecting the temperature changes during the heating period

The room-temperature resistance of the filament is obtainable by gradually decreasing the heating current. For tungsten, the R_{293}/R_{273} ratio equals 1.095. From the determined value of R_{273} and data in Table 16.1, one calculates the resistance of the filament at temperatures 1500 K, 1600 K and so on up to 2800 K. Below 1500 K, cold-end effects become significant, while above 2800 K a problem appears because of evaporation of tungsten in vacuum. With a gas-filled lamp, the measurements are possible up to 3400 K. The calculated resistances are set at the bridge in turn, and one balances the bridge by changing the heating current. After the bridge is nearly balanced, it is easy to adjust the variable resistor to achieve the balance at the start and at the end of the heating (Fig. 16.3). From this adjustment, the increase in the resistance ΔR is readily available. The values of β are taken from Table 16.1.

Another possibility consists in approximation of the temperature dependence of the resistance of tungsten. In the range 1500–3600 K, the necessary polynomial fits are as follows:

$$T = 170 + 184X - 1.8X^2 + 0.02X^3, \qquad (16.3a)$$

$$X = -0.706 + 0.005T + 4.8 \times 10^{-7}T^2 - 2.56 \times 10^{-11}T^3, \qquad (16.3b)$$

$$\beta = dX/dT = 0.005 + 9.6 \times 10^{-7}T - 7.68 \times 10^{-11}T^2. \qquad (16.3c)$$

In our case, $P - Q = Q$, so that there are several ways to determine these values. The simplest one is to achieve the same temperatures of the filament with a DC current. In this case, the applied power equals the power of heat losses, Q. By changing the period of heating, a passage from an adiabatic regime to a nonadiabatic regime is clearly seen.

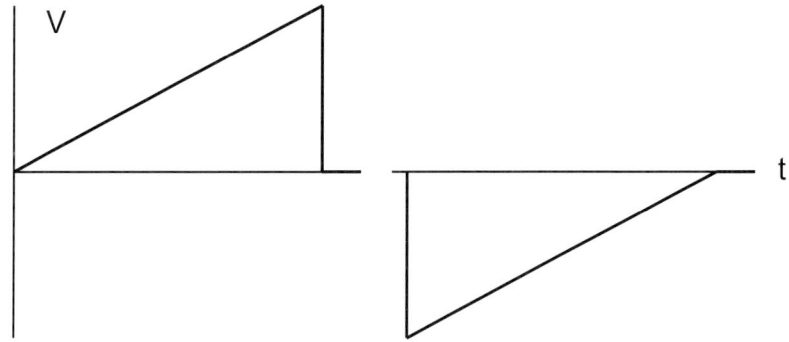

Fig. 16.3. By adjusting the variable resistor of the bridge, one determines the resistance of the filament at the start and at the end of the heating

Table 16.1. Resistance ratio R/R_{273} and the temperature coefficient of resistance $\beta = R'/R_{273}$ for tungsten (Roeser and Wensel 1941)

T (K)	R/R_{273}	$\beta[10^{-5}\ K^{-1}]$	T (K)	R/R_{273}	$\beta[10^{-5}\ K^{-1}]$
1500	7.78	621	2600	15.08	699
1600	8.41	629	2700	15.78	704
1700	9.04	637	2800	16.48	709
1800	9.69	645	2900	17.19	713
1900	10.34	654	3000	17.90	718
2000	11.00	661	3100	18.62	723
2100	11.65	669	3200	19.35	728
2200	12.33	676	3300	20.08	733
2300	13.01	682	3400	20.82	737
2400	13.69	688	3500	21.56	739
2500	14.38	694	3600	22.30	740

16.2 Principles of Modulation Calorimetry

A low-voltage (2.5 V) vacuum light bulb serves for verification of the basic theory of modulation calorimetry (Sect. 2.1). A DC current with a small AC component heats the filament, and a photodiode detects the temperature oscillations in it. The light bulb is fed from a DC source and an oscillator through resistors of sufficiently high resistance. The signal from the photodiode proceeds to the Y-input of an oscilloscope or to a data-acquisition system. The X-input is fed by the voltage across the filament or a resistor placed in series with it (Fig. 16.4). In this experiment, only relative values of the temperature oscillations are determined.

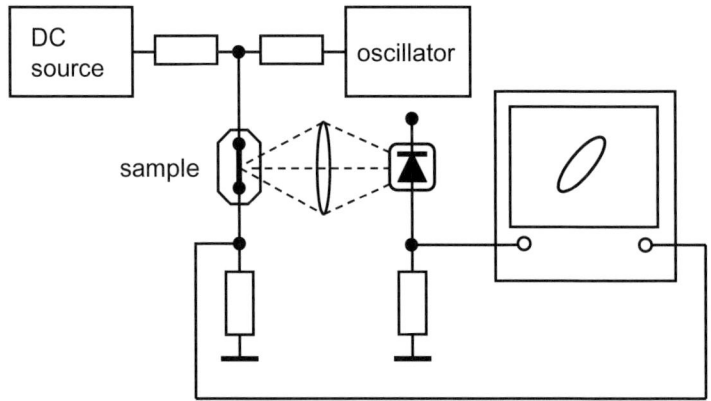

Fig. 16.4. Set-up for observing frequency dependence of the amplitude and the phase of the temperature oscillations in a low-power light bulb heated by a DC current with a small AC component

Using the Lissajous figures (Fig. 16.5), the amplitude Θ_0 and the phase of the temperature oscillations ϕ are measured versus of the modulation frequency. The results are presented as $\Theta_0(\omega)$, $\tan\phi(\omega)$, and a polar diagram $\Theta_0(\phi)$ (Fig. 16.6).

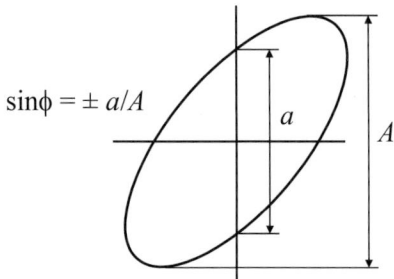

Fig. 16.5. Determination of the phase shift ϕ from the Lissajous figure

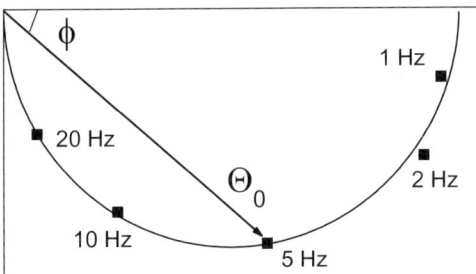

Fig. 16.6. Polar diagram $\Theta_0(\phi)$ from data obtained shows the amplitude and the phase of the temperature oscillations

Another way to measure the phase shift ϕ is the use of an integrating RC circuit at the X-input of the oscilloscope or the data-acquisition system (Fig. 16.7). The phase shift introduced by the circuit is set equal to the phase shift between the AC component of the heating current and the temperature oscillations in the sample, and a straight line is observed on the screen. The parameters of the RC circuit should not depend on the modulation frequency. This is seen from comparison of Eq. (2.3c) with the expression for the output voltage of an integrating RC-circuit: $U_{out} = U_{in}/(1 + \omega^2 R^2 C^2)^{1/2}$. The input voltage U_{in} is analogous to the quantity p/Q' in Eq. (2.3c), while the output voltage U_{out} is analogous to the amplitude of the temperature oscillations in the sample. Therefore, the quantity RC is analogous to mc/Q' and should be independent of the frequency. The measurements are carried out for several mean temperatures deduced from the resistance of the filament.

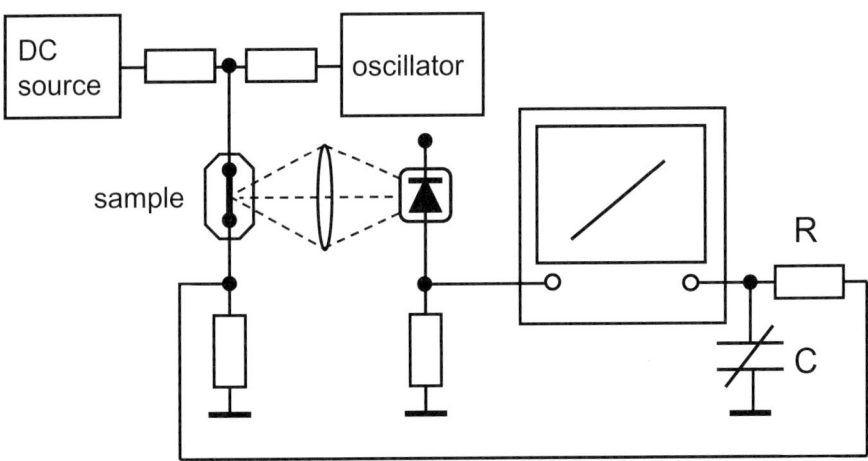

Fig. 16.7. Integrating RC circuit at the second input of an oscilloscope or a data-acquisition system introduces a phase shift equal to that between the AC component of the heating current and the temperature oscillations in the sample. The parameters of the RC circuit do not depend on the modulation frequency

16.3 Third-Harmonic Technique

When a current $I\sin\omega t$ heats a sample, the temperature oscillations in it, under adiabatic conditions, are $\Theta = -\Theta_0\sin2\omega t$. If the sample is fed through a sufficiently high resistance, the voltage across the sample equals

$$V = (I\sin\omega t)\times(R_0 - R'\Theta_0\sin2\omega t)$$

$$= IR_0[\sin\omega t - (\beta\Theta_0/2)\cos\omega t + (\beta\Theta_0/2)\cos3\omega t],$$

(16.4)

where R_0 is the resistance of the sample, and $\beta = R'/R_0$ is the temperature coefficient of resistance. The voltage contains a third-harmonic component directly related to the amplitude of the temperature oscillations in the sample and thus to its heat capacity (Sect. 4.1.2).

The sample, a low-voltage light bulb, is included in a bridge (Fig. 16.8). A variable resistor R balances the in-phase component of the voltage (the first term in the brackets). A variable capacitor C is necessary to balance the out-of-phase component (the second term). The third-harmonic component of the voltage across the sample that equals $V_{3\omega} = (\beta\Theta_0 IR_0/2)\cos3\omega t$ cannot be balanced. It directly relates to the heat capacity of the sample.

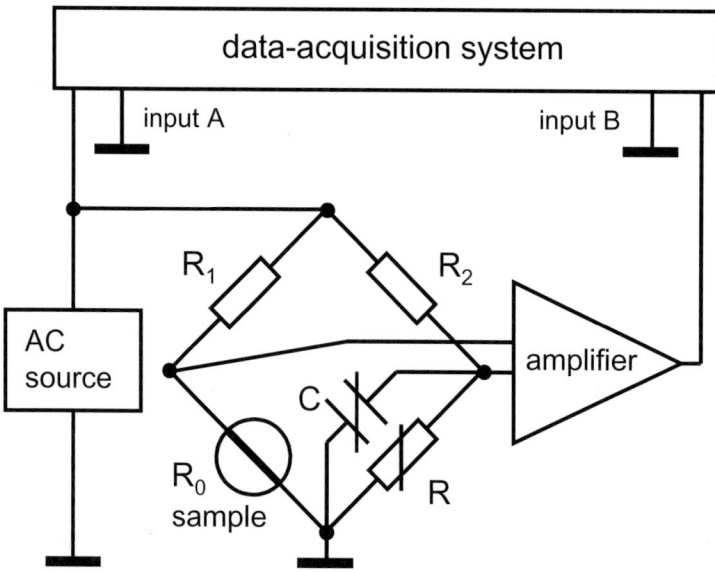

Fig. 16.8. Arrangement for measurements by the third-harmonic technique. The output voltage of the bridge is monitored with an oscilloscope or a data-acquisition system

After balancing the in-phase fundamental-frequency component, the output voltage of the bridge contains the out-of-phase fundamental-frequency component and the third-harmonic voltage (Fig. 16.9). From this output voltage, the amplitude of the temperature oscillations in the sample is also available. It is easy to calculate that its amplitude equals 1.54 times the amplitude of the third-harmonic component. By further balancing with the variable capacitor, only the third-harmonic voltage remains. It should be remembered that the output third-harmonic voltage differs from the voltage drop across the sample given by Eq. (16.4). To obtain the output voltage, one has to consider the resistances of the arms of the bridge. When $R_0 = R_1$, then $R = R_2$, and $V_{out} = \frac{1}{2}V_{3\omega}$.

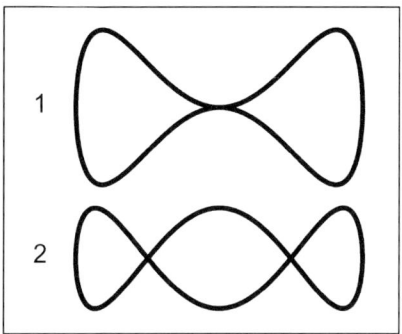

Fig. 16.9. Lissajous figures on the screen of an oscilloscope or a data-acquisition system: 1 – output voltage after balancing the in-phase component of fundamental frequency, 2 – output voltage after completely balancing both components of the fundamental frequency

Under a constant heating current, the modulation frequency is varied in a wide range (0.5–100 Hz) including nonadiabatic conditions (Fig. 16.10). In an adiabatic regime, the amplitude of the third-harmonic component is inversely proportional to the frequency.

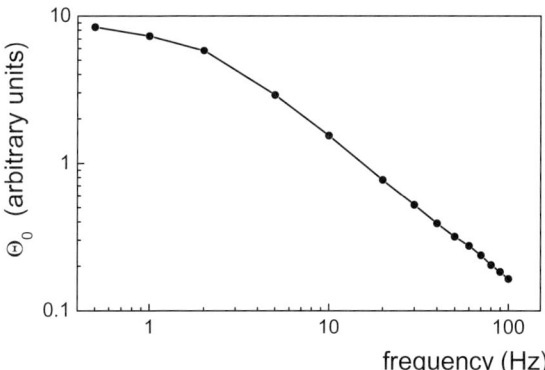

Fig. 16.10. Frequency dependence of the amplitude of temperature oscillations obtained from the third-harmonic signal generated by the sample. The frequency range (0.5–100 Hz) includes both nonadiabatic and adiabatic regimes

16.4 Equivalent-Impedance Method

In this student experiment (Kraftmakher 1994b), a gas-filled light bulb (12 V, 10 W) is included in a bridge circuit and heated by a DC current with a small AC component (Fig. 16.11). A variable capacitor C shunts one arm of the bridge to compensate for the quadrature component caused by the temperature oscillations in the sample (Sect. 4.1.3). An oscilloscope serves as the null indicator of the bridge, and the Lissajous figure is seen on its screen. A selective amplifier is useful to reduce noise and interference. It should be tuned not to shift the phase of the amplified signal.

An ammeter measures the total current feeding the bridge, I_0, so that it is necessary to take into account the distribution of the current between the arms. In our case, $R2/R1 = 10$, and an additional factor $(10/11)^2$ should be included into the expression for specific heat. The modulation frequency, 80 Hz, meets the adiabaticity conditions. From the measurements, one obtains the quantity

$$mc/\beta = 2(10/11)^2 I_0^2 R_{273}/\omega^2 RC \tag{16.5}$$

and plots this ratio as a function of the resistance of the sample (Fig. 16.12). The scatter of the experimental points is less than 0.5%. After measuring the R/R_{273} ratio, one calculates the values of the resistance for the temperatures 1500 K, 1600 K, and so on, up to 3400 K.

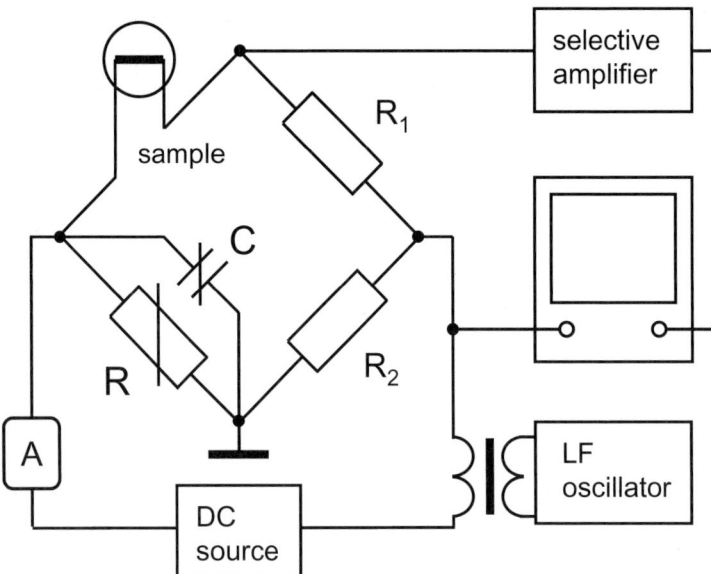

Fig. 16.11. Set-up for experiment employing equivalent-impedance technique (Kraftmakher 1994b). A selective amplifier is useful to reduce noise and interference. It should be tuned not to shift the phase of the signal

Then it is easy to find points on the graph corresponding to these temperatures and to determine the mc/β values. Another possibility is to calculate the resistance of the sample at these temperatures. They are set at the bridge, and the balance is achieved by adjusting the heating current and the capacitor C. It should be remembered that R_{273} in Eq. (16.5) is the resistance of the sample, while R relates to the resistance set at the bridge. In our case, this resistance is ten times larger than that of the sample.

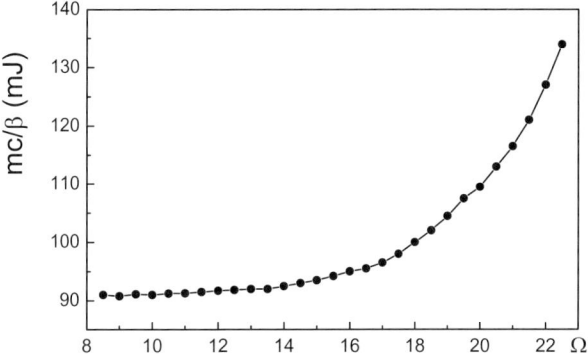

Fig. 16.12. With the equivalent-impedance technique, the quantity mc/β is measured versus electrical resistance of the sample (Kraftmakher 1994b)

Absolute values of the specific heat are available after determination of the mass of the filament. The mass of our filament was 3.5 ± 0.2 mg. The weighing caused an error larger than other errors of the measurements. As an alternative, one can fit the data at a convenient point, where results obtained by various methods, including the equivalent-impedance technique, are in good agreement. For example, the specific heat of tungsten at 2000 K equals 31.5 ± 0.5 J.mol^{-1}.K^{-1}.

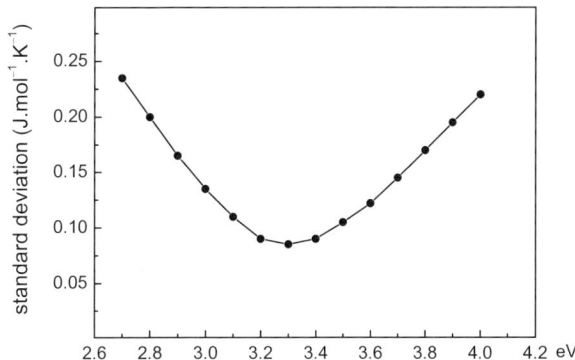

Fig. 16.13. Plot of standard deviation versus accepted value of H_F shows the most probable value of the formation enthalpy and its uncertainty

Accepting the vacancy origin of the non-linear increase in specific heat, it is easy to evaluate the parameters of vacancy formation (Sect. 13.1). The plot of $\ln T^2 \Delta C$ versus $1/T$ is a straight line with a slope $-H_F/k_B$. After the formation enthalpy is determined, the equilibrium vacancy concentration becomes available. A more rigorous determination of these values consists in fitting the experimental data by an equation taking into account vacancy formation. Various values of H_F should be tried. A plot of the standard deviation versus the assumed H_F shows the most probable formation enthalpy and its uncertainty (Fig. 16.13).

16.5 Determination of Boltzmann's Constant

Thermal noise is an example of fluctuation phenomena, which play an important role in modern physics and technology (Sect. 10.1). Fluctuation phenomena are a rare subject of student experiments because measurements of electrical fluctuations seem to be too complicated. Papers describing student experiments in this field could not disprove this opinion. However, measurements of electrical fluctuations are quite accessible to undergraduate students (Kraftmakher 1995).

To quantitatively study electrical fluctuations, one has to solve three problems:

- To find an amplifier of proper sensitivity and frequency band.
- To determine the frequency response of the amplifier.
- To measure correctly the mean square of the fluctuation voltage.

The appropriate frequency band is 10 kHz to 1 MHz. Low frequencies are to be excluded to avoid $1/f$-noise and low-frequency interference. High frequencies are also undesirable because the frequency band of the noise to be measured is restricted by the time constant of the input circuit. This time constant is governed by the resistance of the source and the capacitance of the wiring and of the input of the amplifier.

The best tool to measure noise voltages is a 'true RMS' voltmeter. This term means that RMS values of a voltage are measured regardless of its waveform. One can manufacture a simple and inexpensive device for such measurements, though of less accuracy than commercial voltmeters. To determine the frequency response of the amplifier, a high-frequency oscillator is necessary. It should be remembered that the mean square of the sum of two or more independent (uncorrelated) noise voltages equals the sum of the mean squares of the voltages.

Fundamental electrical fluctuations take the form of small random variations of voltage and current. Even when a resistor carries no current, a small fluctuating voltage appears between its terminals. The noise originates from thermal motion of the charge carriers and is referred to as thermal noise or Johnson noise. This voltage can be amplified, observed with an oscilloscope, and measured. The fluctuation voltage across a resistor R, in a narrow frequency interval Δf, obeys the Nyquist formula:

$$<\Delta V^2> = 4k_B TR\Delta f. \tag{16.6}$$

This expression is valid when $hf \ll k_B T$, where h is Planck's constant. The spectral density of the noise, $\langle \Delta V^2 \rangle / \Delta f$, is independent of frequency (the so-called 'white noise'). When the sample possesses a complex impedance, its real part enters the formula. A proof of Nyquist's theorem is given in many textbooks.

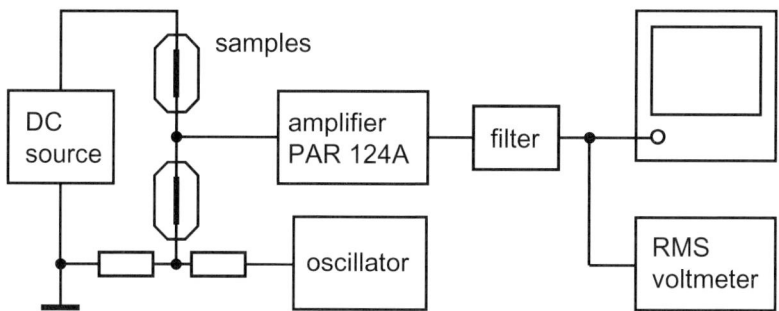

Fig. 16.14. Arrangement for observation of thermal noise (Kraftmakher 1995). The thermal noise generated by the light bulbs varies in a wide range due to changes in temperature and resistance of the samples

The experiment involves two sources of thermal noise. First, tungsten filaments of high-resistance vacuum light bulbs (28 V, 40 mA) generate the noise. To demonstrate the nature of thermal noise, it is useful to vary the temperature of the resistor in a wide range, and the filament of a light bulb provides excellent opportunity for this purpose. To avoid superfluous evaluations, two similar light bulbs are employed. They are connected in series to a DC source and in parallel to the input of an amplifier (Fig. 16.14). Hence, the noise source is considered to be a resistor of a half of the resistance of one filament. An amplifier, PAR 124A, operating in the selective mode (the range 10 μV, f_0 = 50 kHz, Q = 2 or 5), amplifies the noise. Values of Q correspond to selectivity of the amplifier set by a switch.

The output voltage of the amplifier proceeds to the input of an oscilloscope, Kenwood CS-4025, through a low pass filter (R = 1 kΩ, C = 2200 pF). The filter reduces the signal at high frequencies. The oscilloscope has an output terminal of one amplification channel. A 'true RMS' voltmeter, Keithley 196 digital multimeter, measures the output voltage. The inherent filter of the voltmeter suppresses fluctuations in the readings. The mean square of the amplified noise is

$$\langle V^2 \rangle = 4k_B RT \int_0^\infty G^2 df$$
$$= 4k_B RT G_0^2 \int_0^\infty (G/G_0)^2 df = 4k_B RT G_0^2 B, \qquad (16.7)$$

where G is the frequency-dependent gain of the amplifier, G_0 is the maximum gain, and $B = \int_0^\infty (G/G_0)^2 df$ is called the noise equivalent bandwidth.

An oscillator serves to determine the frequency response of the amplifier. These measurements are performed with a low gain of the amplifier. From numerical integration, the noise equivalent bandwidth appeared to be 33.3 kHz for Q = 2 and 16.2 kHz for Q = 5. The internal attenuators of the oscillator (–20 dB and –40 dB) help to determine the maximum total gain. The resistance of the filaments is available from the voltage drop across them and the heating current. In the range 400–2500 K, the absolute temperature of the filaments obeys the relation

$$T = 103 + 207X - 1.8X^2, \tag{16.8}$$

where $X = R/R_{273}$. This expression fits data recommended by Roeser and Wensel (1941) and by Kohl (1962). The $<V^2>$ values obtained for the two frequency bands are plotted versus RT (Fig. 16.15). The plots are straight lines with slopes $4k_B G_0^2 B$. Due to the inherent noise of the amplifier, the lines cross the X-axis at a negative value of RT. The inherent noise of an amplifier is specified by the so-called equivalent noise resistance. This means a resistance, which at room temperature generates thermal noise equal to the inherent noise of the amplifier attributed to its input. The equivalent noise resistance of our amplifier is 1.47 kΩ.

Second, thermal noise of a usual resistor, 2.4 kΩ, is measured at room temperature and at liquid nitrogen temperature. A metal box containing the probe can be immersed in a Dewar filled with liquid nitrogen. The measurements are performed at only two but well known temperatures. The resistance of the probe is measured at both temperatures. The gain and the frequency band of the amplifier are the same as when measuring the thermal noise generated by the light bulbs, and all the data can be presented together. All the experimental points are close to the straight lines. Measurements of thermal noise are thus possible even by means of an amplifier with relatively high inherent noise.

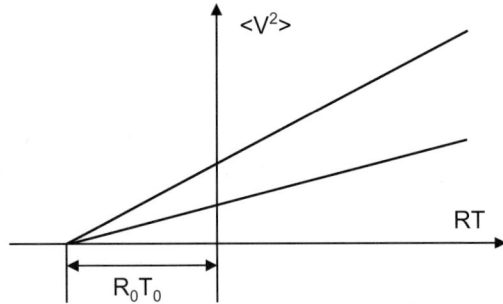

Fig. 16.15. Plots for determination of Boltzmann's constant and of equivalent noise resistance of the amplifier, for two frequency bands of the amplifier. The quantity $R_0 T_0$ corresponds to the inherent noise of the amplifier

An additional part of the experiment may be a determination of the absolute temperature of the melting point of ice. The absolute temperature scale was established with intention to retain the temperature unit of the centigrade scale. Hence, the difference between the absolute temperatures of the normal boiling point of water and of the melting point of ice has to be 100 K (exactly). Denoting the absolute temperature of the ice point as T_0, one obtains

$$R_{100}(T_0 + 100)/R_0 T_0 = <V_{100}^2>/<V_0^2>, \tag{16.9}$$

where $\langle V_0^2 \rangle$ and $\langle V_{100}^2 \rangle$ are the corresponding values of the thermal noise, and R_0 and R_{100} are the resistances of the probe. All the quantities, except T_0, are measured directly.

A simple and inexpensive device was manufactured to measure mean squares of noise voltages (Fig. 16.16). Three resistors and a low-power light bulb (6 V, 50 mA) form a bridge circuit fed by an adjustable DC current. The temperature of the filament is about 1000 K. A DC microammeter (50 µA, 4 kΩ) acts as the balance indicator of the bridge. Before the measurements, the feeding current is adjusted to balance the bridge. Then an AC voltage to be measured is applied to the filament through a capacitor. Due to changes in the temperature and in the resistance of the filament, a DC voltage appears at the output of the bridge. The current measured by the microammeter is therefore a function of the mean square of the AC voltage regardless of its waveform. A source of sine voltage and a usual AC voltmeter serve for the calibration. This calibration is valid for any other waveform of the AC voltage, including noise. The curvature of the calibration line is due to changes in the resistance and the heat loss from the filament. It does not mean that the tool responses to something else than the mean square of the applied AC voltage. As a disadvantage, the sensitivity of the device is insufficient to measure the voltage obtainable at the output of the amplifier employed. An additional amplifier providing output voltages up to 1 V is therefore necessary.

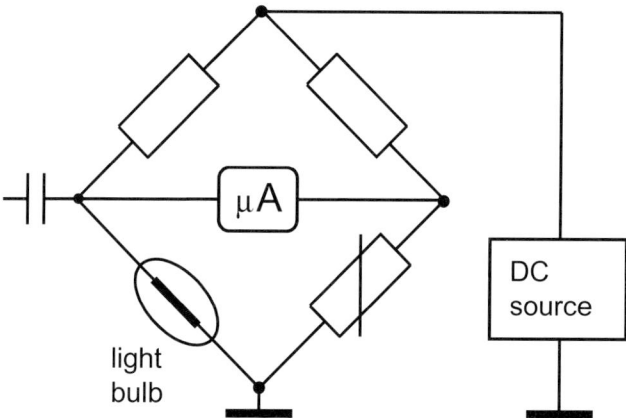

Fig. 16.16. Simple circuit operating as a 'true RMS' voltmeter. The operation rests on the fact that the heat dissipated in the filament of a light bulb depends only on the RMS value of the current, regardless of its waveform

Conclusion

In conclusion, it is worth attracting attention to some points presented in the book.

● The frequency-conversion method (Sect. 4.1.1) may appear to be useful in many cases, where high-frequency temperature oscillations in a sample are measured. In particular, the possibility to extend the frequency band in specific-heat spectroscopy (Sect. 7.4.3) should be examined.

● Measurements of high-temperature specific heat are possible by the use of reference samples, such as tungsten or platinum, having a blackbody model. Such samples provide temperature oscillations of definite amplitude, which can be compared with those in other samples with a blackbody model (Sect. 4.2).

● High-frequency temperature oscillations appeared to be measurable with pyroelectric sensors (Sect. 4.3), and this technique probably deserves more attention.

● High-temperature specific heat of nonconducting materials could be measured by the use of the equivalent-impedance technique or a capillary filled with a material under study (Sect. 6.1).

● All variants of modulation dilatometry are promising (Chap. 8). Nowadays, thermal expansivity is measurable along the thickness of the sample, so that the problem of temperature gradients in the sample would be completely avoided. Furthermore, the measurements are probably possible along different axes of small single-crystal samples (Sect. 8.5).

● Direct measurements of the temperature derivative of resistance could be used in many studies, especially at low temperatures (Sect. 9.1.2).

● Modulation measurements of spectral absorptance would extend the temperature range and enhance the accuracy of data (Sect. 9.3).

● Measurements of temperature oscillations in wire samples through their thermal noise (Chap. 10) would make it possible to more widely use such samples in all the modulation measurements.

● Until now, we have no reliable knowledge of equilibrium point-defect concentrations in metals. To clarify this problem, modulation calorimetry and related techniques would probably be useful (Sect. 13.1).

● It is worth applying the method of direct measurements of the temperature coefficient of specific heat (Sect. 13.2) to studies of phase transitions. In studies of high-temperature superconductors (Sect. 14.3), direct measurements of the temperature coefficient of specific heat might provide more accurate data of the temperature dependence of the specific heat.

• The unusual premelting anomaly in specific heat observed by modulation measurements (Sect. 13.3) deserves further studies, also by means of other techniques.

• It is very desirable to use modern experimental feasibilities for observing temperature fluctuations in small samples at high temperatures (Sect. 13.4) and to obtain further data on isochoric specific heat of metals.

• It is worth applying modulation microcalorimetry to studies of second-order phase transitions in ferro- and antiferromagnets. Measurements on single crystals would provide more accurate values of the critical indices (Sect. 14.2).

• The relaxation phenomenon observed in the specific heat of tungsten and platinum (Sect. 15.3.2) is still waiting for confirmation. It is worth remembering that the scatter of experimental data for $C(X)$ and $\tan\phi$ should increase in a region, where $X = \omega\tau$ becomes close to unity. This offers an additional approach for studying the phenomenon. The same probably relates to other relaxation phenomena in specific heat.

• Further efforts are necessary to clarify the origin of the relaxation phenomenon in high-temperature specific heat of metals (Sect. 15.3). This relates to both modulation and rapid-heating experiments (Sects. 15.3.4 and 15.3.6).

• The rapid-heating technique might solve two important problems: (i) the vacancy contributions to enthalpy, and (ii) the mechanism of melting (Sect. 15.3.5).

• The student experiments described (Chap. 16) seem to be a useful addition to existing laboratory courses.

References

Actis A, Cibrario A, Crovini L (1972), in: Plumb HH (ed) Temperature, Its Measurement and Control in Science and Industry, vol 4. Instrument Society of America, Pittsburgh, pp 355-364
Adamovsky SA, Minakov AA, Schick C (2003) Thermochim Acta 403:55–63
Affortit C, Lallement R (1968) Rev Int Hautes Tempér et Réfract 5:19–26
Akhmatova IA (1965) Dokl Akad Nauk SSSR 162:127–129
Akhmatova IA (1967) Izmer Tekhn N 8:14–17
Akiba A, Yamada H, Matsuo R, Kanke Y, Haeiwa T, Kita E (1998) J Phys Soc Jpn 67:1303–1305
Akimov AI, Kraftmakher YA (1970) Phys Stat Sol 42:K41–42
Akutsu H, Saito K, Yamamura Y, Kikuchi K, Nishikawa H, Ikemoto I, Sorai M (1999) J Phys Soc Jpn 68:1968–1974
Akutsu H, Saito K, Sorai M (2000) Phys Rev B 61:4346–4352
Akutsu N, Ikeda H (1981) J Phys Soc Jpn 50:2865–2871
Albrecht T, Armbruster S, Stühn B, Vogel K, Strobl G (2001) Thermochim Acta 377:159–172
Alkhafaji MT, Migone AD (1992) Phys Rev B 45:5729–5732
Alkhafaji MT, Migone AD (1993) Phys Rev B 48:1761–1764
Alkhafaji MT, Migone AD (1996) Phys Rev B 53:11152–11158
Androsch R, Wunderlich B (1999) Thermochim Acta 333:27–32
Annino A, Grasso F, Musumeci F, Triglia A (1984) Appl Phys A 35:115–118
Anshukova NA, Bugoslavskiy YV, Veselago VG, Golovashkin AI, Ershov OV, Zaytzev IA, Ivanenko OM, Minakov AA, Mitzen KV (1989) Acta Phys Polonica A 76:35–40
Arpaci E, Frohberg MG (1984) Z Metallkunde 75:614–618
Ashman J, Handler P (1969) Phys Rev Lett 23:642–644
Aubin M, Ghamlouch H, Fournier P (1993) Rev Sci Instrum 64:2938–2941
Ausloos M, Benhaddou M, Cloots R (1994) Physica C 235/240:1767–1768
Bachmann R, DiSalvo FJ, Geballe TH, Greene RL, Howard RE, King CN, Kirsch HC, Lee KN, Schwall RE, Thomas H-U, Zubeck RB (1972) Rev Sci Instrum 43:205–214
Baloga JD, Garland CW (1977) Rev Sci Instrum 48:105–110
Baur H, Wunderlich B (1998) J Thermal Analysis 54:437–465
Beaubois F, Claverie T, Marcerou JP, Rouillon JC, Nguyen HT, Garland CW, Haga H (1997) Phys Rev E 56:5566–5574
Bechthold PS, Campagna M, Schober T (1980) Solid State Commun 36:225–231
Bednarz G, Millier B, White MA (1992) Rev Sci Instrum 63:3944–3952
Bednarz G, Geldart DJ W, White MA (1993) Phys Rev B 47:14247–14259
Beiner M, Kahle S, Hempel E, Schröter K, Donth E (1998) Europhys Lett 44:321–327
Bennett SJ (1977) J Phys E 10:525–530
Bentefour EH, Glorieux C, Chirtoc M, Thoen J (2003) J Appl Phys 93:9610–9614

Berger R, Gerber C, Gimzewski JK, Meyer E, Güntherodt HJ (1996)
　　Appl. Phys. Lett 69:40–42
Berret J-F, Meissner M, Mertz B (1992) Z Phys B 87:213–217
Bezemer J, Jongerius RT (1976) Physica C 83:338–346
Bibi I, Jenkins TE (1983) J Phys C 16:L57–60
Bimberg D, Bubenzer A (1981) Appl Phys Lett 38:803–805
Biondi MA (1954) Phys Rev 96:534–535
Biondi MA (1956) Phys Rev 102:964–967
Birge NO (1986) Phys Rev B 34:1631–1642
Birge NO, Nagel SR (1985) Phys Rev Lett 54:2674–2677
Birge NO, Nagel SR (1987) Rev Sci Instrum 58:1464–1470
Birge NO, Dixon PK, Menon N (1997) Thermochim Acta 304/305:51–66
Birmingham JT, Richards PL, Meyer H (1996) J Low Temp Phys 103:183–208
Bittner DN, Bretz M (1985) Phys Rev B 31:1060–1068
Blagonravov LA, Filippov LP, Alekseev VA, Shnerko VN (1983)
　　Inzh–Fiz Zh 44:438–444
Blagonravov LA, Filippov LP, Alekseev VA, Shnerko VN (1984)
　　Teplofiz Vysok Temper 22:177–179
Bleckwedel A, Eichler A (1985) Solid State Commun 56:693–696
Bloemen E, Garland CW (1981) J Physique 42:1299–1302
Bockstahler LI (1925) Phys Rev 25:677–685
Boerio-Goates J, Garland CW, Shashidhar R (1990) Phys Rev A 41:3192–3196
Bohn H, Eichler A (1991) Z Phys B 83:105–111
Bohn K-P, Prahm A, Petersson J, Krüger J K (1997) Thermochim Acta 304/305:283–290
Boller A, Jin Y, Wunderlich B (1994) J Thermal Analysis 42:307–330
Boller A, Schick C, Wunderlich B (1995) Thermochim Acta 266:97–111
Boller A, Ribeiro M, Wunderlich B (1998) J Thermal Analysis 54:545–563
Bonilla A, Garland CW (1974) J Phys Chem Solids 35:871–877
Borkowski CJ, Blalock TV (1974) Rev Sci Instrum 45:151–162
Bouquet F, Wang Y, Wilhelm H, Jaccard D, Junod A (2000)
　　Solid State Commun 113:367–371
Bouvier M, Lethuillier P, Schmitt D (1991) Phys Rev B 43:13137–13144
Boyarskii SV, Novikov II (1981) Teplofiz Vysok Temper 19:201–203
Braun M, Kohlhaas R, Vollmer O (1968) Z angew Physik 25:365–372
Bręczewski T, Piskunowicz P, Jaroma-Weiland G (1984)
　　Acta Phys Polonica A 66:555–560
Brixy H, Hecker R, Oehmen J, Rittinghaus KF, Setiawan W, Zimmermann E (1992),
　　in: Schooley JF (ed) Temperature, Its Measurement and Control in Science and
　　Industry, vol 6. American Institute of Physics, New York, pp 993–996
Brophy JJ, Epstein M, Webb SL (1965) Rev Sci Instrum 36:1803–1806
Bruins DE, Garland CW, Greytak TJ (1975) Rev Sci Instrum 46:1167–1170
Brusetti R, Garoche P, Bechgaard K (1983) J Phys C 16:3535–3545
Buckingham MJ, Edwards C, Lipa JA (1973) Rev Sci Instrum 44:1167–1172
Büchner B, Korpiun P (1987) Appl Phys B 43:29–33
Calzona V, Putti M, Siri AS (1990) Thermochim Acta 162:127–132
Campbell JH, Bretz M (1985) Phys Rev B 32:2861–2869
Carpentier L, Bustin O, Descamps M (2002) J Phys D 35:402–408
Carrington A, Mackenzie AP, Tyler A (1996) Phys Rev B 54:R3788–3791

Carrington A, Marcenat C, Bouquet F, Colson D, Bertinotti A, Marucco JF,
    Hammann J (1997) Phys Rev B 55:R8674–8677
Carslaw HS, Jaeger JC (1959) Conduction of Heat in Solids. Clarendon Press, Oxford
Castro M, Burriel R (1995a) Thermochim Acta 269/270:523–535
Castro M, Burriel R (1995b) Thermochim Acta 269/270:537–552
Castro M, Puértolas JA (1997) Thermochim Acta 304/305:291–301
Castro M, Puértolas JA (2003) Thermochim Acta 402:159–168
Celasco M, Fiorillo F, Mazzetti P (1976) Phys Rev Lett 36:38–42
Cezairliyan A (1971) J Res Nat Bur Stand A 75:565–571
Cezairliyan A (1983) Int J Thermophys 4:159–171
Cezairliyan A (1984), in: Maglić KD, Cezairliyan A, Peletsky VE (eds) Compendium of
    Thermophysical Property Measurement Methods, vol 1. Plenum, New York,
    pp 643–668
Cezairliyan A (1988), in: Ho CY (ed) Specific Heat of Solids. Hemisphere, New York,
    pp 323–353
Cezairliyan A (1992), in: Maglić KD, Cezairliyan A, Peletsky VE (eds) Compendium of
    Thermophysical Property Measurement Methods, vol 2. Plenum, New York,
    pp 483–517
Cezairliyan A, McClure JL (1971) J Res Nat Bur Stand A 75:283–290
Cezairliyan A, Righini F (1996) Metrologia 33:299–306
Cezairliyan A, Morse MS, Berman HA, Beckett CW (1970)
    J Res Nat Bur Stand A 74:65–92
Cezairliyan A, Gathers GR, Malvezzi AM, Miiller AP, Righini F, Shaner JW (1990)
    Int J Thermophys 11:819–833
Cezairliyan A, Krishnan S, McClure JL (1996) Int J Thermophys 17:1455–1473
Cezairliyan A, Krishnan S, Basak D, McClure JL (1998) Int J Thermophys 19:1267–1276
Chae HB, Bretz M (1989) J Low Temp Phys 76:199–223
Chaikin PM, Kwak JF (1975) Rev Sci Instrum 46:218–220
Chan MHW, Migone AD, Miner KD, Li ZR (1984) Phys Rev B 30:2681–2694
Chan T, Garland CW, Nguyen HT (1995) Phys Rev E 52:5000–5003
Charalambous M, Riou O, Gandit P, Billon B, Lejay P, Chaussy J, Hardy WN, Bonn DA,
    Liang R (1999) Phys Rev Lett 83:2042–2045
Chaussy J, Gandit P, Bret JL, Terki F (1992) Rev Sci Instrum 63:3953–3958
Chekhovskoi VY (1981) Metallofizika 3:116–119
Chekhovskoi VY (1984), in: Maglić KD, Cezairliyan A, Peletsky VE (eds)
    Compendium of Thermophysical Property Measurement Methods, vol 1.
    Plenum, New York, pp 555–589
Chekhovskoi V Y (1992), in: Maglić KD, Cezairliyan A, Peletsky VE (eds) Compendium of
    Thermophysical Property Measurement Methods, vol 2. Plenum, New York,
    pp 457–481
Chekhovskoi VY, Petrov VA (1970) Sov Phys Solid State 12:2473–2474
Chen X, Perel AS, Brooks JS, Guertin RP, Hinks DG (1993) J Appl Phys 73:1886–1891
Cheung KM, Ullman FG (1974) Phys Rev B 10:4760–4764
Chirtoc M, Bentefour EH, Glorieux C, Thoen J (2001) Thermochim Acta 377:105–112
Christofides C (1993) Critical Reviews in Solid State and Materials Sciences 18:113–174
Chu CW (1974) Phys Rev Lett 33:1283–1286
Chu CW, Knapp GS (1973) Phys Lett A 46:33–35
Chu CW, Testardi LR (1974) Phys Rev Lett 32:766–769

Chu CW, Vieland LJ (1974) J Low Temp Phys 17:25–29
Chui TCP, Swanson DR, Adriaans MJ, Nissen JA, Lipa JA (1992)
    Phys Rev Lett 69:3005–3008
Chung M, Wang Y, Brill JW, Xiang X-D, Mostovoy R, Hou JG, Zettl A (1992)
    Phys Rev B 45:13831–13833
Chung M, Figueroa E, Kuo Y-K, Wang Y, Brill JW, Burgin T, Montgomery LK (1993a)
    Phys Rev B 48:9256–9263
Chung M, Wang Y, Brill JW, Burgin T, Montgomery LK (1993b) Synth Met 56:2755–2760
Chung M, Verebelyi DT, Schneider CW, Nevitt MV, Skove MJ, Payne JE, Marone M,
    Kostić P (1996) Physica C 265:301–308
Connelly DL, Loomis JS, Mapother DE (1971) Phys Rev B 3:924–934
Corbino OM (1910) Phys Z 11:413–417
Corbino OM (1911) Phys Z 12:292–295
Coufal H (1984) Appl Phys Lett 44:59–61
Craven RA, Meyer SF (1977) Phys Rev B 16:4583–4593
Craven RA, Salamon MB, DePasquali G, Herman RM, Stucky G, Schultz A (1974)
    Phys Rev Lett 32:769–772
Craven RA, Di Salvo FJ, Hsu FSL (1978a) Solid State Commun 25:39–42
Craven RA, Tsuei CC, Stephens R (1978b) Phys Rev B 17:2206–2211
Crovini L, Actis A, Galleano R (1992), in: Schooley JF (ed) Temperature, Its
    Measurement and Control in Science and Industry, vol 6. AIP, New York, pp 47–52
Csáthy GA, Tulimieri D, Yoon J, Chan MHW (1998) Phys Rev Lett 80:4482–4485
Danley RL (2003) Thermochim Acta 402:91–98
Das P, Ema K, Garland CW (1989a) Liquid Crystals 4:205–208
Das P, Ema K, Garland CW, Shashidhar R (1989b) Liquid Crystals 4:581–589
Dawes DG, Coles BR (1979) J Phys F 9:L215–220
Day P, Hahn I, Chui TCP, Harter AW, Rowe D, Lipa JA (1997)
    J Low Temp Phys 107:359–370
del Cerro J, Martín-Olalla JM, Romero FJ (2003) Thermochim Acta 401:149–158
Deligiannis K, Billon B, Chaussy J, Charalambous M, Liang R, Bonn D, Hardy WN (2000)
    Physica C 332:360–364
Demuer A, Marcenat C, Thomasson J, Calemczuk R, Salce B, Lejay P, Braithwaite D,
    Flouquet J (2000) J Low Temp Phys 120:245–257
Demuer A, Jaccard D, Sheikin I, Raymond S, Salce B, Thomasson J, Braithwaite D,
    Flouquet J (2001) J Phys: Condens Matter 13:9335–9347
Denlinger DW, Abarra EN, Kimberly Allen, Rooney PW, Messer MT, Watson SK,
    Hellman F (1994) Rev Sci Instrum 65:946–959
Derman AS, Bogorodskii OV (1970) Izv Akad Nauk SSSR (Ser Fiz) 34:1215–1216
Devoille L, Salce B, Thomasson J, Marcenat C, Calemczuk R, Ziman T, Cépas O,
    Dhalenne G, Revcolevschi A (2002) J Phys: Condens Matter 14:2569–2575
Diosa JE, Vargas RA, Mina E, Toruano E, Mellander B-E (2000a)
    Phys Stat Sol B 220:641–646
Diosa JE, González-Montero G, Vargas RA (2000b) Phys Stat Sol B 220:647–650
Diosa JE, Aparicio GM, Vargas RA, Jurado JF (2000c) Phys Stat Sol B 220:651–654
Diosa JE, Fernández ME, Vargas RA (2001) Phys Stat Solidi B 227:465–468
Ditmars DA (1984), in: Maglić KD, Cezairliyan A, Peletsky VE (eds) Compendium of
    Thermophysical Property Measurement Methods, vol 1. Plenum, New York,
    pp 527–553

Dixon GS, Black SG, Butler CT, Jain AK (1982) Anal Biochem 121:55–61
Dixon PK (1990) Phys Rev B 42:8179–8186
Dixon PK, Nagel SR (1988) Phys Rev Lett 61:341–344
Dobrosavljević AS, Maglić KD (1989) High Temp – High Press 21:411–421
Dobrosavljević AS, Maglić KD (1991) High Temp – High Press 23:129–133
Dobrosavljević AS, Maglić KD, Perović NL (1989) High Temp – High Press 21:317–324
Donth E, Korus J, Hempel E, Beiner M (1997) Thermochim Acta 304/305:239–249
Dumrongrattana S, Nounesis G, Huang CC (1986) Phys Rev A 33:2181–2183
Dutzi J, Buckel W (1984) Z Phys B 55:99–102
Edsinger RE, Schooley JF (1991) Int J Thermophys 12:665–677
Efremov MY, Schiettekatte F, Zhang M, Olson EA, Kwan AT, Berry RS, Allen LH (2000) Phys Rev Lett 85:3560–3563
Efremov MY, Olson EA, Zhang M, Allen LH (2003) Thermochim Acta 403:37–41
Egry I (2000) High Temp – High Press 32:127–134
Egry I, Diefenbach A, Dreier W, Piller J (2001) Int J Thermophys 22:569–578
Eichler A, Gey W (1979) Rev Sci Instrum 50:1445–1452
Eichler A, Bohn H, Gey W (1980) Z Phys B 38:21–25
Eichler A, Cieslik J, Gey W (1981) Physica B 108:1005–1006
Ema K (1983) J Phys Soc Jpn 52:2798–2809
Ema K, Yao H (1997) Thermochim Acta 304/305:157–163
Ema K, Yao H (1998) Phys Rev E 57:6677–6684
Ema K, Hamano K, Kurihara K, Hatta I (1977) J Phys Soc Jpn 43:1954–1961
Ema K, Hamano K, Ikeda Y (1979) J Phys Soc Jpn 46:345–346
Ema K, Garland CW, Sigaud G, Nguyen HT (1989a) Phys Rev A 39:1369–1375
Ema K, Nounesis G, Garland CW, Shashidhar R (1989b) Phys Rev A 39:2599–2608
Ema K, Uematsu T, Sugata A, Yao H (1993) Jpn J Appl Phys (Part 1) 32:1846–1850
Ema K, Watanabe J, Takagi A, Yao H (1995) Phys Rev E 52:1216–1219
Ema K, Ogawa M, Takagi A, Yao H (1996a) Phys Rev E 54:R25–28
Ema K, Takagi A, Yao H (1996b) Phys Rev E 53:R3036–3039
Ema K, Yao H, Fukuda A, Takanishi Y, Takezoe H (1996c) Phys Rev E 54:4450–4453
Ema K, Takagi A, Yao H (1997) Phys Rev E 55:508–513
Eno HF, Tyler EH, Luo HL (1977) J Low Temp Phys 28:443–448
Ershov OV, Minakov AA, Veselago VG (1984) Sov Phys Sold State 26:1527–1528
Evans-Lutterodt KW, Chung JW, Ocko BM, Birgeneau RJ, Chiang C, Garland CW, Chin E, Goodby J, Nguyen HT (1987) Phys Rev A 36:1387–1395
Fanton JT, Kino GS (1987) Appl Phys Lett 51:66–68
Farrant SP, Gough CE (1975) Phys Rev Lett 34:943–946
Fecht H-J, Johnson WL (1991) Rev Sci Instrum 62:1299–1303
Fecht H-J, Wunderlich RK (1994) Mater Sci Eng A 178:61–64
Feder R, Charbnau HP (1966) Phys Rev 149:464–471
Felner I, Schmitt D, Barbara B (1997) Physica B 229:153–166
Feng YP, Jin A, Finotello D, Gillis KA, Chan MHW, Greedan JE (1988) Phys Rev B 38:7041–7044
Fermi E (1937) Nuova Antologia 72:313–316
Filippov LP (1960) Inzh-Fiz Zh 3 N 7:121–123
Filippov LP (1966) Int J Heat Mass Transfer 9:681–691
Filippov LP (1967) Measurement of Thermal Properties of Solid and Liquid Metals at High Temperatures, in Russian. University Press, Moscow

Filippov LP (1984) Measurement of Thermophysical Properties by Methods of Periodic Heating, in Russian. Energoatomizdat, Moscow
Filippov LP, Makarenko IN (1968) High Temperature 6:143–149
Filippov LP, Yurchak RP (1965) High Temperature 3:837–844
Filippov LP, Tugareva HA, Markina LI (1964) Inzh-Fiz Zh 7 N 6:3–7
Filippov LP, Blagonravov LA, Alekseev VA (1976) High Temp – High Press 8:658–659
Finotello D, Iannacchione GS (1995) Int J Mod Phys B 9:2247–2283
Finotello D, Gillis K A, Wong A, Chan MHW (1988) Phys Rev Lett 61:1954–1957
Finotello D, Qian S, Iannacchione GS (1997) Thermochim Acta 304/305:303–316
Fisher ME, Langer JS (1968) Phys Rev Lett 20:665–668
Flerov IN, Burriel R, Gorev MV, Isla P, Voronov VN (2003) Phys Solid State 45:167–170
Florian R, Pelzl J, Rosenberg M, Vargas H, Wernhardt R (1978) Phys Stat Sol A 48:K35–38
Fominaya F, Fournier T, Gandit P, Chaussy J (1997a) Rev Sci Instrum 68:4191–4195
Fominaya F, Villain J, Gandit P, Chaussy J, Caneschi A (1997b) Phys Rev Lett 79:1126–1129
Fominaya F, Gandit P, Gaudin G, Chaussy J, Sessoli R, Sangregorio C (1999a) J Magn Magn Mater 195:L253–255
Fominaya F, Villain J, Fournier T, Gandit P, Chaussy J, Fort A, Caneschi A (1999b) Phys Rev B 59:519–528
Fontes E, Lee WK, Heiney PA, Nounesis G, Garland CW, Riera A, McCauley JP, Smith AB (1990) J Chem Phys 92:3917–3929
Forster R, Gmelin E (1996) Rev Sci Instrum 67:4246–4255
Fortune NA, Brooks JS, Graf MJ, Montambaux G, Chiang LY, Perenboom JAAJ, Althof D (1990) Phys Rev Lett 64:2054–2027
Fortune NA, Murata K, Ishibashi M, Tokumoto M, Kinoshita N, Anzai H (1991) Solid State Commun 77:265–269
Fortune NA, Murata K, Ikeda K, Takahashi T (1992) Phys Rev Lett 68:2933–2936
Fraile-Rodríguez A, Ruiz-Larrea I, López-Echarri A, Tello MJ (2001) Thermochim Acta 377:131–140
Freeman RH, Bass J (1970) Rev Sci Instrum 41:1171–1174
Frenkel JI (1926) Z Phys 35:652–669
Fridman VY (1983) Inzh-Fiz Zh 44:986–988
Fritsch G, Lüscher E (1983) Phil Mag A 48:21–29
Fritsch G, Lachner R, Diletti H, Lüscher E (1982) Phil Mag A 46:829–839
Fritsch G, Diletti H, Lüscher E (1984) Phil Mag A 50:545–558
Frohberg MG (1999) Thermochim Acta 337:7–17
Fröchte B, Khan Y, Kneller E (1990) Rev Sci Instrum 61:1954–1957
Gallardo MC, Burriel R, Romero FJ, Gutiérrez FJ, Salje EKH (2002) J Phys: Condens Matter 14:1881–1886
Gangopadhyay TK, Henderson PJ (1999) Meas Sci Technol 10:R129–138
Garfield NJ, Patel M (1998) Rev Sci Instrum 69:2186–2187
Garfield NJ, Howson MA, Overend N (1998) Rev Sci Instrum 69:2045–2049
Garfield NJ, Howson MA, Yang G, Abell S (1999) Physica C 321:1–11
Garland CW (1985) Thermochim Acta 88:127–142
Garland CW (1996) Mol Cryst Liq Cryst A 288:25–31
Garland CW, Baloga JD (1977) Phys Rev B 16:331–339
Garland CW, Huster ME (1987) Phys Rev A 35:2365–2368

Garland CW, Kasting GB, Lushington KJ (1979) Phys Rev Lett 43:1420–1423
Garland CW, Meichle M, Ocko BM, Kortan AR, Safinya CR, Yu LJ, Litster JD, Birgeneau RJ (1983) Phys Rev A 27:3234–3240
Garland CW, Nounesis G, Stine KJ (1989a) Phys Rev A 39:4919–4922
Garland CW, Nounesis G, Stine KJ, Heppke G (1989b) J Physique 50:2291–2301
Garnier PR (1971) Phys Lett A 35:413–414
Garnier PR, Salamon MB (1971) Phys Rev Lett 27:1523–1526
Garoche P, Bigot J (1983) Phys Rev B 28:6886–6895
Garoche P, Johnson WL (1981) Solid State Commun 39:403–406
Garoche P, Manuel P, Veyssié JJ, Molinié P (1978) J Low Temp Phys 30:323–336
Garoche P, Brusetti R, Jérome D, Bechgaard K (1982) J Physique Lettres 43:L147–152
Garrison JB, Lawson AW 1949 Rev Sci Instrum 20:785–794
Gathers GR (1986) Rep Progr Phys 49:341–396
Geer R, Huang CC, Pindak R, Goodby JW (1989) Phys Rev Lett 63:540–543
Geer R, Stoebe T, Huang CC, Pindak R, Srajer G, Goodby JW, Cheng M, Ho JT, Hui SW (1991a) Phys Rev Lett 66:1322–1325
Geer R, Stoebe T, Pitchford T, Huang CC (1991b) Rev Sci Instrum 62:415–421
Geer R, Stoebe T, Huang CC (1993) Phys Rev E 48:408–427
Geraghty P, Wixom M, Francis AH (1984) J Appl Phys 55:2780–2785
Gerlich D, Abeles B, Miller RE 1965 J Appl Phys 36:76–79
Gesi K (1992) J Phys Soc Jpn 61:1225–1231
Gesi K, Osaka T (1995) Solid State Commun 95:639–642
Ghiron K, Salamon MB, Veal BW, Paulikas AP, Downey JW (1992) Phys Rev B 46:5837–5840
Gibson BC, Ginsberg DM, Tai PCL (1979) Phys Rev B 19:1409–1419
Gill PS, Sauerbrunn SR, Reading M (1993) J Thermal Analysis 40:931–939
Ginsberg DM, Inderhees SE, Salamon MB, Goldenfeld N, Rice JP, Pazol BG (1988) Physica C 153/155:1082–1085
Glass AM (1968) Phys Rev 172:564–571
Glazkov SY (1985) Int J Thermophys 6:421–426
Glazkov SY (1987) High Temperature 25:51–57
Glazkov SY (1988) Teplofiz Vysok Temper 26:501–503
Glazkov SY, Kraftmakher YA (1983) High Temperature 21:591–594
Glazkov SY, Kraftmakher YA (1986) High Temp – High Press 18:465–470
Glorieux C, Thoen J (1994) J Physique IV 4:C7-271–274
Glorieux C, Caerels J, Thoen J (1994) J Physique IV 4:C7-267–270
Glorieux C, Thoen J, Bednarz G, White MA, Geldart DJW (1995) Phys Rev B 52:12770–12778
Glukhikh LK, Efremova RI, Kuskova NV, Matizen EV (1966), in: Novikov II, Strelkov PG (eds) High-Temperature Studies, in Russian. Nauka, Novosibirsk, pp 75–88
Gmelin E (1987) Thermochim Acta 110:183–208
Gmelin E (1997) Thermochim Acta 304/305:1–26
Gmelin E (1999) J Therm Anal Calorim 56:655–671
Gmelin E, Fischer R, Stitzinger R (1998) Thermochim Acta 310:1–17
Gobrecht H, Hamann K, Willers G (1971) J Phys E 4:21–23
González C, Pérez Alcázar GA, Zamora LE, Tabares JA, Greneche J-M (2002) J Phys: Condens Matter 14:6531–6542
Goto T, Yoshizawa M, Tamaki A, Fujimura T (1982) J Phys C 15:3041–3051

Goto T, Li JH, Hirai T, Maeda Y, Kato R, Maesono A (1997) Int J Thermophys 18:569–577
Graebner JE (1989) Rev Sci Instrum 60:1123–1128
Graebner JE, Schneemeyer LF, Thomas JK (1989) Phys Rev B 39:9682–9684
Graebner JE, Haddon RC, Chichester SV, Glarum SH (1990) Phys Rev B 41:4808–4810
Graf MJ, Fortune NA, Brooks JS, Smith JL, Fisk Z (1989) Phys Rev B 40:9358–9361
Greene RL, King CN, Zubeck RB, Hauser JJ (1972) Phys Rev B 6:3297–3305
Griesheimer RN (1947), in: Montgomery CG (ed) Technique of Microwave Measurements. McGraw-Hill, New York, pp 79–220
Gulish OK, Polandov IN, Kuyumchev AA (1983) Sov Phys Solid State 25:1217–1219
Haeiwa T, Kita E, Siratori K, Kohn K, Tasaki A (1988) J Phys Soc Jpn 57:3381–3390
Haga H, Garland CW (1997a) Liquid Crystals 23:645–652
Haga H, Garland CW (1997b) Phys Rev E 56:3044–3052
Haga H, Garland CW (1998) Phys Rev E 57:603–609
Haga H, Nozaki R, Shiozaki Y, Ema K (1995a) J Phys Soc Jpn 64:4258–4264
Haga H, Onodera A, Shiozaki Y, Ema K, Sakata H (1995b) J Phys Soc Jpn 64:822–829
Haga H, Kutnjak Z, Iannacchione GS, Qian S, Finotello D, Garland CW (1997) Phys Rev E 56:1808–1818
Halvorson JJ, Wimber RT (1972) J Appl Phys 43:2519–2522
Handler P, Mapother DE, Rayl M 1967 Phys Rev Lett 19:356–358
Hatori J, Komukae M, Osaka T, Makita Y (1996) J Phys Soc Jpn 65:1960–1962
Hatta I (1979) Rev Sci Instrum 50:292–295
Hatta I (1994) Jpn J Appl Phys (Part 2) 33:L686–688
Hatta I (1996) Thermochim Acta 272:49–52
Hatta I, Ikeda H (1980) J Phys Soc Jpn 48:77–85
Hatta I, Ikushima A (1971) Phys Lett A 37:207–208
Hatta I, Ikushima A (1972) Phys Lett A 40:235–236
Hatta I, Ikushima A (1973) J Phys Chem Solids 34:57–66
Hatta I, Ikushima A (1976) J Phys Soc Jpn 41:558–564
Hatta I, Katayama N (1998) J Thermal Analysis 54:577–584
Hatta I, Kobayashi KLI (1977) Solid State Commun 22:775–777
Hatta I, Minakov AA (1999) Thermochim Acta 330:39–44
Hatta I, Muramatsu S (1996) Jpn J Appl Phys (Part 2) 35, L858–860
Hatta I, Nakayama S (1998) Thermochim Acta 318:21–27
Hatta I, Rehwald W (1977) J Phys C 10:2075–2081
Hatta I, Shiroishi Y, Müller KA, Berlinger W (1977) Phys Rev B 16:1138–1145
Hatta I, Matsuda T, Doi H, Nagasawa H, Ishiguro T, Kagoshima S (1978) Solid State Commun 27:479–481
Hatta I, Nakayama T, Matsuda T (1980) Phys Stat Sol A 62:243–248
Hatta I, Suzuki K, Imaizumi S (1983) J Phys Soc Jpn 52:2790–2797
Hatta I, Imaizumi S, Akutsu Y (1984) J Phys Soc Jpn 53:882–888
Hatta I, Sasuga Y, Kato R, Maesono A (1985a) Rev Sci Instrum 56:1643–1647
Hatta I, Matsuura M, Yao H, Gouhara K, Kato N (1985b) Thermochim Acta 88:143–148
Hatta I, Ichikawa H, Todoki M (1995) Thermochim Acta 267:83–94
Hellenthal W, Ostholt H (1970) Z angew Physik 28:313–316
Hempstead RD, Mochel JM (1973) Phys Rev B 7:287–299
Henning PF, Brooks JS, Crow JE, Tanaka Y, Kinoshita T, Kinoshita N, Tokumoto M, Anzai H (1995) Solid State Commun 95:691–694
Hensel A, Schick C (1997) Thermochim Acta 304/305:229–237

Hensel A, Dobbertin J, Schawe JEK, Boller A, Schick C (1996)
    J Thermal Analysis 46:935–954
Hiraka H, Endoh Y (1999) J Phys Soc Jpn 68:36–38
Hirayama T, Nakagawa M, Oda Y (2000) Solid State Commun 113:121–124
Hirotsu S, Miyamota M, Ema K (1983) J Phys C 16:L661–666
Hixson RS, Winkler MA (1990) Int J Thermophys 11:709–718
Hixson RS, Winkler MA (1992) Int J Thermophys 13:477–487
Holland LR (1963) J Appl Phys 34:2350–2357
Holland LR, Smith RC (1966) J Appl Phys 37:4528–4536
Holmes AT, Demuer A, Jaccard D (2003) Acta Phys Polonica B 34:567–570
Howling DH, Mendoza E, Zimmerman JE (1955) Proc Roy Soc A 229:86–109
Howson MA, Salamon MB, Friedmann TA, Inderhees SE, Rice JP, Ginsberg DM,
    Ghiron KM (1989) J Phys: Condens Matter 1:465–471
Howson MA, Salamon MB, Friedmann TA, Rice JP, Ginsberg D (1990)
    Phys Rev B 41:300–306
Höhne GWH (1997a) Thermochim Acta 304/305:121–123
Höhne GWH (1997b) Thermochim Acta 304/305:209–218
Höhne GWH (1999a) Thermochim Acta 330:45–54
Höhne GWH (1999b) Thermochim Acta 330:93–99
Höhne GWH, Merzlyakov M, Schick C (2002) Thermochim Acta 391:51–67
Hsu L-S (1994) Phys Lett A 184:476–480
Hu X, Tan TB, Li Y, Wilde G, Perepezko JH (1999) J Non-Cryst Solids 260:228–234
Huang CC, Viner JM (1982) Phys Rev A 25:3385–3388
Huang CC, Goldman AM, Toth LE (1980) Solid State Commun 33:581–584
Huang CC, Viner JM, Pindak R, Goodby JW (1981) Phys Rev Lett 46:1289–1292
Huth H, Beiner M, Weyer S, Merzlyakov M, Schick C, Donth E (2001)
    Thermochim Acta 377:113–124
Hwang GH, Shieh JH, Ho JC, Ku HC (1992) Physica C 201:171–175
Iannacchione GS, Finotello D (1992) Phys Rev Lett 69:2094–2097
Iannacchione G, Finotello D (1993) Liquid Crystals 14:1135–1142
Iannacchione GS, Finotello D (1994) Phys Rev E 50:4780–4795
Iannacchione G, Gorecka E, Pyzuk W, Kumar S, Finotello D (1995a)
    Phys Rev E 51:3346–3349
Iannacchione GS, Qian S, Crawford GP, Keast SS, Neubert ME, Doane JW, Finotello D,
    Steele LM, Sokol PE, Zumer S (1995b) Mol Cryst Liq Cryst 262:13–23
Iannacchione GS, Crawford GP, Qian S, Doane JW, Finotello D, Zumer S (1996)
    Phys Rev E 53:2402–2411
Iannacchione GS, Qian S, Finotello D, Aliev FM (1997) Phys Rev E 56:554–561
Iannacchione GS, Garland CW, Mang JT, Rieker TP (1998a) Phys Rev E 58:5966–5981
Iannacchione GS, Garland CW, Mieczkowski J, Gorecka E (1998b) Phys Rev E 58:595–601
Ikeda H (1977) J Phys C 10:L469–472
Ikeda H (1986) J Phys C 19:L811–816
Ikeda H, Hatta I, Tanaka M (1976) J Phys Soc Jpn 40:334–339
Ikeda H, Okamura N, Kato K, Ikushima A (1978) J Phys C 11:L231–235
Ikeda H, Abe T, Hatta I (1981) J Phys Soc Jpn 50:1488–1494
Ikeda S, Ishikawa Y (1979) Jpn J Appl Phys 18:1367–1372
Ikeda S, Ishikawa Y (1980) J Phys Soc Jpn 49:950–956
Imaizumi S, Garland CW (1987) J Phys Soc Jpn 56:3887–3892

Imaizumi S, Garland CW (1989) J Phys Soc Jpn 58:597–601
Imaizumi S, Hatta I (1984) J Phys Soc Jpn 53:4476–4487
Imaizumi S, Matsuda T, Hatta I (1979) J Phys Soc Jpn 47:1643–1646
Imaizumi S, Hatta I, Matsuda T (1981) J Phys Soc Jpn 50:276–280
Imaizumi S, Suzuki K, Hatta I (1983) Rev Sci Instrum 54:1180–1185
Inaba S, Oda S, Morinaga K (2002) J Non–Cryst Solids 325:42–49
Inaba S, Oda S, Morinaga K (2003) J Non–Cryst Solids 306:258–266
Inada T, Kawaji H, Atake T, Saito Y (1990) Thermochim Acta 163:219–224
Inderhees SE, Salamon MB, Friedmann TA, Ginsberg DM (1987)
    Phys Rev B 36:2401–2403
Inderhees SE, Salamon MB, Goldenfeld N, Rice JP, Pazol BG, Ginsberg DM, Liu JZ,
    Crabtree GW (1988) Phys Rev Lett 60:1178–1180
Inderhees SE, Salamon MB, Rice JP, Ginsberg DM (1991) Phys Rev Lett 66:232–235
Inoue R, Enoki T, Tsujikawa I (1982) J Phys Soc Jpn 51:3592–3600
Inoue M, Muneta Y, Negishi H, Sasaki M (1986) J Low Temp Phys 63:235–245
Irokawa K, Komukae M, Osaka T, Makita Y (1994) J Phys Soc Jpn 63:1162–1171
Ishikawa M, Nakazawa Y, Takabatake T, Kishi A, Kato R, Maesono A (1988)
    Solid State Commun 66:201–204
Itskevich ES, Kraidenov VF, Syzranov VS (1978) Cryogenics 18:281–284
Izawa T, Tajima K, Yamamoto Y, Fujii M, Fujimaru O, Shinoda Y (1996)
    J Phys Soc Jpn 65:2640–2644
Jackson JJ, Koehler JS (1960) Bull Amer Phys Soc 5:154
Jacobs SF, Bradford JN, Berthold JW (1970) Appl Optics 9:2477–2480
James JD, Spittle JA, Brown SGR, Evans RW (2001) Meas Sci Technol 12:R1–15
Jeong YH (1997) Thermochim Acta 304/305:67–98
Jeong YH, Moon IK (1995) Phys Rev B 52:6381–6385
Jeong YH, Bae DJ, Kwon TW, Moon IK (1991) J Appl Phys 70:6166–6168
Jin AJ, Veum M, Stoebe T, Chou CF, Ho JT, Hui SW, Surendranath V, Huang CC (1995)
    Phys Rev Lett 74:1863–1866
Jin XC, Hor PH, Wu MK, Chu CW (1984) Rev Sci Instrum 55:993–995
Johansen TH (1987) High Temp – High Press 19:77–87
Johansen TH, Feder J, Jøssang T (1986) Rev Sci Instrum 57:1168–1174
Johnson DL, Hayes CF, deHoff RJ, Schantz CA (1978) Phys Rev B 18:4902–4912
Johnson JB (1928) Phys Rev 32:97–109
Jones KJ, Kinshott I, Reading M, Lacey AA, Nikolopoulos C, Pollock HM (1997)
    Thermochim Acta 304/305:187–199
Jones RC (1953), in: Advances in Electronics, vol 5. Academic Press, New York, pp 1–96
Jonsson UG, Andersson O (1998) Meas Sci Technol 9:1873–1885
Jung DH, Kwon TW, Bae DJ, Moon IK, Jeong YH (1992) Meas Sci Technol 3:475–484
Jung DH, Moon IK, Jeong YH (2002) Thermochim Acta 391:7–12
Jung DH, Moon IK, Jeong YH, Lee SH (2003) Thermochim Acta 403:83–88
Jurado JF, Ortiz E, Vargas RA (1997) Meas Sci Technol 8:1151–1155
Kagan DN (1984), in: Maglić KD, Cezairliyan A, Peletsky VE (eds) Compendium of
    Thermophysical Property Measurement Methods, vol 1. Plenum, New York,
    pp 461–526
Kamilov IK, Abdulvagidov SB, Shakhshaev GM, Aliev KK, Batdalov AB (1995)
    Int J Thermophys 16:821–829

Kamper RA (1972), in: Plumb HH (ed) Temperature, Its Measurement and Control in
    Science and Industry, vol 4. Instrument Society of America, Pittsburgh, pp 349–354
Kanda E, Yoshizawa M, Yamakami T, Fujimura T (1982) J Phys C 15:6823–6831
Kanel' OM, Kraftmakher YA (1966) Sov Phys Solid State 8:232–233
Kaschnitz E, Pottlacher G, Jäger H (1992) Int J Thermophys 13:699–710
Kasting GB, Lushington KJ, Garland CW (1980) Phys Rev B 22:321–331
Kato H, Nara K, Okaji M (2001) Cryogenics 41:373–378
Kato R, Maeda Y, Maesono A, Tye RP (1999) High Temp – High Press 31:23–28
Katsumoto S, Kobayashi S, Urayama H, Yamochi H, Saito G (1988)
    J Phys Soc Jpn 57:3672–3673
Kawai M, Tahira K, Kitagawa K, Miyakawa T (1978) Appl Phys Lett 33:9–10
Kawai M, Miyakawa T, Tako T (1984) Jpn J Appl Phys 23:1202–1208
Kawaji H, Atake T, Saito Y (1989) J Phys Chem Solids 50:215–220
Kämpf G, Buckel W (1977) Z Phys B 27:315–319
Kämpf G, Selisky H, Buckel W (1981) Physica B 108:1263–1264
Kenny TW, Richards PL (1990a) Rev Sci Instrum 61:822–829
Kenny TW, Richards PL (1990b) Phys Rev Lett 64:2386–2389
Kettler W, Wernhardt R, Rosenberg M (1982) J Appl Phys 53:8248–8250
Kettler W, Kaul SN, Rosenberg M (1984) Phys Rev B 29:6950–6956
Kettler WH, Wernhardt R, Rosenberg M (1986) Rev Sci Instrum 57:3053–3058
Kim HK, Zhang QM, Chan MHW (1986a) Phys Rev Lett 56:1579–1582
Kim HK, Zhang QM, Chan MHW (1986b) Phys Rev B 34:4699–4709
Kim K-S, Kim J-H, Lee J-K, Jarng S-S (1997) J Mat Sci Lett 16:1753–1756
Kimball MO, Mehta S, Gasparini FM (2000) J Low Temp Phys 121:29–51
Kirby RK (1991) Int J Thermophys 12:679–685
Kirillin VA, Sheindlin AE, Chekhovskoi VY, Zhukova IA (1967)
    High Temperature 5:1016–1017
Kirsch T, Eichler A, Morin P, Welp U (1992) Z Phys B 86:83–86
Kishi A, Kato R, Azumi T, Okamoto H, Maesono A, Ishikawa M, Hatta I,
    Ikushima A (1988) Thermochim Acta 133:39–42
Kittel C (1988) Physics Today, May, 93
Kluin J-E, Hehenkamp T (1991) Phys Rev B 44:11598–11608
Kohl WH (1962) Materials and Techniques for Electron Tubes.
    Chapman and Hall, London, p 277
Korn D, Mürer W (1977) Z Phys B 27:309–314
Korostoff E (1962) J Appl Phys 33:2078–2079
Korus J, Beiner M, Busse K, Kahle S, Unger R, Donth E (1997)
    Thermochim Acta 304/305:99–110
Kraev OA (1967) High Temperature 5:727–730
Kraftmakher YA (1962) Zh Prikl Mekhan Tekhn Fiz N 5:176–180
Kraftmakher YA (1963a) Zh Prikl Mekhan Tekhn Fiz N 2:158–160
Kraftmakher YA (1963b) Sov Phys Solid State 5:696–697
Kraftmakher YA (1964) Sov Phys Solid State 6:396–398
Kraftmakher YA (1966a) Sov Phys Solid State 8:1048–1049
Kraftmakher YA (1966b) J Appl Mech Tech Phys 7:100
Kraftmakher YA (1966c) in: Novikov II, Strelkov PG (eds) High-Temperature Studies,
    in Russian. Nauka, Novosibirsk, pp 5–54
Kraftmakher YA (1967a) Sov Phys Solid State 9:1199–1200

Kraftmakher YA (1967b) Sov Phys Solid State 9:1197–1198
Kraftmakher YA (1967c) Sov Phys Solid State 9:1458
Kraftmakher YA (1967d) Zh Prikl Mekhan Tekhn Fiz N 4:143–144
Kraftmakher YA (1971a) Sov Phys Solid State 13:2918–2919
Kraftmakher YA (1971b) Phys Stat Sol B 48:K39–43
Kraftmakher YA (1972) Sov Phys Solid State 14:325–327
Kraftmakher YA (1973a) High Temp – High Press 5:433–454
Kraftmakher YA (1973b) High Temp – High Press 5:645–656
Kraftmakher YA (1974), in: Vasu KI, Raman KS, Sastry DH, Prasad YVRK (eds) Defect Interactions in Solids. Indian Institute of Science, Bangalore, pp 64–70
Kraftmakher YA (1977) Scripta Metall 11:1033–1038
Kraftmakher YA (1978), in: 6th European Conf Thermophys Properties – Research and Applications, Dubrovnik, paper 101
Kraftmakher YA (1981) Teplofiz Vysok Temp 19:656–658
Kraftmakher YA (1984), in: Maglić KD, Cezairliyan A, Peletsky VE (eds) Compendium of Thermophysical Property Measurement Methods, vol 1. Plenum, New York, pp 591–641
Kraftmakher YA (1985) Sov Phys Solid State 27:141–142
Kraftmakher YA (1990) Phys Lett A 149:284–286
Kraftmakher YA (1991) Phys Lett A 154:43–44
Kraftmakher YA (1992), in: Maglić KD, Cezairliyan A, Peletsky VE (eds) Compendium of Thermophysical Property Measurement Methods, vol 2. Plenum, New York, pp 409–436
Kraftmakher Y (1994a) High Temp – High Press 26:497–505
Kraftmakher Y (1994b) Eur J Phys 15:329–334
Kraftmakher Y (1995) Amer J Phys 63:932–935
Kraftmakher Y (1996) Phil Mag A 74:811–822
Kraftmakher Y (1997) Defect Diffusion Forum 143/147:37–42
Kraftmakher Y (1998) High Temp – High Press 30:449–455
Kraftmakher Y (2000) Lecture Notes on Equilibrium Point Defects and Thermophysical Properties of Metals. World Scientific, Singapore
Kraftmakher Y (2001a) Defect Diffusion Forum 194/199:17–22
Kraftmakher Y (2001b) Defect Diffusion Forum 194/199:23–28
Kraftmakher YA, Cheremisina IM (1965) Zh Prikl Mekhan Tekhn Fiz N 2:114–115
Kraftmakher YA, Cherepanov VY (1978) High Temperature 16:557–559
Kraftmakher YA, Cherevko AG (1972a) Prib Tekhn Eksper N 4:150–151
Kraftmakher YA, Cherevko AG (1972b) Phys Stat Sol A 14:K35–38
Kraftmakher YA, Cherevko AG (1974) Phys Stat Sol A 25:691–695
Kraftmakher YA, Cherevko AG (1975) High Temp – High Press 7:283–286
Kraftmakher YA, Krylov SD (1980a) High Temperature 18:261–264
Kraftmakher YA, Krylov SD (1980b) Sov Phys Solid State 22:1845–46
Kraftmakher YA, Lanina EB (1965) Sov Phys Solid State 7:92–95
Kraftmakher YA, Nezhentsev VP (1971), in: Novikov II, Strelkov PG (eds) Solid State Physics and Thermodynamics, in Russian. Nauka, Novosibirsk, pp 233–237
Kraftmakher YA, Pinegina TY (1970) Phys Stat Sol 42:K151–152
Kraftmakher YA, Pinegina TY (1971) Sov Phys Solid State 13:2345–2346
Kraftmakher YA, Pinegina TY (1974) Sov Phys Solid State 16:78–81
Kraftmakher YA, Pinegina TY (1978) Phys Stat Sol A 47:K81–83

Kraftmakher YA, Romashina TY (1965) Sov Phys Solid State 7:2040–2041
Kraftmakher YA, Romashina TY (1966) Sov Phys Solid State 8:1562–1563
Kraftmakher YA, Romashina TY (1967) Sov Phys Solid State 9:1459–1460
Kraftmakher YA, Shestopal VO (1965) Zh Prikl Mekhan Tekhn Fiz N 4:170–171
Kraftmakher YA, Strelkov PG (1960) Zh Prikl Mekhan Tekhn Fiz N 3:194–197
Kraftmakher YA, Strelkov PG (1962) Sov Phys Solid State 4:1662–1664
Kraftmakher YA, Strelkov PG (1966a) Sov Phys Solid State 8:460–462
Kraftmakher YA, Strelkov PG (1966b) Sov Phys Solid State 8:838–841
Kraftmakher YA, Strelkov PG (1970), in: Seeger A, Schumacher D, Schilling W, Diehl J (eds) Vacancies and Interstitials in Metals. North-Holland, Amsterdam, pp 59–78
Kraftmakher YA, Sushakova GG (1972) Phys Stat Sol B 53:K73–76
Kraftmakher YA, Sushakova GG (1974) Sov Phys Solid State 16:82–84
Kraftmakher YA, Tarasenko AP (1987) J Eng Phys 53:787–791
Kraftmakher YA, Tonaevskii VL (1972) Phys Stat Sol A 9:573–579
Kramer W, Nölting J (1972) Acta Metall 20:1353–1359
Kratz W, Kahle HG, Paul W, (1996) J Magn Magn Mater 161: 249–254
Krauss G, Buckel W (1975) Z Phys B 20:147–153
Krüger JK, Bohn K-P, Le Coutre A, Mesquida P (1998) Meas Sci Technol 9:1866–1872
Kuo Y-K, Figueroa E, Brill JW (1995) Solid State Commun 94:385–389
Kuo Y-K, Powell DK, Brill JW (1996) Solid State Commun 98:1027–1031
Kutnjak Z, Garland CW (1997) Phys Rev E 55:488–495
Lacey AA, Nikolopoulos C, Reading M (1997) J Thermal Analysisysis 50:279–333
Lai SL, Ramanath G, Allen LH, Infante P, Ma Z (1995) Appl Phys Lett 67:1229–1231
Landau LD, Lifshitz EM (1980) Statistical Physics. Pergamon, London
Lannin JS, Eno HF, Luo HL (1978) Solid State Commun 25:81–84
Lægreid T, Fossheim K, Trætteberg O, Sandvold E, Julsrud S, (1988) Physica C 153/155:1026–1027
Lægreid T, Tuset P, Nes O-M, Slaski M, Fossheim K (1989) Physica C 162/164:490–491
Lebedev SV, Savvatimskii AI (1974) Sov Phys Uspekhi 27:749–771
Lederman FL, Salamon MB (1974) Solid State Commun 15:1373–1376
Lederman FL, Salamon MB, Shacklette LW (1974) Phys Rev B 9:2981–2988
Lederman FL, Salamon MB, Peisl H (1976) Solid State Commun 19:147–150
Leedertz JA (1970) J Phys E 3:214–218
LeGrange JD, Mochel JM (1980) Phys Rev Lett 45:35–38
LeGrange JD, Mochel JM (1981) Phys Rev A 23:3215–3223
Lerchner J, Wolf A, Wolf G (1999) J Therm Anal Calorim 67:241–251
Lerchner J, Wolf G, Auguet C, Torra V (2002) Thermochim Acta 382:65–76
Lewis EAS (1970) Phys Rev B 1:4368–4377
Leyser H, Schulte A, Doster W, Petry W, (1995) Phys Rev E 51:5899–5904
Li Y, Ng SC, Lu ZP, Feng YP, Lu K (1998) Phil Mag Lett 78:37–44
Lien SC, Huang CC, Goodby JW (1984) Phys Rev A 29:1371–1374
Lin S, Li Lu, Zhang D, Duan HM, Kiehl W, Hermann AM (1993) Phys Rev B 47:8324–8326
Liu HY, Huang CC, Bahr C, Heppke G (1988) Phys Rev Lett 61:345–348
Loponen MT, Dynes RC, Narayanamurti V, Garno JP (1982) Phys Rev B 25:1161–1173
Lowe AJ, Regan S, Howson MA (1991) Phys Rev B 44:9757–9759
Lowenthal GC (1963) Austral J Phys 16:47–67
Løkberg OJ, Malmo JT, Slettemoen GÅ (1985) Appl Optics 24:3167–3172

Lushington KJ, Garland CW (1980) J Chem Phys 72:5752–5759
Lushington KJ, Kasting GB, Garland CW (1980a) J Physique Lettres 41:L419–422
Lushington KJ, Kasting GB, Garland CW (1980b) Phys Rev B 22:2569–2572
Luyt AS, Vosloo HCM, Reading M (1998) Thermochim Acta 320:135–140
Machado FLA, Clark WG (1988) Rev Sci Instrum 59:1176–1181
Machado FLA, Clark WG, Azevedo LJ, Yang DP, Hines WA, Budnick JI,
    Quan MX (1987a) Solid State Commun 61:145–149
Machado FLA, Clark WG, Yang DP, Hines WA, Azevedo LJ, Giessen BC, Quan MX
    (1987b) Solid State Commun 61:691–695
Maesono A, Tye R P (1998) High Temp – High Press 30, 695–700
Maglić KD (2003) Int J Thermophys 24:489–500
Maglić KD, Dobrosavljević AS (1992) Int J Thermophys 13:3–16
Maglić KD, Pavičić DZ (2001) Int J Thermophys 22:1833–1841
Maglić KD, Perović NL, Vuković GS, Zeković LP (1994) Int J Thermophys 15:963–972
Maglić KD, Dobrosavljević AS, Perović NL, Stanimirović AM, Vuković GS (1995/96)
    High Temp – High Press 27/28:389–402
Maglić KD, Perović NL, Vuković GS (1997) High Temp – High Press 29:97–102
Mahmood R, Lewis M, Johnson D, Surrendranath V (1988) Phys Rev A 38:4299–4303
Majumdar A, Lai J, Chandrachood M, Nakabeppu O, Wu Y, Shi Z (1995)
    Rev Sci Instrum 66:3584–3592
Makarenko IN, Trukhanova LN, Filippov LP (1970a) High Temperature 8:416–418
Makarenko IN, Trukhanova LN, Filippov LP (1970b) High Temperature 8:628–631
Mandelbrot BB (1989) Physics Today, January, 71–73
Mandelis A, Zver MM (1985) J Appl Phys 57:4421–4430
Mandelis A, Care F, Chan KK, Miranda LCM (1985) Appl Phys A 38:117–122
Mangelschots I, Andersen NH, Lebech B, Wisniewski A, Jacobsen CS (1992)
    Physica C 203:369–377
Manuel P, Veyssié JJ (1972) Phys Lett A 41:235–236
Manuel P, Veyssié JJ (1973) Solid State Commun 13:1819–1823
Manuel P, Veyssié JJ (1976) Phys Rev B 14:78–88
Manuel P, Niedoba H, Veyssié JJ (1972) Rev Phys Appliquée 7:107–116
Marinelli Massimo, Murtas F, Mecozzi M G, Zammit U, Pizzoferrato R, Scudieri F,
    Martellucci S, Marinelli Marco (1990) Appl Phys A 51:387–393
Marinelli M, Zammit U, Mercuri F, Pizzoferrato R (1992) J Appl Phys 72:1096–1100
Marinelli M, Mercuri F, Zammit U, Gusev V (1994a) Appl Phys Lett 65:2663–2665
Marinelli M, Mercuri F, Zammit U, Pizzoferrato R, Scudieri F, Dadarlat D (1994b)
    J Physique IV 4:C7-261–266
Marinelli M, Mercuri F, Foglietta S, Belanger DP (1996) Phys Rev B 54:4087–4092
Marinelli M, Mercuri F, Zammit U, Scudieri F (1998) Int J Thermophys 19:595–601
Marone MJ, Payne JE (1997) Rev Sci Instrum 68:4516–4520
Maszkiewicz M (1978) Phys Stat Sol A 47:K77–80
Maszkiewicz M, Mrygoń B, Wentowska K (1979) Phys Stat Sol A 54:111–115
Matsumoto T, Ono A (2001) Meas Sci Technol 12:2095–2102
Matsuura M, Yao H, Gouhara K, Hatta I, Kato N (1985) J Phys Soc Jpn 54:625–629
McNeill DJ (1962) J Appl Phys 33:597–600
Mehta S, Gasparini FM (1997) Phys Rev Lett 78:2596–2599
Mehta S, Gasparini FM (1998) J Low Temp Phys 110:287–292
Mehta S, Kimball MO, Gasparini FM (1999) J Low Temp Phys 114:467–521

Meichle M, Garland CW (1983) Phys Rev A 27:2624–2631
Meissner M, Spitzmann K (1981) Phys Rev Lett 46:265–268
Melero JJ, Burriel R (1996) J Magn Magn Mater 157/158:651–652
Melero JJ, Bartolomé J, Burriel R, Aleksandrova IP, Primak S (1995)
    Solid State Commun 95:201–206
Menczel JD, Judovits (1998) J Therm Anal Calorim 54:419–436
Menges H, v. Löhneysen H (1991) J Low Temp Phys 84:237–260
Menon N (1996) J Chem Phys 105:5246–5257
Mercuri F, Marinelli M, Zammit U, Pizzoferrato R, Scudieri F (1994)
    J Physique IV 4:C7-253–256
Mertig M, Pompe G, Hegenbarth E (1984) Solid State Commun 49:369–372
Merzlyakov M (2003) Thermochim Acta 403:65–81
Merzlyakov M, Schick C (1999a) Thermochim Acta 330:55–64
Merzlyakov M, Schick C (1999b) Thermochim Acta 330:65–73
Merzlyakov M, Schick C (2000) J Therm Anal Calorim 61: 649–659
Merzlyakov M, Schick C (2001a) Thermochim Acta 377:193–204
Merzlyakov M, Schick C (2001b) Thermochim Acta 380:5–12
Merzlyakov M, Höhne GWH, Schick C (2002) Thermochim Acta 391:69–80
Mesquida P, le Coutre A, Krüger JK (1999) Thermochim Acta 330:137–144
Migone AD, Li ZR, Chan MHW (1984) Phys Rev Lett 53:810–813
Miiller AP, Cezairliyan A (1982) Int J Thermophys 3:259–288
Miiller AP, Cezairliyan A (1985) Int J Thermophys 6:695–704
Miiller AP, Cezairliyan A (1988) Int J Thermophys 9:195–203
Miiller AP, Cezairliyan A (1990) Int J Thermophys 11:619–628
Miiller AP, Cezairliyan A (1991) Int J Thermophys 12:643–656
Milatz JMW, Van der Velden HA 1943 Physica 10:369–380
Milošević ND, Vuković GS, Pavičić DZ, Maglić KD (1999)
    Int J Thermophys 20:1129–1136
Minakov AA (2000) Thermochim Acta 345:3–12
Minakov AA, Ershov OV (1994) Cryogenics 34:461–464 (Supplement)
Minakov AA, Bugoslavsky YV, Schick C (1999) Thermochim Acta 342:7–18
Minakov AA, Adamovsky SA, Schick C (2001) Thermochim Acta 377: 173–182
Minakov AA, Adamovsky SA, Schick C (2003) Thermochim Acta 403:89–103
Miyazaki A, Sakata K, Komukae M, Osaka T, Makita Y (1989a)
    J Phys Soc Jpn 58:3635–3641
Miyazaki A, Ikeda T, Osaka T, Komukae M, Makita Y (1989b)
    J Phys Soc Jpn 58:4496–4500
Mizuno H, Nagano Y, Tashiro K, Kobayashi M (1992) J Chem Phys 96:3234–3239
Molz E, Wong APY, Chan MHW, Beamish JR (1993) Phys Rev B 48:5741–5750
Monazam ER, Maloney DJ, Lawson LO (1989) Rev Sci Instrum 60:3460–3465
Moon IK, Jeong YH (1996) Rev Sci Instrum 67:3553–3556
Moon IK, Jeong YH (2001) Thermochim Acta 377:51–61
Moon IK, Jeong YH, Kwun SI (1996) Rev Sci Instrum 67:29–35
Moon I, Androsch R, Wunderlich B (2000a) Thermochim Acta 357/358:285–291
Moon IK, Jung DH, Lee K-B, Jeong YH (2000b) Appl Phys Lett 76:2451–2453
Morilov VV, Ivliev AD (1995) High Temperature 33:367–372
Mosig K, Wolff J, Kluin J-E, Hehenkamp T (1992) J Phys: Cond Matter 4:1447–1458

Murayama S, Morita Y, Hoshi K, Onodera A, Obi Y (1995)
    J Magn Magn Mater 140/144:309–310
Murtazaev AK, Abdulvagidov SB, Aliev AM, Musaev OK (2001)
    Phys Solid State 43: 1103–1107
Müller J, Lang M, Helfrich R, Steglich F, Sasaki T (2002) Phys Rev B 65:140509-1–4
Müller J, Lang M, Helfrich R, Steglich F, Sasaki T (2003) Synth Met 133/134:235–237
Nagano H, Nakanishi T, Yao H, Ema K (1995a) Phys Rev E 52:4244–4250
Nagano H, Yao H, Ema K (1995b) Phys Rev E 51:3363–3367
Nahm K, Kim CK, Mittag M, Jeong YH (1995) J Appl Phys 78:3980–3982
Nakagawa Y, Schäfer R, Güntherodt H-J (1998) Appl Phys Lett 73:2296–2298
Nakazawa Y, Sato A, Seki M, Saito K, Hiraki K, Takahashi T, Kanoda K, Sorai M (2003)
    Phys Rev B 68:085112-1–8
Nes O-M, Castro M, Slaski M, Lægreid T, Fossheim K, Motohira N, Kitazawa K (1991)
    Supercond Sci Technol 4:S388–390 (Supplement)
Nishikawa M, Saruyama Y (1995) Thermochim Acta 267:75–81
Nounesis G, Huang CC, Goodby JW (1986) Phys Rev Lett 56:1712–1715
Nyquist H (1928) Phys Rev 32:110–113
Ogawa S, Yamadaya T (1974) Phys Lett A 47:213–214
Ogura H, Shimizu T, Motoyama H, Ochiai M, Chiba A (1992)
    Jpn J Appl Phys (Part 1) 31:835–839
Ohmatsu K, Suematsu H, Suzuki M (1983) Synth Met 6:135–140
Ohsawa J, Nishinaga T, Uchiyama S (1978) Jpn J Appl Phys 17:1059–1065
Okazaki N, Hasegawa T, Kishio K, Kitazawa K, Kishi A, Ikeda Y, Takano M, Oda K,
    Kitaguchi H, Takada J, Miura Y (1990) Phys Rev B 41:4296–4301
Onodera A, Strukov BA, Belov AA, Taraskin SA, Haga H, Yamashita H, Uesu Y (1993)
    J Phys Soc Jpn 62:4311–4315
O'Reilly KAQ, Cantor B (1996) Proc Roy Soc (London) A 452:2141–2160
Oussena M, Gagnon R, Wang Y, Aubin M (1992) Phys Rev B 46:528–531
Overend N, Howson MA, Lawrie ID (1994) Phys Rev Lett 72:3238–3241
Overend N, Howson MA, Lawrie ID, Abell S, Hirst PJ, Chen Changkang, Chowdhury S,
    Hodby JW, Inderhees SE, Salamon MB (1996) Phys Rev B 54:9499–9508
Ozawa T, Kanari K (2000) J Therm Anal Calorim 59:257–270
Papp E (1984) Z Phys B 55:17–22
Park SH, Jeong Y-H, Lee K-B, Kwon SJ (1997) Phys Rev B 56:67–70
Park T, Salamon MB, Jung CU, Park M-S, Kim K, Lee S-I (2002)
    Phys Rev B 66:134515-1–5
Park T, Salamon MB, Choi EM, Kim HJ, Lee S-I (2003) Phys Rev Lett 90:177001-1–4
Pavičić DZ, Maglić KD (2002) Int J Thermophys 23:1319–1325
Pepper MG, Brown JB (1979) J Phys E 12:31–34
Pérez J, Blasco J, García J, Castro M, Stankiewicz J, Sánchez MC, Sánchez RD (1999)
    J Magn Magn Mater 196/197:541–542
Perović NL, Maglić KD, Vuković GS (1996) Int J Thermophys 17:1047–1055
Phelps RB, Birmingham JT, Richards PL (1993) J Low Temp Phys 92:107–125
Pichon C, Le Liboux M, Fournier D, Boccara AC (1979) Appl Phys Lett 35:435–437
Pitchford T, Huang CC, Pindak R, Goodby JW (1986) Phys Rev Lett 57:1239–1242
Pochapsky TE (1953) J Chem Phys 21:1539–1540
Polandov IN, Chernenko VA, Novik VK (1981) High Temp – High Press 13:399–406
Pottlacher G, Kaschnitz E, Jäger H (1991) J Phys: Condens Matter 3:5783–5792

Pottlacher G, Kaschnitz E, Jäger H (1993) J Non-Cryst Solids 156/158:374–378
Powell DK, Miebach T, Cheng S-L, Montgomery LK, Brill JW (1997) Solid State Commun 104:95–99
Powell DK, Starkey KP, Shaw G, Sushko YV, Montgomery LK, Brill JW (2001) Solid State Commun 119:637–640
Priestley EB (1974), in: Priestley EB, Wojtowicz PJ, Ping Sheng (eds) Introduction to Liquid Crystals, Plenum, New York, pp 1–13
Puértolas JA, Castro M, De la Fuente MR, Pérez Jubindo MA, Dreyfus H, Guillon D, González Y (1996) Mol Cryst Liq Cryst 287:69–82
Pursey H, Pyatt E (1959) J Sci Instrum 36:260–264
Qian S, Finotello D (1997) Mol Cryst Liq Cryst 304:519–524
Qian S, Iannacchione GS, Finotello D, Steele L M, Sokol P E (1995) Mol Cryst Liq Cryst 265:395–402
Qian S, Iannacchione GS, Finotello D (1996) Phys Rev E 53:R4291–4294
Qian S, Iannacchione GS, Finotello D (1998) Phys Rev E 57:4305–4315
Rajeswari M, Raychaudhuri AK (1993) Phys Rev B 47:3036–3046
Rao NAHK, Goldman AM (1981) J Low Temp Phys 42:253–276
Rasor NS, McClelland JD (1960a) Rev Sci Instrum 31:595–604
Rasor NS, McClelland JD (1960b) J Phys Chem Solids 15:17–26
Reading M (1997) Thermochim Acta 292:179–187
Reading M (2001) J Therm Anal Calorim 64:7–14
Reading M, Elliott D, Hill VL (1993) J Thermal Analysis 40:949–955
Reading M, Luget A, Wilson R (1994) Thermochim Acta 238:295–307
Regan S, Lowe AJ, Howson MA (1991) J Phys: Condens Matter 3:9245–9248
Resel R, Gratz E, Burkov AT, Nakama T, Higa M, Yagasaki K (1996) Rev Sci Instrum 67:1970–1975
Ribeiro M, Grolier J-PE (1999) J Therm Anal Calorim 57:253–263
Richmond J C (1984), in: Maglić KD, Cezairliyan A, Peletsky VE (eds) Compendium of Thermophysical Property Measurement Methods, vol 1. Plenum, New York, pp 709–768
Rieger P, Baumann F (1991) J Phys: Condens Matter 3:2309–2317
Righini F, Rosso A (1983) Int J Thermophys 4:173–181
Righini F, Roberts RB, Rosso A (1985) Int J Thermophys 6:681–693
Righini F, Roberts RB, Rosso A, Cresto P C (1986a) High Temp – High Press 18:561–571
Righini F, Roberts RB, Rosso A (1986b) High Temp – High Press 18:573–583
Righini F, Spišiak J, Bussolino GC, Rosso A (1993) High Temp – High Press 25:193–203
Righini F, Spišiak J, Bussolino GC, Gualano M (1999) Int J Thermophys 20:1107–1116
Righini F, Bussolino GC, Spišiak J (2000) Thermochim Acta 347:93–102
Riou O, Gandit P, Charalambous M, Chaussy J (1997) Rev Sci Instrum 68:1501–1509
Robinson DS, Salamon MB (1982) Phys Rev Lett 48:156–159
Roeser WF, Wensel HT (1941), in: Temperature, Its Measurement and Control in Science and Industry. Reinhold, New York, pp 1293–1323
Rogez J (1998) Revue de Metallurgie – Cahiers d'Informations Techniques 95:1047–1057
Ronchi C, Heinz W, Musella M, Selfslag R, Sheindlin M (1999) Int J Thermophys 20:987–996
Rosencwaig A, Gersho A (1976) J Appl Phys 47:64–69
Rosenthal LA (1961) Rev Sci Instrum 32:1033–1036
Rosenthal LA (1965) Rev Sci Instrum 36:1179–1182

Rotter M, Müller H, Gratz E, Doerr M, Loewenhaupt M (1998)
    Rev Sci Instrum 69:2742–2746
Rudyi AS (1993) Int J Thermophys 14:159–172
Sabbah R, An Xu-wu, Chickos JS, Planas Leitão ML, Roux MV, Torres LA (1999)
    Thermochim Acta 331:93-204
Saito K, Kamio H, Kikuchi K, Ikemoto I (1990) Thermochim Acta 163:241–248
Saito K, Akutsu H, Sorai M (1999a) Solid State Commun 111:471–475
Saito K, Yamamura Y, Akutsu H, Takeda M, Asaoka H, Nishikawa H, Ikemoto I, Sorai M
    (1999b) J Phys Soc Jpn 68:1277–1285
Saito K, Sato A, Kikuchi K, Nishikawa H, Ikemoto I, Sorai M (2000)
    J Phys Soc Jpn 69:3602–3606
Saito K, Akutsu H, Kikuchi K, Nishikawa H, Ikemoto I, Sorai M (2001a)
    J Phys Soc Jpn 70:1635–1641
Saito K, Sato A, Bhattacharjee A, Sorai M (2001b) Solid State Commun 120:129–132
Salamon MB (1970) Phys Rev B 2:214–220
Salamon MB (1973) Solid State Commun 13:1741–1745
Salamon MB, Chun SH (2003) Phys Rev B 68:014441-1–8
Salamon MB, Hatta I (1971) Phys Lett A 36:85–86
Salamon MB, Ikeda H (1973) Phys Rev B 7:2017–2024
Salamon MB, Simons DS (1973) Phys Rev B 7:229–232
Salamon MB, Simons DS, Garnier PR (1969) Solid State Commun 7:1035–1038
Salamon MB, Garnier PR, Golding B, Buehler E (1974) J Phys Chem Solids 35:851–859
Salamon MB, Bray JW, DePasquali G, Craven RA, Stucky G, Schultz A (1975)
    Phys Rev B 11:619–622
Salamon MB, Inderhees SE, Rice JP, Pazol BG, Ginsberg DM, Goldenfeld N (1988)
    Phys Rev B 38:885–888
Salamon MB, Inderhees SE, Rice JP, Ginsberg DM (1990) Physica A 168:283–290
Salamon MB, Shi J, Overend N, Howson MA (1993) Phys Rev B 47:5520–5523
Sandvold E, Fossheim K (1986) J Phys C 19:1481–1489
Saruyama Y (1992) J Thermal Analysis 38:1827–1833
Saruyama Y (1997) Thermochim Acta 304/305:171–178
Saruyama Y (1998) J Thermal Analysis 54:687–693
Saruyama Y (2000) J Therm Anal Calorim 59:271–278
Saruyama Y (2001) Thermochim Acta 377:151–158
Sato A, Akutsu H, Saito K, Sorai M (2001) Synt Met 120:1035–1036
Sato M, Fujishita H, Hoshino S (1983) J Phys C 16:L417–421
Schaefer H-E, Schmid G (1989) J Phys: Condens Matter 1:SA49–54 (Supplement A)
Schantz CA, Johnson DL (1978) Phys Rev A 17:1504–1512
Schawe JEK (1995) Thermochim Acta 261:183–194
Schawe JEK (1996) Thermochim Acta 271:127–140
Schawe JEK (1997) Thermochim Acta 304/305:111–119
Schick C, Jonsson U, Vassiliev T, Minakov A, Schawe J, Scherrenberg R, Lőrinczy D
    (2000a) Thermochim Acta 347:53–61
Schick C, Merzlyakov M, Minakov A, Wurm A (2000b)
    J Therm Anal Calorim 59:279–288
Schick C, Wurm A, Merzlyakov M, Minakov A, Marand H (2001)
    J Therm Anal Calorim 64:549–555
Schmidt U, Vollmer O, Kohlhaas R (1970) Z Naturforsch A 25:1258–1254

Schmiedeshoff GM, Fortune NA, Brooks JS, Stewart GR (1987)
    Rev Sci Instrum 58:1743–1745
Schnelle W, Gmelin E (1995) Thermochim Acta 269/270:27–32
Schottky (1918) Ann Physik 57:541–567
Schoubs E, Mondelaers H, Thoen J (1994) J Physique IV 4:C7-257–260
Schowalter LJ, Salamon MB, Tsuei CC, Craven RA (1977)
    Solid State Commun 24:525–529
Schwartz P (1971) Phys Rev B 4:920–928
Seidman DN, Balluffi RW (1965) Phys Rev 139:A1824–1840
Sekine T, Uchinokura K, Iimura H, Yoshizaki R, Matsuura E (1984)
    Solid State Commun 51:187–189
Sekine T, Kuroe H, Makimura C, Tanokura Y, Takeuchi T (1995) Synth Met 70:1383–1384
Senchenko VN, Sheindlin MA (1987) High Temperature 25:364–368
Seville AH (1974) Phys Stat Sol A 21:649–658
Seydel U, Fischer U (1978) J Phys F 8:1397–1404
Shacklette LW (1974) Phys Rev B 9:3789–3792
Shang H-T, Salamon MB (1980) Phys Rev B 22:4401–4411
Shang H-T, Huang C-C, Salamon MB (1978) J Appl Phys 49:1366–1368
Shepard RL, Carroll RM, Falter DD, Blalock TV, Roberts MJ (1992), in: Schooley (ed)
    Temperature, Its Measurement and Control in Science and Industry, vol 6.
    AIP, New York, pp 997–1002
Shepherd JP (1985) Rev Sci Instrum 56:273–277
Shestopal VO (1965) Sov Phys Solid State 7:2798–2799
Shore FJ, Williamson RS (1966) Rev Sci Instrum 37:787–788
Simon SL, McKenna GB (1997) J Chem Phys 107:8678–8685
Simons DS, Salamon MB (1971) Phys Rev Lett 26:750–752
Simons DS, Salamon MB (1974) Phys Rev B 10:4680–4686
Skelskey D, Van den Sype J (1970) J Appl Phys 41:4750–4751
Skelskey DA, Van den Sype J (1974) Solid State Commun 15:1257–1262
Slaski M, Lægreid T, Nes O-M, Gjølmesli S, Tuset P, Fossheim K, Pajaczkowska A (1989)
    Physica C 162/164:492–493
Smaardyk JE, Mochel JM (1978) Rev Sci Instrum 49:988–993
Smith KK, Bigler PW (1922) Phys Rev 19:268–270
Smith RC (1966) J Appl Phys 37:4860–4865
Smith RC, Holland LR (1966) J Appl Phys 37:4866–4869
Sohn M, Baumann F (1996) J Phys: Condens Matter 8:6857–6872
Soulen RJ, Fogle WE, Colwell JH (1992), in: Schooley (ed) Temperature, Its Measurement
    and Control in Science and Industry, vol 6. AIP, New York, pp 983–988
Spišiak J, Righini F, Bussolino GC (2001) Int J Thermophys 22:1241–1251
Stanimirović A, Vuković G, Maglić K (1999) Int J Thermophys 20:325–332
Steele LM, Finotello D (1992) J Low Temp Phys 89:645–648
Steele LM, Finotello D (1994) Physica B 194/196:637–638
Steele LM, Yeager CJ, Finotello D (1993) Phys Rev Lett 71:3673–3676
Steinmetz N, Menges H, Dutzi J, v. Löhneysen H, Goldacker W (1989)
    Phys Rev B 39:2838–2841
Stewart GR (1983) Rev Sci Instrum 54:1–11
Stine KJ, Garland CW (1989) Phys Rev A 39:1482–1485
Stoebe T, Huang CC, Goodby JW (1992) Phys Rev Lett 68:2944–2947

Stokka S, Fossheim K (1982a) J Phys E 15:123–127
Stokka S, Fossheim K (1982b) J Phys C 15:1161–1176
Stokka S, Samulionis V (1981) Phys Stat Sol A 67:K89–92
Stokka S, Fossheim K, Ziolkiewicz S (1981) Phys Rev B 24:2807–2811
Stokka S, Fossheim K, Johansen T, Feder J (1982) J Phys C 15:3053–3058
Storm L (1970) Z angew Physik 28:331–333
Strukov BA, Onodera A, Taraskin SA, Shnaidshtein IV, Red'kin BS, Haga H (1996) Ferroelectrics 185:181–184
Suematsu H, Suzuki M, Ikeda H (1980) J Phys Soc Jpn 49:835–836
Suga H (2000) Thermochim Acta 355:69–82
Suga H (2001) Thermochim Acta 377:35–49
Sugimoto N, Matsuda T, Hatta I (1981) J Phys Soc Jpn 50:1555–1559
Sukhovei KS (1967) Sov Phys Solid State 9:2893–2895
Sullivan P, Seidel G (1966) Annales Academiæ Scientiarum Fennicæ A, Physica N 210:58–62
Sullivan P, Seidel G (1967) Phys Lett A 25:229–230
Sullivan PF, Seidel G (1968) Phys Rev 173:679–685
Suska J, Tschirnich J (1999) Meas Sci Technol 10:N55–59
Suzuki M, Ikeda H (1978) J Phys C 11:3679–3685
Suzuki T, Tsuboi T (1977) J Phys Soc Jpn 43:444–450
Suzuki T, Tsuboi T, Takaki H (1982) Jpn J Appl Phys 21:368–372
Tanaka M, Akimitsu J, Inada Y, Kimizuka N, Shindo I, Siratori K (1982) Solid State Commun 44:687–690
Takase K, Koyano M, Katoh Y, Sasaki M, Inoue M (1994) J Low Temp Phys 97:335–345
Tanasijczuk OS, Oja T (1978) Rev Sci Instrum 49:1545–1548
Tashiro K, Saito M, Tsuzuku T (1985) Synth Met 12:63–69
Tashiro K, Ozawa N, Sugihara K, Tsuzuku T (1990) J Phys Soc Jpn 59:4022–4028
Terki F, Gandit P, Chaussy J (1992) Phys Rev B 46:922–929
Thoen J, Glorieux C (1997) Thermochim Acta 304/305:137–150
Tong HM, Hsuen HKD, Saenger KL, Su GW (1991) Rev Sci Instrum 62:422–430
Trost W, Differt K, Maier K, Seeger A (1986), in: Janot C, Petry W, Richter D, Springer T (eds) Atomic Transport and Defects in Metals by Neutron Scattering. Springer, Berlin, pp 219–234
Tsuboi T, Suzuki T (1977) J Phys Soc Jpn 42:437–444
Tsuchiya Y (1991) J Phys: Condens Matter 3:3163–3172
Tsuchiya Y (1993) J Non-Cryst Solids 156/158:704–707
Tsuchiya Y (1995) J Phys Soc Jpn 64:159–163
Tura V, Mitoseriu L, Papusoi C, Osaka T, Okuyama M (1998) Jpn J Appl Phys (Part 1) 37:1950–1954
Uchida K, Yao H, Ema K (1997) Phys Rev E 56:661–666
Van den Sype J (1970) Phys Stat Sol 39:659–664
Van der Ziel A (1958) Solid State Physical Electronics. Macmillan, London
Varchenko AA, Kraftmakher YA (1973) Phys Stat Sol A 20:387–393
Varchenko AA, Kraftmakher YA, Pinegina TY (1978) High Temperature 16:720–723
Vargas RA, Diosa JE (1997) Solid State Commun 103:511–513
Vargas R, Salamon MB, Flynn CP (1976) Phys Rev Lett 37:1550–1553
Vargas RA, Salamon MB, Flynn CP (1977) Phys Rev B 17:269–281
Vargas RA, Chacón M, Tróchez J C, Palacios I (1989) Phys Lett A 139:81–84

Vargas RA, Diosa JE, Torijano E (1995) Solid State Commun 95:191–193
Vassilev T, Velinov T, Avramov I, Surnev S (1995) Appl Phys A 61:129–134
Velichkov IV (1992) Cryogenics 32:285–290
Viner JM, Huang CC (1981) Solid State Commun 39:789–791
Viner JM, Lamey D, Huang CC, Pindak R, Goodby JW (1983) Phys Rev A 28:2433–2441
Viswanathan R, Johnston DC (1975) J Phys Chem Solids 36:1093–1096
Viswanathan R, Wu CT, Luo HL, Webb GW (1974) Solid State Commun 14:1051–1054
Viswanathan R, Lawson AC, Pande CS (1976) J Phys Chem Solids 37:341–343
Vuković GS, Perović NL, Maglić KD (1996) Int J Thermophys 17:1057–1067
Wagner C, Schottky W (1930) Z phys Chemie 11:163–210
Wagner T, Kasap SO (1996) Phil Mag B 74:667–680
Waite TR, Craig RS, Wallace WE (1956) Phys Rev 104:1240–1241
Wang JK, Campbell JH (1988) Rev Sci Instrum 59:2031–2035
Wang JK, Campbell JH, Tsui DC, Cho AY (1988) Phys Rev B 38:6174–6184
Wang JK, Tsui DC, Santos M, Shayegan M (1992) Phys Rev B 45:4384–4389
Wantenaar GHJ, Campbell SJ, Chaplin DH, Wilson GVH (1977)
    J Phys E 10:825–828
Wen X, Garland CW, Wand MD (1990) Phys Rev A 42:6087–6092
Wen X, Garland CW, Shashidhar R, Barois P (1992) Phys Rev B 45:5131–5139
Weyer S, Hensel A, Korus J, Donth E, Schick C (1997a)
    Thermochim Acta 304/305:251–255
Weyer S, Hensel A, Schick C (1997b) Thermochim Acta 304/305:267–275
White DR, Galleano R, Actis A, Brixy H, DeGroot M, Dubbeldam J, Reesink AL, Edler F,
    Sakurai H, Shepard RL, Gallop JC (1996) Metrologia 33:325–335
White GK, Minges ML (1997) Int J Thermophys 18:1269–1327
Wilde G (2002) J Non-Cryst Solids 307/310:853–862
Wilhelm H, Jaccard D (2002a) Phys Rev B 66:064428-1–9
Wilhelm H, Jaccard D (2002b) J Phys: Condens Matter 14:10683–10687
Williams CC, Wickramasinghe HK (1986) Appl Phys Lett 49:1587–1589
Winter W, Höhne GWH (2003) Thermochim Acta 403:43–53
Wu L, Garland CW, Shashidhar R (1993) J Chem Phys 98:4309–4310
Wu L, Zhou B, Garland CW, Bellini T, Schaefer DW (1995) Phys Rev E 51:2157–2165
Wunderlich B (1997) J Thermal Analysis 48:207–224
Wunderlich B, Okazaki I (1997) J Thermal Analysis 49:57–70
Wunderlich B, Jin Y, Boller A (1994) Thermochim Acta 238:277–293
Wunderlich B, Boller A, Okazaki I, Kreitmeier S (1996a) Thermochim Acta 283:143–155
Wunderlich B, Boller A, Okazaki I, Kreitmeier S (1996b) J Thermal Analysis 47:1013–1026
Wunderlich B, Boller A, Okazaki I, Ishikiriyama K (1997) Thermochim Acta 305:125–136
Wunderlich B, Okazaki I, Ishikiriyama K, Boller A (1998) Thermochim Acta 324:77–85
Wunderlich B, Boller A, Okazaki I, Ishikiriyama K, Chen W, Pyda M, Pak J, Moon I,
    Androsch R (1999) Thermochim Acta 330:21–38
Wundrlich B, Androsch R, Pyda M, Kwon YK (2000) Thermochim Acta 348:181–190
Wunderlich B, Pyda M, Pak J, Androsch R (2001) Thermochim Acta 377:9–33
Wunderlich RK, Fecht H-J (1993) J Non–Cryst Solids 156/158:421–424
Wunderlich RK, Fecht H-J (1996) Int J Thermophys 17:1203–1216
Wunderlich RK, Fecht H-J, Willnecker R (1993) Appl Phys Lett 62:3111–3113
Wunderlich RK, Lee DS, Johnson WL, Fecht H-J (1997) Phys Rev B 55:26–29
Wunderlich RK, Ettl C, Fecht H-J (2001) Int J Thermophys 22:579–591

Xu X-Q, Hagen SJ, Jiang W, Peng JL, Li ZY, Greene RL (1992) Phys Rev B 45:7356–7359
Xu X-Q, Peng JL, Li ZY, Ju HL, Greene RL (1993) Phys Rev B 48:1112–1118
Yakunkin M M (1983) High Temperature 21:848–853
Yang G, Migone AD, Johnson KW (1991) Rev Sci Instrum 62:1836–1839
Yao H, Hatta I (1988) Jpn J Appl Phys (Part 2) 27:L121–122
Yao H, Hatta I (1995) Thermochim Acta 266:301–308
Yao H, Ema K, Garland CW (1998) Rev Sci Instrum 69:172–178
Yao H, Ema K, Hatta I (1999) Jpn J Appl Phys (Part 1) 38:945–950
Yeager CJ, Steele LM, Finotello D (1994a) Phys Rev B 49:9782–9793
Yeager CJ, Steele LM, Finotello D (1994b) Physica B 194/196:639–640
Yeager CJ, Steele LM, Finotello D (1995) Phys Rev B 51:15274–15280
Yi W, Lu L, Zhang DL, Pan ZW, Xie SS (1999) Phys Rev B 59:R9015–9018
Yoon J, Chan MHW (1997) Phys Rev Lett 78:4801–4804
Yoshizawa M, Fujimura T, Goto T, Kamiyoshi K-I (1983) J Phys C 16:131–142
Yoshizawa M, Suzuki T, Goto T, Yamakami T, Fujimura T, Nakajima T, Yamauchi H (1984) J Phys Soc Jpn 53:261–269
Yu RC, Naughton MJ, Yan X, Chaikin PM, Holtzberg F, Greene RL, Stuart J, Davies P (1988) Phys Rev B 37:7963–7966
Zaitseva GG, Kraftmakher YA (1965) Zh Prikl Mekhan Tekhn Fiz N 3:117
Zally GD, Mochel JM (1971) Phys Rev Lett 27:1710–1712
Zally GD, Mochel JM (1972) Phys Rev B 6:4142–4150
Zammit U, Marinelli M, Pizzoferrato R, Scudieri F, Martellucci S (1988) J Phys E 21:935–937
Zammit U, Marinelli M, Pizzoferrato R, Scudieri F, Martellucci S (1990) Phys Rev A 41:1153–1155
Zhang M, Efremov MY, Schiettekatte F, Olson EA, Kwan AT, Lai SL, Wisleder T, Greene JE, Allen LH (2000) Phys Rev B 62:10548–10557
Zhang M, Efremov MY, Olson EA, Zhang ZS, Allen LH (2002) Appl Phys Lett 81:3801–3803
Zhang S, Migone AD (1988) Phys Rev B 38:12039–12042
Zhou B, Buan J, Huang CC, Waszczak JV, Schneemeyer LF (1991) Phys Rev B 44:10408–10410
Zhou B, Iannacchione GS, Garland CW (1997a) Liquid Crystals 22:335–339
Zhou B, Iannacchione GS, Garland CW, Bellini T (1997b) Phys Rev E 55:2962–2968
Zinov'ev OS, Lebedev SV (1976) High Temperature 14:73–75
Zoller P, Dillinger J R (1969) Phys Lett A 28:682–683
Zwikker C (1928) Z Phys 52:668–677

Index

Active thermal shield 81
Adiabaticity 16
Bath modulation 104
Biological materials 110
Blackbody models 43
Boltzmann's constant 159, 254
Bridge circuit 40, 252
Calorimeter ACC-1 98
Calorimetry in space 85
Complex specific heat 228
Confined systems 222
Correlation technique 164, 202
Critical index 211
Electron-bombardment heating 29, 182
Equilibration times 235, 242
Equilibrium vacancies 193
Equivalent-impedance technique 37, 252
Equivalent noise resistance 256
Estimates of errors 179
Ferro- and antiferroelectrics 211
Ferro- and antiferromagnets 211
Freestanding thin films 102
Frequency conversion 35, 198
Glass transitions 118, 230
Heat losses 15
Heat transfer coefficient 15, 80
High pressures 91
Induction heating 30, 84
Integrated-circuit calorimeter 71, 107
Isochoric specific heat 201
Levitation calorimetry 57
Liquid crystals 107
Lock-in detection 50, 171
MDSC 124
Modulated-light heating 26, 96, 98, 152
Modulation dilatometry 135
Molten metals 81
Nanocalorimetry 71, 104
Noise thermometry 159
Nonadiabatic regime 22, 77

Nonconducting materials 82, 142
Noncontact calorimetry 84
Nyquist's formula 159, 254
Organic conductors 226
Peltier heating 31
Phase transitions 207
Photoacoustic techniques 113
Photoelectric detectors 42, 170
Polar diagram 17, 249
Potentiometer circuit 41
Premelting anomaly 199
Pulse calorimetry 58, 245
Pyroelectric sensors 45, 230
Rapid-heating experiments 67, 238, 240
Relaxation phenomena 227
Resistance thermometers 49
Resistive heaters 25
Shot effect 162
Solid electrolytes 209
Specific-heat ratio 204
Specific-heat spectroscopy 115
Spectral absorptance 153
Spectral emissivity 60
Superconductors 217
Supplementary-current technique 33
Temperature coefficient of specific heat 196
Temperature derivative of resistance 147
Temperature fluctuations 201
TEMPUS 85
Thermal coupling 19
Thermal diffusivity 96, 114
Thermal effusivity 116
Thermal expansivity 136
Thermal noise 159, 254
Thermocouples 47
Thermopower 151
Thin films 101, 222
Third-harmonic method 35, 117, 250
Vacancy equilibration 236
Vacancy-related enthalpy 239